THEMES IN OLD WORLD ZOOARCHAEOLOGY

THEMES IN OLD WORLD ZOOARCHAEOLOGY

FROM THE MEDITERRANEAN TO THE ATLANTIC

Edited by

Umberto Albarella, Cleia Detry, Sónia Gabriel, Catarina Ginja, Ana Elisabete Pires and João Pedro Tereso

OXBOW | books
Oxford & Philadelphia

Published in the United Kingdom in 2021 by
OXBOW BOOKS
The Old Music Hall, 106–108 Cowley Road, Oxford, OX4 1JE

and in the United States by
OXBOW BOOKS
1950 Lawrence Road, Havertown, PA 19083

Hardback Edition: ISBN 978-1-78925-534-8
Digital Edition: ISBN 978-1-78925-535-5 (epub)

A CIP record for this book is available from the British Library

Library of Congress Control Number: 2021931166

Printed in Malta by Melita Press

Typeset by Versatile PreMedia Services (P) Ltd.

For a complete list of Oxbow titles, please contact:

UNITED KINGDOM
Oxbow Books
Telephone (01865) 241249
Email: oxbow@oxbowbooks.com
www.oxbowbooks.com

UNITED STATES OF AMERICA
Oxbow Books
Telephone (610) 853-9131, Fax (610) 853-9146
Email: queries@casemateacademic.com
www.casemateacademic.com/oxbow

Oxbow Books is part of the Casemate Group

Front cover: Hyena: pixabay.com/tommileew; Launceston castle, Cornwall, England built by Robert the Count of Mortain after 1068: Shutterstock.com/studiogeorge; A Soay Ram: Shutterstock.com/Stephen John David Wain
Back cover: Photo of Simon Davis by Ana Elisabete Pires

Contents

Historic times

Contributors

Umberto Albarella
University of Sheffield (UK)
u.albarella@sheffield.ac.uk

Ana Margarida Arruda
UNIARQ, FLUL, Universidade de Lisboa (Portugal)
a.m.arruda@letras.ulisboa.pt

Polydora Baker
Historic England (UK)
polydora.baker@HistoricEngland.org.uk

Evelyne Browaeys
University of Sheffield (UK)
edmbrowaeys1@sheffield.ac.uk

Annie Brown
University of Manitoba, Winnipeg (Canada)
brown.annie6@gmail.com

Ana Caessa
Câmara Municipal de Lisboa (Portugal)
anacaessa68@gmail.com

João Luís Cardoso
Universidade Aberta, Lisboa (Portugal)
cardoso18@netvisao.pt

Tânia Casimiro
HTC/IAP Universidade Nova de Lisboa (Portugal)
tmcasimiro@fcsh.unl.pt

Orr Comay
Tel Aviv University (Israel)
orrcomay@tauex.tau.ac.il

Tamar Dayan
Tel Aviv University (Israel)
dayant@tauex.tau.ac.il

Pierre de Miroschedji
Centre National de la Recherche Scientifique, Nanterre (France)
miroschedji@mac.com

Cleia Detry
UNIARQ, FLUL, Universidade de Lisboa (Portugal)
cdetry@letras.ul.pt

Lídia Fernandes
Museu de Lisboa – Teatro Romano, EGEAC/CML (Portugal)
lidiafernandes@egeac.pt

Sónia Gabriel
LARC-DGPC, Laboratório de Arqueociências, Lisboa (Portugal)
gabriel.sonia@gmail.com

Catarina Ginja
CIBIO-InBIO, Universidade do Porto (Portugal)
catarinaginja@cibio.up.pt

Haskel J. Greenfield
University of Manitoba, Winnipeg (Canada)
Haskel.Greenfield@umanitoba.ca

Sílvia Guimarães
Institución Milà y Fontanals de Humanidades, CSIC, Barcelona (Spain)
CIBIO-InBIO, Universidade do Porto (Portugal)
silvia.guimaraes@cibio.up.pt

Angelos Hadjikoumis
University of Sheffield (UK)
a.hadjikoumis@sheffield.ac.uk

Salima Ikram
American University in Cairo (Egypt)
salima@aucegypt.edu

Leif Jonsson
Gothenburg Museum of Natural History (Sweden)
leif.jonsson@osteology.se

Laura Llorente-Rodríguez
Leiden Universiteit (The Netherlands)
l.llorente.rodriguez@arch.leidenuniv.nl

Lluís Lloveras
Universitat de Barcelona (Spain)
lluislloveras@gmail.com

Marco Masseti
University of Florence (Italy)
marco.masseti@unifi.it

Arturo Morales-Muñiz
Universidad Autónoma de Madrid (Spain)
arturo.morales@uam.es

Marta Moreno-García
Instituto de Historia, CCHS-CSIC, Madrid (Spain)
marta.moreno@cchs.csic.es

Nuno Mota
Câmara Municipal de Lisboa (Portugal)
nuno.miguel.mota@cm-lisboa.pt

Ariadna Nieto-Espinet
Institución Milà y Fontanals de Humanidades, CSIC, Barcelona (Spain)
arinietoespinet@gmail.com

Ana Elisabete Pires
CIBIO-InBIO, Universidade do Porto (Portugal)
aepires@cibio.up.pt

Eufrasia Roselló-Izquierdo
Universidad Autónoma de Madrid (Spain)
eufrasia.rosello@uam.es

Ana Beatriz Santos
UNIARQ, FLUL, Universidade de Lisboa (Portugal)
asantos5@campus.ul.pt

João Pedro Tereso
CIBIO-InBIO, Universidade do Porto (Portugal)
jptereso@gmail.com

Angela Trentacoste
Institute of Archaeology, University of Oxford (UK)
a.trentacoste@gmail.com

Silvia Valenzuela-Lamas
Institución Milà y Fontanals de Humanidades, CSIC, Barcelona (Spain)
svalenzuela@imf.csic.es

João Zilhão
Institució Catalana de Recerca i
Estudis Avançats (ICREA)
Universitat de Barcelona (Spain)
UNIARQ, Universidade de Lisboa (Portugal)
joao.zilhao@ub.edu

To Simon Davis,
zooarchaeologist, woodworker, internationalist, friend

Photo by Paulo Gaspar Ferreira

1

From the Mediterranean to the Atlantic: Simon Davis' exceptional contribution to the world of zooarchaeology

Umberto Albarella

Introduction

This book intends to showcase recent developments in the zooarchaeology of Europe and Western Asia, as well as honour the life and career of Simon Davis – zooarchaeologist and human being extraordinaire. It represents the culmination of a project that started with the organisation of the conference 'From the Mediterranean to the Atlantic: Themes in Old World Zooarchaeology', which was held on 1–3 June 2017 at the National Museum of Natural History and Science in Lisbon, Portugal (Fig. 1.1).

The conference was organised by Cleia Detry, Umberto Albarella, Sónia Gabriel, Catarina Ginja, Ana Elisabete Pires and João Tereso, who have also taken on the editorial task of preparing the publication of these proceedings. The impetus for the organisation of the conference came from the consideration that Simon Davis' seminal book *The Archaeology of Animals*, originally published in 1987 (Davis 1987), was approaching its 30th birthday. The conference was a friendly and informal but academically rigorous event, which attracted researchers from all over the world, many of

Figure 1.1. Participants to the conference 'From the Mediterranean to the Atlantic: Themes in Old World Zooarchaeology', held on 1–3 June 2017 at the National Museum of Natural History and Science in Lisbon, Portugal. Published with permission of Cleia Detry.

them Simon's longstanding associates. Some of Simon's old friends and colleagues, who could not make the conference, sent messages of congratulations, which were publicly read at the conference. Not all speakers could write a chapter for the book, but others who were not with us in Lisbon managed to contribute in a written format.

This Introduction aims to discuss Simon Davis' contribution to the world of archaeology, as well as briefly introduce the papers included in this book. Before we proceed, it is, however, necessary to add a brief clarification about the title of the book. The expression 'Old World' is here used because of its historical value. Though geographically vague, it does serve us well as a broad definition of the research region which has represented Simon's main area of activity. In no way is 'Old World' supposed to suggest some kind of supremacy of this part of the world over others. The term dates back to the period of global explorations; that such explorations eventually led to colonialism and exploitation represents a phenomenon that, in its shameful legacy, is better to be remembered and discussed than hidden under the carpet. The expression 'From the Mediterranean to the Atlantic' is especially significant for us, as it symbolises a pathway in Simon's research that was also reflected in his life. From an early focus on the Eastern Mediterranean, Simon's research eventually 'migrated' west towards the European Atlantic, as is exemplified by his work (and life) first in England and then Portugal. However, the East was neither forgotten nor abandoned, though Simon's main research focus in the last 30 years has primarily, but not exclusively, been on western Europe.

Figure 1.2. Simon Davis (then one year old) featured on the cover of the magazine Nursery World. *Published with permission of Simon Davis.*

Simon Davis

Simon Davis has been a genuine pioneer in the world of zooarchaeology and his work has had a long-standing influence on the development of the discipline. A refined intellectual – his expertise in archaeology and biology goes hand in hand with his interest in literature, history, linguistics, art, music, politics and world cultures – Simon is, nonetheless, especially known for his humility, informality and affability. The academic inspiration and friendship he has provided to so many people over the years have made him a much-loved personality in the research world and it is thus unsurprising that the conference we organised in his honour saw a massive outpour of affection and respect. Simon epitomises a lack of arrogance – while possessing an encyclopaedic knowledge, he has never shied away from humble jobs, such as spending endless hours preparing skeletons for the reference collection, making cabinets for his cherished bones, or even cleaning bench surfaces. Simon's sense of hospitality is also legendary and he could easily invite you for dinner on a first encounter and start preparing a delicious fish soup while, at the same time, discussing subjects such as Persian music, Orwell's essays, the Salazar's regime, or, of course, our relationship with animals.

Born in Brighton, on the southern coast of Britain (Fig. 1.2), Simon spent his childhood in the company of a younger sister, Sara, who eventually became a very well-known illustrator and artist. Simon and Sara have remained close friends for all their lives, united by a very special bond of solidarity. After a not entirely happy experience in public school, Simon moved to London for his higher education and completed a BSc in Zoology at the University College London in 1971. These exposures to the discipline of biology and a metropolitan lifestyle will remain defining elements of his life. Brighton was for him a bourgeoisie outpost of largely Daily Telegraph[1] readers, while London was hustle and bustle, experienced at its best in working-class areas with lively markets – the kind of atmospheres Simon made his own and that eventually attracted him to the Middle East and Southern Europe. Once, while in his flat in Southampton Row, where many long hours were spent playing music, cooking food and conversing about everything, Simon told me that he would consider no other place to live in Britain but London. If he had to leave London, then he would rather go abroad, preferably to the south. This eventually proved to be prophetic.

In fact, well before the time of that conversation, Simon had already spent several years living outside his home country. Following the completion of his first degree, he moved to Jerusalem to study for his master's degree, which represented his first proper foray into zooarchaeology. For his master's dissertation, he studied the Kebaran animal bone assemblage from Ein Gev I (Israel). He completed his degree in 1973 and the ensuing publication was rapidly produced (Davis 1974) – those were the days! He remained in Jerusalem to move on to PhD level study and during those years he cemented his expertise as a zooarchaeologist. He had access to many Pleistocene and Holocene Israeli assemblages, which led to an impressive corpus of published material (*e.g.* Davis 1976; 1977; 1980a; 1980b; 1980c; 1981; 1982; 1983a; Davis and Valla 1978; Davis, Goring-Morris and Gopher 1982). This remarkable collection of publications shows how, still in his twenties, he was already a fully formed researcher, investigating research questions that would remain at the core of his work in the following decades. Prominent among those themes were the origins of animal domestication, the dynamics of body size variation, and the morphometric separation of different taxa. Methodologically, he had already developed innovative biometric approaches and, well ahead of his time, he had realised the importance of a selective approach to the recording of animal bone assemblages. In doing so, he was working in parallel with his friend John Watson – the inventor of the 'diagnostic zone' approach in zooarchaeology (Watson 1979). John was a researcher with whom Simon had a natural affinity, which explains their long-standing research collaboration (*e.g.* Watson and Davis 2019; 2020). While in Jerusalem, Simon also had his first teaching experiences. Although he never held a full teaching job, he learned to love the interaction with students especially in a lab setting, where he could make his teaching lively and largely hands-on, which suited his laid-back personality.

In the late seventies, Simon also had the opportunity to work in the Iranian Kurdistan (Davis 1984a), shortly before the Islamic revolution (Fig. 1.3). This was an important formative experience that he would frequently refer to in later years. In 1979 his PhD was completed, and his Israeli experience had come to its natural conclusion. There were, however, many returns to Israel to continue research activities – the animal bones from the site Hatoula (Davis, Lernau and Pichon 1994) represented one of his most important projects – but also to visit old friends.

In England, a post-doctoral fellowship at the University College London (UCL) provided him with the opportunity to be back in his beloved London. The research concerned the study of island populations of mice from Britain and Faroe, a topic that opened up what was going to become a long-lasting interest in the biology and archaeology of islands. The research was duly published (Davis 1983b), reinforcing a pattern that has characterised Simon's career throughout his life – he has always delivered.

The early eighties also saw the beginning of one of Simon's most important projects, which would last for many years. He joined the French CNRS Cypriot mission

Figure 1.3. Mahidasht, Iranian Kurdistan 1978. Photo taken in the garden of the chief of the Sanjabi tribe of Kurds, Asghar Sanjabi, who is seated on the left. Standing in the centre is Louis Levene, who led the Mahidasht archaeological expedition. Simon is seated on the right. Published with permission of Simon Davis.

directed by Alain Le Brun and Odile Daune Le Brun, excavating the Neolithic site of Khirokitia. The animal bone assemblage was huge and most interesting, and kept Simon occupied for a long time. From England, he visited Cyprus almost every year and developed a great affection for the island and in particular for the little village of Khirokitia, where he thought he would one day buy a house and retire – so far, an unfulfilled dream! The animal bones provided important information about the early colonisation of the island, the extinction of the native megafauna (Fig. 1.4), and the introduction of domestic and wild mammals from the main continent (Davis 1984b). Regular fieldwork meant that, although the core research questions had been addressed in the first publication, incremental knowledge was being accumulated over the course of the later eighties and nineties. Simon, therefore, found himself in the good position of being able to publish regular accounts of the progress of his work. The initial 1984 publication was followed by *Some more animal remains from the Aceramic Neolithic of Cyprus* published in 1989 (Davis 1989), which the late Tony Legge – a friend and eminent zooarchaeologist – defined as 'the most boring title he had ever heard of'! Simon responded in tune by publishing his 1989 update with the title *Even more bones from Khirokitia!* (Davis 1994). Although the fieldwork was completed several years ago, the Khirokitia project represents the most enduring of

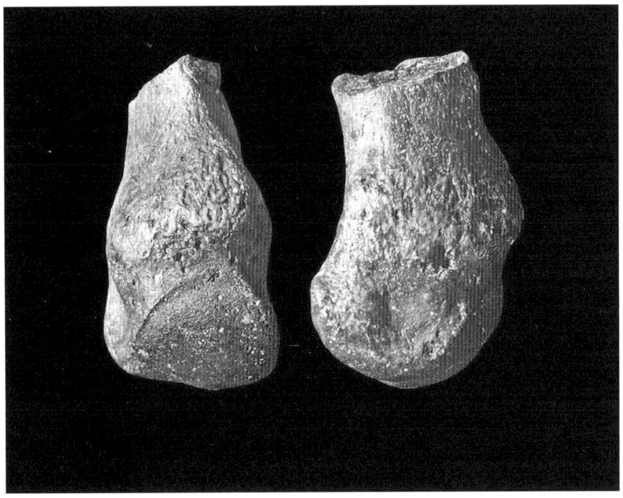

Figure 1.4. Distal metacarpal of a dwarf hippo from the Aceramic Neolithic site of Cape Andreas Kastros (Cyprus). Published with permission of Simon Davis.

Figure 1.5. The book cover of the Korean edition of Simon Davis' The Archaeology of Animals *(originally published in 1987). Published with permission of Simon Davis.*

Simon's projects and the one that he has arguably remained most attached to. He also produced a more synthetic work integrating data from many years of excavation and comparing them with the mainland evidence (Davis 2003).

Most of the eighties saw Simon busy with teaching and demonstration jobs at the University of London and fieldwork in Israel and Cyprus but, although life in London was vibrant and exciting, permanent employment was hard to come by. Freelance work provided much freedom and opportunity for research but limited financial reward and, to pay the bills, at some point Simon even had to work for Marks and Spencer (a job in a British multinational retailer that, unsurprisingly, he found very boring!). However, there were more exciting opportunities too. One of these was being commissioned by the publisher Batsford to work on a zooarchaeology textbook. It was the late Juliet Clutton-Brock, a famous zooarchaeologist and natural historian, who recommended Simon – still a young researcher – for the job. Although a few zooarchaeology manuals existed (*e.g.* Cornwall 1956; Ryder 1968; Chaplin 1971) these were largely old-fashioned books. *The Archaeology of Animals* (Davis 1987), however, managed to combine a modern approach to zooarchaeology with an accessible style, which made the book hugely popular. One of its best features was careful attention to graphic design and engaging illustrations – aspects that will characterise the entirety of Simon's publication output. The book was soon translated into Spanish and in much more recent years – to Simon's delight – a Korean edition also came to light (Fig. 1.5). Ever the internationalist, Simon cherished the opportunity for his work to be known on a global scale.

The late eighties finally brought stable employment, as Simon was hired as a zooarchaeologist by English Heritage, an agency of the British Government in charge of the preservation of the historic environment (now called Historic England). For the following ten years he would work at the Ancient Monuments Laboratory (AML) at Fortress House, Savile Row (Fig. 1.6), a street known for tailor-making for wealthy people – a far cry from Simon's casual and down-to-earth style! The English Heritage job provided him with the opportunity to work with his old friend Sebastian Payne, with whom he shared many ideas about zooarchaeology approaches. The two would form a strong partnership, which produced some highly significant work, such as the important study of the Irthlingborough Bronze Age Barrow (Northamptonshire), which demonstrated the ritual use of cattle skulls in a totemic–like fashion (Davis and Payne 1993). Simon and Sebastian also agreed about the importance of establishing good reference collections of animal skeletons to enhance the reliability of zooarchaeological identifications. They worked hard in improving both the accessibility and the completeness of the pre-existing English Heritage collection, eventually making it one of the best in the world (Davis and Payne 1992; Davis, Corke and Payne 1998) (Fig. 1.7).

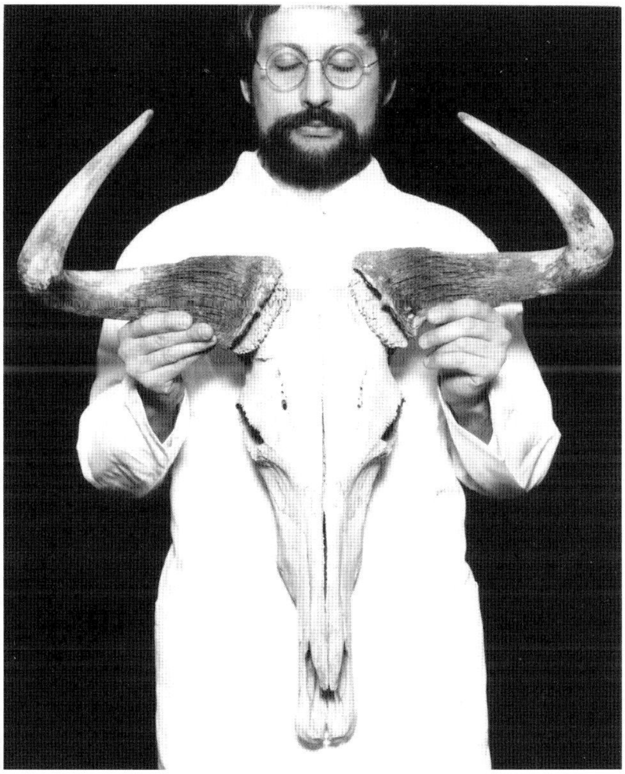

Figure 1.6. Simon Davis holding a hartebeest skull at the English Heritage Ancient Monuments Laboratory (AML) in 1985. Although he would not be fully employed by English Heritage until 1988, he did some contract work at the AML in earlier years, when this photo was taken. Published with permission of Simon Davis.

Figure 1.7. Simon Davis and Polydora Baker at the Ancient Monuments Laboratory c.1993. The English Heritage animal bone reference collection, which Simon contributed substantially to develop, is in good view. Published with permission of Simon Davis and Polydora Baker.

Although Simon continued his work in the Eastern Mediterranean, taking some unpaid leave every year, his English Heritage years also provided him with the opportunity to focus his attention on the archaeology of his own country. There was, of course, much to explore and Simon studied a formidable number of animal bone assemblages, despite being somewhat burdened by unwelcome administrative duties as well as the more interesting distractions of reference collection building. Having spent most of his career exploring prehistory, he was sufficiently open-minded to start investigating the historical periods with great enthusiasm. One topic he became especially interested in was the onset of the so-called Agricultural Revolution in Early Modern England, which led to a substantial body of work on the topic (*e.g.* Davis 1997). Simon was intrigued by the historian Eric Kerridge's suggestion that the Agricultural Revolution had started earlier than proposed by traditional scholarship (Kerridge 1967) – perhaps as early as the 16th century, having its roots in the post-Black Death medieval period (*i.e.* late 14th–15th century). He decided to seek parallel evidence in the archaeological record. The integration of archaeological and historical evidence became an increasing priority in Simon's work, leading to a collaboration with a historian to investigate Early Modern husbandry innovations (Davis and Beckett 1999) and a paper synthesising useful husbandry information collected from medieval and Early Modern agricultural textbooks (Davis 2002a). A fundamental stimulus to the study of the Agricultural Revolution in England was provided by the study of the animal bone assemblage from Launceston Castle in Cornwall (Albarella and Davis 1996; Albarella, Davis and Smith 2006) (Fig. 1.8), which started in 1992, and also cemented our work partnership – a collaboration that also evolved into a great friendship that was going to continue unabated for the following 30 years. The animal bone assemblage from Launceston was of exceptional importance, indicating significant changes in agriculture already occurring in the 16th–early 17th century (Kerridge was vindicated!), and also providing valuable information

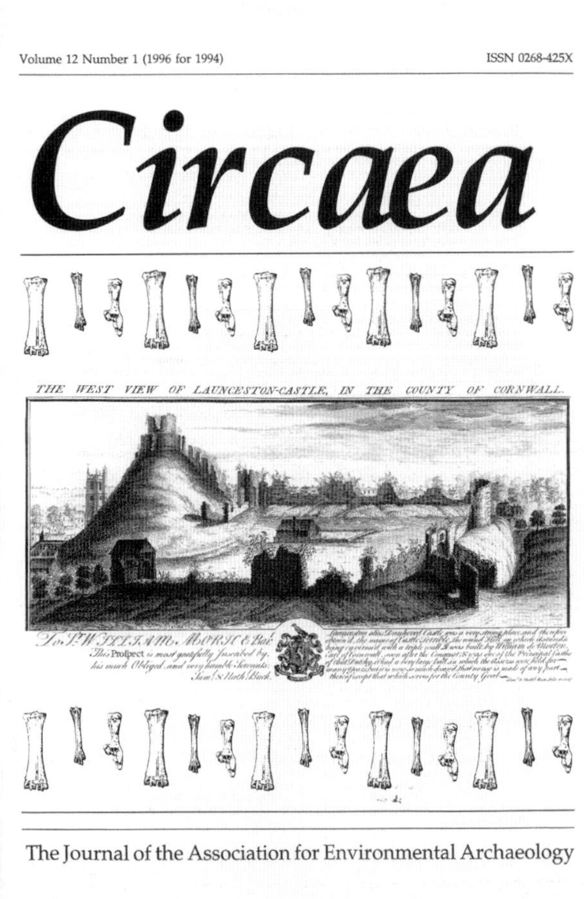

Figure 1.8. The front cover of the Circaea *monograph, dedicated to the study of the mammal and bird bones from Launceston Castle (Cornwall, UK). The cover was designed by Simon's sister, Sara Midda. Published with permission of Simon Davis.*

on food consumption, social status and trade in medieval and post-medieval times. It always remained an inspiring project for Simon, and it motivated some of his future work. Working at the other end of the social spectrum, the study of the animal bones from the deserted medieval village of West Cotton (Northamptonshire) represented another important project, which contributed to an understanding of rural medieval life in a context dominated by urban data (Albarella and Davis 2010). From this project, Simon was especially struck by the notion that there was geographic variation in the size of medieval livestock, with animals from central regions (such as Northamptonshire) generally larger than those from peripheral areas (such as Cornwall) (Davis 1997). This explained why at Launceston Castle there was a very sharp increase in size in the 16th century, as the medieval livestock was rather diminutive in that region.

Not all Simon's work focused on ecology, economy and diet, as the investigation of the above-mentioned ritual site of Irthlingborough demonstrates. He also became interested in ritual offerings in classical temples, mainly in connection to his analysis of the material from the Temple of Apollo at Kourion (Cyprus), where he found a bias in the body side of offered sheep remains (Davis 1996a; 2008a).

In addition to site-based projects, in the nineties Simon also continued working on his methodological projects, producing two especially important papers dealing with the morphometric variability of sheep bones of known breed, sex and age (Davis 1996b; 2000). But there was more to the nineties than academic work. In 1993, during one of his return trips to Khirokitia, Simon met a young French artist and illustrator, Cathy Douzil, who was eventually going to become his wife and the mother of his daughter, Miriam. Cathy was also going to make an important contribution to Simon's work, illustrating some of his books and articles with delightful sketches of animals and fulfilling Simon's aspiration to make his publications aesthetically pleasing (Fig. 1.9). The early nineties were fun and exciting times, full of vibrancy and endless conversations covering the most disparate subjects – academic and not. In 1991 Simon and myself also met Marco Masseti (contributing to this volume), who was visiting London and, in Simon's inimitable style, was invited for dinner on the spot. Marco was going to become a long-standing friend of both of us and his frequent visits to London provided the opportunity for gatherings of various other colleagues and friends – Caroline Grigson, Dale Serjeantson, Adrian Lister, Louise Martin, Polydora Baker, and several others would often be part of the company. There were also occasional visits from other colleagues from abroad, such as John Speth, John Watson and Sandi Olsen. Due to Simon's sense of hospitality and its central location in London, the flat on Southampton Row was in the nineties a hotspot of zooarchaeological conversations!

Gradually, and especially so in the latter part of the nineties, the job at English Heritage became increasingly bureaucratic and managerial, and therefore less appealing to Simon, who started thinking about an alternative. The final deciding factor was represented by the move of the Ancient Monuments Laboratory from central London to Fort Cumberland, a beautiful but rather isolated outpost near Portsmouth on the British south coast. Moving there was hugely unappealing to Simon, used as he was to his metropolitan life. Fluent in French and Hebrew, and with some knowledge of Greek and Italian, Simon's natural inclination was to look for jobs in the south and south–eastern Mediterranean area, but an interesting opportunity rather materialised in Lisbon, Portugal, mainly promoted by the visionary Portuguese archaeologist João Zilhão (contributor to this volume). At the time, João had an important role in the management of Portuguese archaeology and it was his dream to create a multidisciplinary – government-funded – archaeology laboratory in Lisbon, where different scientific applications could find a home and provide an integrated approach to

Figure 1.9. One of the many delightful papercuts that Cathy Douzil created to illustrate Simon's publications. Published with permission of Cathy Douzil.

Figure 1.10. From left to right, Simon Davis, Cleia Detry and the author on the beach of Melides (Portugal) in April 2011. Author's own photograph.

the study of past human societies. Despite the pioneering work of João Cardoso (also a contributor to this volume), Portuguese zooarchaeology was in need of development and could benefit from an influx of new energy and expertise. This was a good challenge for Simon, and Lisbon looked like the ideal city for him and Cathy to settle in – and so it proved to be. In 1999 Simon got the job and, within a few years, he created a first-rate zooarchaeology laboratory with a superb reference collection – previous expertise acquired in Jerusalem and London helped substantially! In this job, he collaborated with Marta Moreno-García (contributor), who was appointed at the same time as Simon, and Carlos 'Maico' Pimenta. The team got going on many interesting research projects (*e.g.* Moreno-García, Davis and Pimenta 2003; Moreno-García *et al.* 2003; Davis and Moreno García 2007). Soon, several young researchers became interested in zooarchaeology and sought Simon's help and advice. Sónia Gabriel contributed substantially to the building up of the reference collection but also developed a specialisation in the study of fish bones, an area that was new to Simon and proved to be especially valuable. Cleia Detry was at the time a young PhD student and started working closely with Simon. They became great friends but also a close-knit team, working together on many projects, during and after the completion of Cleia's PhD (*e.g.* Albarella *et al.* 2005; Davis *et al.* 2012; Davis and Detry 2013; Davis *et al.* 2018) (Fig. 1.10).

There have been many highlights in Simon's 'Portuguese phase', but the study of two archaeological sites stand out. One is the assemblage from Caldeirão cave (Davis 2002b; Davis, Robert and Zilhão 2007), which provided him with the opportunity to go back to his Pleistocene work, which had been frozen since the Israeli years. The other substantial project concerned the study of the animal bones from the Roman and medieval site of the Alcáçova de Santarém, which was published as a monograph of the *Trabalhos de Arqueologia* (Davis 2005a). The wonderful diachronic sequence provided by the site allowed Simon to explore some of the research questions he had investigated in England, but in a different geographic and cultural context. He became especially interested in Muslim Iberia and the economic and cultural legacy of *Al-Andalus*. He would continue exploring the subject by using evidence from other areas, but still comparing them with Santarém, which remained an important yardstick (*e.g.* Davis 2008b; 2017).

In Lisbon, Simon also developed a strong interest in exploring the potential of genetics for the study of the past human-animal relationship (*e.g.* Pereira *et al.* 2006; Pereira *et al.* 2009). He maintained strong links with the Paris Jacques Monod laboratory directed by Eva-Maria Geigl and Thierry Grange. Eva-Maria invited him to give a keynote speech at the session on palaeogenetics and zooarchaeology she organised for the 2010 conference of the International Council of Archaeozoology (ICAZ) in Paris. Simon's main collaborations in this field have, however, been based in Portugal itself, where he has established a regular and close research connection with Catarina Ginja and Ana Elisabete Pires. He was especially impressed by the work carried out by Emma Svensson, which she had originally presented at the ICAZ conference in Mexico

City in 2006. Emma had been able to sex, on a genetic basis, cattle metapodials from archaeological sites, and this opened up important possibilities for zooarchaeology. The size changes Simon was interested in are affected by sex variation and being able to sex some of the metapodials provided a valuable opportunity to refine the biometric analysis. The collaboration between Simon and Emma led to an important reconsideration of size changes in Britain and Portugal, a project in which Catarina and Ana Elisabete were also involved (Davis *et al.* 2012).

While in Lisbon, Simon also revisited some of his long-standing research interests. Especially prominent among these was his work on the origins of animal domestication. In 1982, he had published an important paper on the subject, linking the beginning of farming with possible climatic change and demographic pressure (Davis 1982). However, in the new millennium, faced with a plethora of contributions to the domestication debate, few people remembered that paper and it was useful for Simon to revisit it based on new evidence and ideas that had meanwhile emerged. This process materialised itself in a new article on the theme published in the *Journal of Archaeological Science* (Davis 2005b). Although a substantial part of the work in Portugal focused on the historical periods, access to the Neolithic animal bone assemblage from Lameiras, Sintra, provided Simon with the opportunity to reconsider issues concerning the introduction of domestic animals that he had originally investigated in the Eastern Mediterranean (*cf.* Davis and Simões 2016; Davis, Gabriel and Simões 2018). A particularly significant project of the last few years, partly carried out with John Watson, has been his work on the size of rabbit bones and their association with climatic fluctuations; once again, this took him back to some of the research themes he had explored in his younger years (Davis 1981). Simon was especially pleased to publish a paper on this subject in the *Biological Journal of the Linnean Society* (Davis 2019), which had hosted his seminal work on equid differentiations almost four decades earlier (Davis 1980a).

At the time we organised the conference in his honour Simon had not yet retired, but he has now and this is providing him with more time to focus on his beloved woodwork, which has been more than a hobby for him in the last few decades; in fact, it is almost a parallel profession. However, those close to him know that Simon is notoriously forgetful, and he often seems to be unaware of his retired status and gets going with yet another zooarchaeology project (to the delight of the research community!). For instance, before the pandemic overturned our lives, he had started working regularly on Greek material (Fig. 1.11), in yet another of his 'migrations' from the Atlantic to the Mediterranean. He simply cannot stay away from the region which saw his first steps as a zooarchaeologist – the west is firmly in his 'bones' but the east remains in his 'heart'!

Figure 1.11. Simon in the village of Alyki on the island of Paros, Greece in 2014 or 2015. He is sorting modern bones, mainly goats, collected on the beach of the uninhabited island of Despotiko. Published with permission of Simon Davis.

This book

The chapters included in this book are authored by many contributors associated with Simon Davis' life and research. Most importantly, however, they address research questions that are close to Simon's interests and topical in the development of modern-day zooarchaeology.

An archaeological problem that Simon found himself investigating in Israel and Portugal has been the nature of bone deposition in Palaeolithic sites and, in particular, whether the accumulating agent was human or not. Luis Lloveras' paper explores the interplay between scavenger, non-human predator, and human predator activities. It proposes an approach to the exploration of this problem, through evidence produced by actualistic studies. The Pleistocene is also the subject of João Luís Cardoso's paper discussing the nature of hyaena's presence at Furninha Cave

(Portugal). The correct taxonomic identification is evaluated especially based on the history of the research at the site and in the wider area. In this paper there are echoes of Simon's work at Caldeirão Cave.

The Levantine connection is provided by two chapters. Tamar Dayan has been a close colleague of Simon for many years and she is contributing, alongside her collaborator Orr Comay, with a paper evaluating the potential of micromammals for palaeoenvironmental reconstructions. The case study they use is represented by the Palaeolithic site of Qesem Cave (Israel). Although the research questions that are addressed are different, the work on micromammals may be taken as a tribute to Simon's work on mice. Haskel Greenfield, Annie Brown, and Pierre de Miroschedji explore later times in the occupation of the southern Levant. Here, the connection with Simon's work is very direct as they provide an elaboration on Simon's original study of the animal bone assemblage from the Bronze Age site of Tel Yarmuth (Israel). The material is used to develop considerations about the connection between food consumption, butchery, and the origin of metallurgy, therefore providing an original contribution to the initial hypotheses concerning the nature of animal husbandry development in the area.

The Eastern Mediterranean is also represented by Angelos Hadjikoumis' methodological chapter on the scale of caprine husbandry in Cyprus. Angelos had the privilege of having Simon as his PhD external examiner and the two are linked by multiple research connections: Angelos' doctoral research at the University of Sheffield (UK) was based on Iberia but, like Simon, his research interests also spanned between the Eastern and Western Mediterranean (and the Atlantic). His paper also contributes to the range of methodological approaches on display in the book, as it is largely based on ethnoarchaeological research.

Salima Ikram's contribution on food consumption at the palace of King Amenhotep III at Malqata (Egypt) generates other interesting connections with Simon's work – though in a different space and time. Like in the case of Simon's investigation of English medieval castles, Salima explores the extent to which animal bones can provide us an insight into social status and dietary choice. There is a religious connection too, provided by the presence of a temple, which is parallel to yet another project Simon was involved with – his study of the animal remains from the Temple of Apollo in Cyprus.

The study of castles is even more explicitly referred to in Polydora Baker's chapter discussing the impact that the analysis of the Launceston Castle assemblage has had on our understanding of Cornish archaeology as well as the broader medieval and post-medieval research agenda in England. Polydora currently holds the job that used to be Simon's at Historic England (then English Heritage) and it is very valuable to have her contribution as a representation of Simon's 'English years'.

The remaining contributions (about half of the book) are focused on Iberia, which is fitting enough given the location of the conference and the 20 years of research Simon has dedicated to the area. We have been especially pleased to have João Zilhão, the man who brought Simon to Portugal, as part of this project. His paper also explores an area very close to Simon's heart – the introduction of caprine husbandry in Europe. Marta Moreno-García, who contributed to the creation of the zooarchaeology laboratory of the Institute of Portuguese Archaeology (IPA) from the very beginning, also deals with sheep and goat husbandry in Iberia, but for a later period (Iron Age). Her approach is mainly based on ageing and morphometry – areas to which Simon has provided a fundamental contribution. Ariadna Nieto-Espinet, Angela Trentacoste, Sílvia Guimarães, and Silvia Valenzuela-Lamas's paper is more diachronic and deals with cattle rather than caprines, but it has some similarities in approach, focused as it is on morphometry – probably the most prominent approach in Simon's opus. Nieto-Espinet *et al.* link morphometry with different geographic areas in Iberia, therefore providing us with the opportunity to understand potential connections between different human populations living in the region.

As mentioned, Simon has been very interested in the legacy of the Muslim occupation on Iberian history. The topic is explored by Simon's old friend, Marco Masseti, who investigates the introduction of new livestock breeds, as well as domestic and wild animal species during the *Al-Andalus* phase. Moving forward in time, the chapter by Cleia Detry, Ana Beatriz Santos, Tânia Casimiro, Ana Caessa and Nuno Mota focuses on the post-medieval period in Lisbon, a historical phase which has been central to Simon's interest in the last 30 years. He has done some biometrical work on the Carnide assemblage himself. Using archaeobotanical information to provide insights into animal husbandry and feed, João Tereso and Lídia Fernandes explore a historical event Simon has always shown a great interest in – the Lisbon earthquake of 1755.

Ana Arruda's chapter on the use of astragali is significant in the economy of this book for two main reasons. One is that it touches upon the symbolic and ritual use of bones, a theme that is only marginally explored in other chapters. Second is that it demonstrates yet another connection between East and West (Mediterranean and Atlantic), as astragali have been studied for their use as gaming and divination pieces especially in the Eastern Mediterranean, as Ana rightly points out, while this paper rather focuses our attention on the western side of the Mediterranean basin.

Finally, we have two papers on birds and fish, animals that Simon has always had a great interest in, without specialising in their study (though he can certainly identify bird bones!). The research topics that these papers investigate are, however, consistent with Simon's interests. The falconry

and M. Moreno-García (eds) *Paleoecologia Humana e Arqueociências. Um Programa Multidisciplinar para a Arqueologia sob a Tutela da Cultura. Trabalhos de Arqueologia* 29, 191–234.

Moreno-García, M., Pimenta, C., Davis, S.J.M. and Gabriel, S. (2003b) A osteoteca: uma ferramenta de trabalho. In J.E. Mateus and M. Moreno–García (eds) *Paleoecologia Humana e Arqueociências. Um Programa Multidisciplinar para a Arqueologia sob a Tutela da Cultura. Trabalhos de Arqueologia* 29, 235–261.

Pereira, F., Davis, S.J.M., Pereira, L., McEvoy, B., Bradley, D.G. and Amorim, A. (2006) Genetic signatures of a Mediterranean influence in Iberian Peninsula sheep husbandry. *Molecular Biology and Evolution* 23(7), 140–146.

Pereira, F., Queirós, S., Gusmão, L., Nijman, I.J., Cuppen, E., Lenstra, J.A., Econogene Consortium, Davis, S.J.M., Nejmeddine, F. and Amorim, A. (2009) Tracing the History of Goat Pastoralism: New Clues from Mitochondrial and Y Chromosome DNA in North Africa. *Molecular Biology and Evolution* 26(12), 2765–2773.

Ryder, M.L. (1968) *Animal Bones in Archaeology: A Book of Notes and Drawings for Beginners*. Oxford, Blackwell Scientific.

Watson, J.P.N. (1979) The estimation of the relative frequencies of mammalian species: Khirokitia 1972. *Journal of Archaeological Science* 6, 127–137.

Watson, J. and Davis, S. (2019) Shape differences in the pelvis of the rabbit, Oryctolagus cuniculus (L.), and their genetic associations. hal-01918838v2.

Watson, J. and Davis, S. (2020) On the relation between number of bones and number of taxa in zooarchaeology. hal-02964845.

2

Taphonomy of carnivores: understanding archaeological small prey accumulations

Lluís Lloveras

A large number of carnivore species, mostly terrestrial mammals and raptors, predate on small prey. These predators, whether through predation or scavenging, are recurrent sources of bone accumulations. Because carnivore nests or dens often occupied the same caves and rock shelters as human settlements, their prey remains are accumulated in the same spaces, often intermingled.

The importance of small prey in the diet of hunter-gatherers during prehistory is an important topic under discussion. This is a relevant subject, as information on the subsistence behaviours of human groups can also inform technology, communication, demographic trends and mobility. However, to understand the acquisition and consumption of small prey during different moments in prehistory, actualistic taphonomic studies that critically assess assemblage formation processes are needed. The comparison of bone modification patterns observed at archaeological sites to actualistic data allows us to suggest what predators, humans or others, were likely to have been responsible for the modifications registered.

Here, we undertake a review of the taphonomic actualistic studies conducted during recent decades and we highlight the importance of their contribution to understand human past behaviour.

Keywords: Terrestrial carnivores, raptors, small prey, taphonomy, prehistory

Introduction

This study deals with carnivores and small prey. Animals that eat other animals, whether through predation or scavenging, are considered carnivores. The term 'carnivores' often refers to the mammals of the Order Carnivora, which includes some 230 species, most of which fit the previous definition. However, considering that 'carnivore' means meat eater, many other taxa of predators and scavengers including different species of birds, reptiles, insects, fishes or other mammals not included in the order Carnivora are also carnivores (Ullrey 2004; Boitani and Powell 2012). Nevertheless, here we are not interested in all animal carnivore species but in those taxa that are able to accumulate small prey skeletal remains that may become part of the archaeological record. Why small prey? Because the main goal of the taphonomic actulistic studies which we are talking about in this paper is to understand archaeological small prey accumulations. When we talk about small prey we make reference to those small-sized animal species that could have been exploited by humans in the past. This mainly includes vertebrates; small mammals (rabbits and hares, hedgehogs, squirrels or marmots, among others), different species of birds, reptiles (turtles, snakes, lizards) and fishes. Furthermore, other prey including different species of invertebrates such as shellfish, snails and crustaceous are also taken into account.

Traditionally, small prey occurrence was undervalued in archaeological research because it was considered of little importance to human populations. The reason for this was precisely its small size. However, this vision changed and as far as subsistence is concerned, from the end of the Pleistocene innovative adaptations associated with anatomically modern humans were identified in southern Europe and the Levant, manifested especially as a diversification of resources. This was termed the Broad Spectrum Revolution (Flannery 1969) which is characterised by an expansion of dietary breadth, resulting in the procurement of a larger number of species, particularly through the greater exploitation of small game (Pérez Ripoll 2001; Aura *et al.* 2002; Hockett and Haws 2002; Stiner and Munro 2002; among others). More recently, different studies also showed evidence of different types of small-prey use prior to the arrival of anatomically modern humans in Europe (Blasco 2008; Sanchis Serra and Fernández Peris 2008; Cochard *et al.* 2012; Blasco *et al.* 2016; Lloveras *et al.* 2018c; among others). At present, the importance of these small animals in different moments of prehistory is a topic that is still under discussion. However, it seems clear that despite being small, small prey animals may have been very abundant and they could have been hunted in large quantities. For this reason, its substantial introduction into the human diet in some specific moments may have been an important advantage for hunter-gatherer groups' subsistence, facilitating population growth and territorial expansion (Villaverde *et al.* 1996; Aura *et al.* 2002; Hockett and Haws 2002).

A good example of small prey importance is the case of the rabbit (*Oryctolagus cuniculus*) in western Mediterranean regions and especially in the Iberian Peninsula. In most Palaeolithic and Epipalaeolithic sites from these regions huge numbers of rabbit remains have been recovered (Fa *et al.* 2013). Rabbits are a gregarious species that construct warrens where they live in large groups (up to 100 rabbits per warren) and they could have been 'easily' mass collected, making important contributions to human subsistence in the past. However, to understand the importance of rabbits and other small prey in the human diet first we need to know which is the agent responsible for their accumulations on archaeological sites. As these animals are an important source of food for a large number of non-human predators, the origin of fossil small prey remains is a key issue to solve first. Among carnivores, different species of small and medium-sized terrestrial mammals and diurnal and nocturnal raptors are the best candidates, because they normally prey on small animals and because most of them are regular visitors to caves and rock shelters where prey leftovers and pellets or scats may accumulate and become part of the archaeological record. Therefore, to understand human subsistence activities in the past it is essential to establish how a given faunal assemblage accumulated. In the last decades, after the analytical methodology developed by Andrews (1990), many systematic actualistic studies on modern small prey remains originating from different carnivores taxa activities have been conducted to identify the corresponding taphonomic signatures. In the present work we undertake a review of the patterns described in taphonomic actualistic studies conducted during recent decades and we will highlight the importance of their contribution to understand human past behaviour. We do not intend to make an exhaustive review of all neotaphonomic studies developed, which would be impossible in the space of this book chapter. We will focus principally on some of the prey that has been most deeply studied, such as rabbits and birds, and in the most important carnivores responsible for their accumulations.

Carnivores and small prey assemblages

Actualistic studies have shown that small prey assemblages can be very diverse. This diversity is mainly based on the behaviour of predators and it is evidenced by different features that when analysed will help to identify the predator responsible for an accumulation. Among these features we would like to highlight the composition of the faunal spectrum, the age of death of prey and the modifications produced on bones during feeding activities.

Composition of the small prey spectrum

The different taxa of carnivores able to accumulate small prey remains normally differ from the list of species hunted. For instance, among terrestrial mammalian carnivores from the Iberian Peninsula, the Iberian lynx (*Lynx pardinus*) is a very specialised predator that mainly preys on rabbits (85 to 100% of its diet) normally complemented with a few bird species (Delibes 1980). By contrast, the red fox (*Vulpes vulpes*) is a generalist or opportunistic predator that can feed on a wide prey and food spectrum, depending on supply. Their diet consists mainly of rabbits, small rodents, birds and other small game, fruit and vegetables, insects and worms (Rau *et al.* 1987). In the case of the European wildcat (*Felis silvestris silvestris*), diet studies showed that throughout its range, rabbits and small rodents are the wildcat's primary prey; however, birds and reptiles are also consumed (Lozano *et al.* 2006). Differently, the Eurasian otter (*Lutra lutra*) diet is mostly based on fishes and crustaceans complemented with amphibians, reptiles, birds, insects and some small rodents (Adrian and Delibes 1987). Also in the Iberian Peninsula, among raptors, the golden eagle (*Aquila chrysaetos*) has been reported to feed upon medium sized birds and mammals, mostly hares, rabbits, grouse, crows, partridges, foxes, young deer, marmots or squirrels (Lloveras *et al.* 2018a). However, studies about Bonelli's eagle (*Aquila fasciata*) have shown that rabbits, partridges and pigeons used to be the dominant prey (Lloveras *et al.* 2014a).

These differences in the list of taxa present in an accumulation are an important variable to take into consideration that will help to characterise the different predators. An exhaustive knowledge of modern predators' diet and ecological niches, together with taphonomic analysis where the whole small taxa recovered is considered, will be very helpful to detect the predator/s involved.

Age at death of the small prey involved

The age in small mammals is estimated taking into account the long bones epiphyseal fusion and teeth eruption and wear. In birds, the maturity of bones is estimated using the relative porosity of the bone. It is assumed that a clearly porous texture belongs to an immature bird, with an age from hatchling to nearly fledgling (Serjeantson, 2009). From neotaphonomic studies it can be checked how the age at death of some of the prey involved in an accumulation differs depending on the carnivore responsible. This mainly occurs with the larger mammal prey such as rabbits, hares or small carnivores. In the case of the rabbits, for example, the percentage of adult or immature individuals clearly varies among raptors. Some prefer adults as the golden eagle (about 85%) whilst in others immature rabbits are more abundant, as the Bonelli's eagle (about 40% of adults) (Lloveras *et al.* 2014a; 2018a). Also, in terrestrial carnivores differences are found. Red foxes normally prefer adults (about 85%) but in the Iberian lynx assemblages young and subadults dominate (Lloveras *et al.* 2008a; 2012). In assemblages created by the common genet (*Genetta genetta*) adults are represented by less than 10% (Pérez-García 2007). Also, when large predators such as the Golden eagle prey in medium-size carnivores such as the red fox, young individuals (about 95% of specimens) are normally preferred (Lloveras *et al.* 2018a).

In the case of birds, the age is more indicative of seasonality or related to assemblages produced by species that used cavities in the cave walls as breeding sites and become incorporated into the archaeological record after a natural death.

The age at death of prey is therefore another important source of information that researchers normally use when trying to establish the taphonomic signature of a determined predator.

Modifications produced on bones during feeding activities

Carnivores are heavy consumers of meat that they acquire from prey through predation or scavenging. Prey carcasses can be consumed close to the hunting space, but they can also be transported to be ingested in a quieter place, or to the den or nest to feed their young. Often, terrestrial carnivores accumulate large quantities of small prey bones in the entrance area and inside the den as well as raptors do in the nesting places. Besides, all these carnivores can ingest completely or partially their prey that, after digestion, is discarded in scats or pellets in which bone prey remains can be identified (Andrews and Evans 1983; Andrews 1990). This means that different types of materials are accumulated, that is, non-ingested parts of carcasses and fragmented and ingested bones that become part of the assemblage from the animal scats or pellets (Fig. 2.1). Often, a mixture of both, ingested and non-ingested parts of the animal carcasses are recovered, being one or the other more abundant depending on the area where the remains were recovered. In a fox den, for instance, digested remains are more abundant inside the den and in the latrine zone whilst non-ingested outnumber in the exit of the burrow and the surrounding area (Denys and Cochard 2017).

When these different types of accumulated materials are analysed from a taphonomic point of view it is evident that carnivores are very destructive agents. After their feeding activities the modifications produced on small prey include: (1) modification of the patterns of survivorship of bones; (2) different degrees of breakage; and (3) different types of bone surface alterations. They are all important variables to analyse in order to characterise small prey assemblages created by different carnivores.

Alterations produced by carnivores to small prey skeletal remains

As mentioned above, very different small prey can be found on an assemblage accumulated by a particular carnivore. Differences in relation to size, skeletal composition, bone density, etc. among different type of prey present are considerable. For this reason, the taphonomic pattern of each carnivore normally has been established separately for different types of prey: leporids (rabbits and hares), armadillos, birds, fishes, small rodents, etc. This brings us to a complex scenario of multiple prey and predator taphonomic signatures, difficult to summarise in the limited space of this book chapter. In all cases bone fragments are studied in order to detect different alterations present.

Anatomical representation

Anatomical representation is one of the useful variables that is normally studied because the skeletal elements present frequently vary depending on the predator responsible. The patterns of survivorship of bones may vary as consequence of processes of transport, disarticulation and dispersion of skeletal elements. Not always the whole carcasses are transported to dens or nests. But the bones present also differ as a result of differences in predator feeding behaviour, that is, differences in the parts ingested and destroyed by gnawing and digestion.

To assess the anatomical representation on an assemblage, following the calculus of the number of identified specimens present (NISP), the minimum number of elements (MNE) and minimum number of individuals (MNI), formulae are used (Dodson and Wexlar 1979) to calculate the relative abundance (or percentage of survivorship) for standardising anatomical profiles and comparing the frequencies of several elements by taxa. Also, to evaluate proportions of skeletal elements several indices or ratios are applied to both mammals and birds remains (Ericson, 1987; Andrews 1990; Bochenski, 2005; Lloveras *et al.* 2008a). From the results obtained, different studies have shown how depending on some variables such as the species of predator or the origin of the assemblage, the loss of skeletal elements differs.

In the case of rabbits (Table 2.1, Fig. 2.2), which are one of the most studied prey, comparisons reveal differences

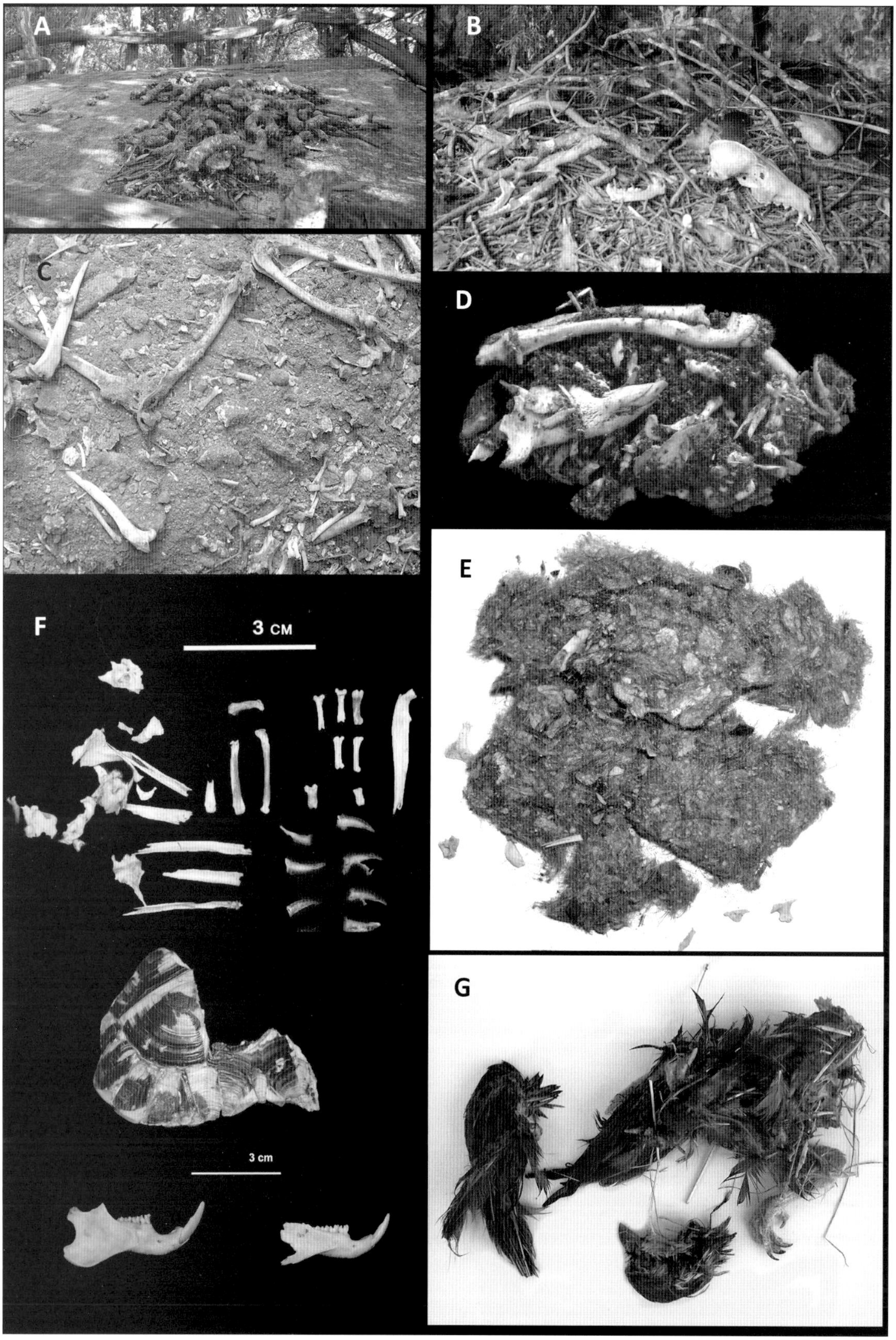

*Figure 2.1 A: Latrine with multiple scats accumulated by a Common genet (*Genetta genetta*). B: Small prey bones accumulated on a Golden eagle (*Aquila chrysaetos*) nest. C: Small prey bones accumulated on an Eagle owl (*Bubo bubo*) nest. D: Eagle owl pellet containing large numbers of ingested small prey remains. E: Disaggregated Iberian lynx scat with many bone prey fragments. F: Ingested and non-ingested small prey remains recovered from a Golden eagle nest. G: Non-ingested remains of a red-legged partridge consumed by an Iberian lynx.*

Table 2.1. Results of anatomical representation, breakage, digestion, teeth/beak/talon marks and age at death obtained on rabbit assemblages accumulated by different types of predators (raptors and terrestrial carnivores).

Rabbit assemblages	*Iberian lynx* Lynx pardinus		*Red fox* Vulpes vulpes		*European wildcat* Felis silvestris		*Geoffroy's cat* Leopardus geoffroyi	*Coyote* Canis latrans	
References	Lloveras *et al.* 2008a	Rodríguez - Hidalgo *et al.* 2013	Lloveras *et al.* *2012*		Lloveras *et al.* 2018b		Alvarez *et al.* 2012	Armstrong 2016	
Origin	Scats	Non-ingested	Scats	Non-ingested	Scats	Non-ingested	Non-ingested	Scats	Non-ingested
N	1522	9564	265	639	87	1457	339	3571	771
Mean RA%	49.2	50.9	25.4	25.6	15.2	75.2	40.2	62.7	
RA% >values	man-teeth-cra	tib-cal-mts	long bones	mts-ast-tib	sc-hu-ra	cr-fe-mts	cr-in-fe	cr-lb	ri-ve
PCRT/CR	+cranial	+postcranial	=	+postcranial	+postcranial	+cranial	+cranial	=	+postcranial
P/D	+proximal	+distal	+proximal	+distal	+proximal	+distal	=	−proximal	+proximal
AN/PO	+forelimb	+hindlimb	+hindlimb	+hindlimb	+forelimb	+hindlimb	+hindlimb	=	+hindlimb
Complete long bones %	2.5	37.6	0	5.4	0	23.7	45.5	0	0
Complete elements %	43	75	12	89.4	11.5	92.3	88.8	25.1	87.3
Average length (in mm)	7.1	17.4	9.3	19.3	5.4	21.5		7.9	15.6
%<10 mm	80	19.7	61	28	98.8	35	–	66.2	35.7
% Digested remains	97.2	–	99.5	–	98.6	–	–	94.4	–
% Digested long bones	100	–	100	–	100	–	–	–	–
Null	2.8	–	0	–	1.4	–	–	6	–
Light	12	–	6	–	1.4	–	–	15	–
Moderate	22	–	26	–	9.6	–	–	30	–
Heavy	43.8	–	43	–	39.7	–	–	34	–
Extreme	19.3	–	25	–	47.9	–	–	14	–
Pits & punctures	0.3	0.9	3	9.5	0	1.2	19.8	–	0.8
Age – % of adults	21.4	–	87	–	–	–	–	–	–

(Continued)

Table 2.1. Results of anatomical representation, breakage, digestion, teeth/beak/talon marks and age at death obtained on rabbit assemblages accumulated by different types of predators (raptors and terrestrial carnivores). (Continued)

Rabbit assemblages	*Bald eagle* Haliaeetus leucocephalus		*Great horned owl* Bubo virginianus		*Eagle owl* Bubo bubo	*S. Imperial eagle* Aquila adalberti	*Egyptian Vulture* Neophron percnopterus	*Bonelli's eagle* Aquila fasciata	*Golden eagle* Aquila chrysaetos
References	Armstrong 2016		Armstrong 2016		Lloveras *et al.* 2009	Lloveras *et al.* 2008b	Lloveras *et al.* 2014b	Lloveras *et al.* 2014a	Lloveras *et al.* 2018a
Origin	Non-ingested	Pellets	Non-ingested	Pellets	Nest	Pellets	Nest	Nest	Nest
N	249	2275	264	1976	1808	824	133	438	1543
Mean RA%		85.3		85.5	45.2	24.4	13.3	28.5	52.5
RA% >values	ri-ve	lb-ve-ri	ri-ve	ve-ri-cr	cal-inn-fem	phal 3-u mol	man-cra-teeth	cr-u mol-inn	inn-cal-tib
PCRT/CR	–	+postcranial	–	+postcranial	+postcranial	+cranial	+cranial	+cranial	+postcranial
P/D	=	+proximal	–	=	+proximal	+distal	–	+proximal	+distal
AN/PO	+hindlimb	+forelimb	+hindlimb	+forelimb	+hindlimb	+hindlimb	–	+hindlimb	+hindlimb
Complete long bones %	92.3	0	98.7	0	14.6	0	–	51.7	45.3
Complete elements %	76.8	47.1	85.6	24.1	53.9	27	81.2	74.7	68.2
Average length (in mm)	29.7	7	31.7	9.1	14.5	8.4	52.5	19.7	23.4
%<10 mm	24.6	77.8	32.8	55.6	49	73	7.8	54.9	44.7
% Digested remains	–	96	–	72.5	68.8	98	0.8	31.2	32
% Digested long bones	–		–		88.9	100	–	31	50.3
Null	–	4	–	27	31.2	2	99.2	68.8	68
Light	–	18	–	42	40.2	18.2	0	2.3	1.4
Moderate	–	32	–	22	19.8	46.8	0.8	7.9	4.3
Heavy	–	34	–	8	8	27.4	0	14.4	8.1
Extreme	–	12	–	1	0.7	5.6	0	6.5	18.2
Pits & punctures	0.6	–	2.3	–	2	0.5	7.5	0.8	1.1
Age – % of adults	–	–	–	–	50	–	100	41.4	83.5

among most predators. These differences are evident from different parameters. The average values of relative abundance for instance, show how in scat assemblages values vary from 15.2% for the wildcat to 62.7% for coyotes, indicating that wildcats produce much greater bone loss on this type of assemblages. By contrast, in non-ingested remains samples the relative abundance average in wildcat accumulations reach the highest value while the lowest is recorded for the red fox. In nest and pellets samples great differences are also recorded, values varying from 13.3% in Egyptian vulture nests to 85.5% in great horned owl assemblages.

Regarding the most frequent elements, coyote, Iberian lynx and red fox assemblages of non-ingested remains are characterised by low frequencies of cranial remains whilst these remains dominate wildcat and Geoffroy's cat samples. Wildcats differ from other predators because all skeletal elements, with the exception of the scapula, are better represented in non-ingested remains accumulations. When considering the scats samples of all these carnivores, the coyote assemblage displays higher percentages of most skeletal elements whilst the wildcat sample stands out for presenting the highest values of scapulae (Lloveras *et al.* 2008a; 2012; 2018b; Alvarez *et al.* 2012; Rodríguez-Hidalgo *et al.* 2013; Armstrong 2016).

From all nest assemblages (Table 2.1, Fig. 2.2), rabbit remains are best preserved in bald eagle and great horned owl accumulations. In Bonelli's eagle assemblages' innominates were the best represented element whilst tibiae, femora and calcanei were rare. Comparisons with the taphonomic signature of Spanish imperial eagle pellet samples also show some differences. This eagle tends to accumulate larger numbers of third phalanges and cranial remains. Differences are also found with the European eagle owl, in which the anatomical profile is characterised by lower percentages of cranial remains (especially teeth) and third phalanges (Sanchis 2000; Cochard 2004; Lloveras *et al.* 2008b; 2009; 2014a; 2014b; 2018a; Armstrong 2016).

In the case of birds (Table 2.2), the average values of relative abundance tend to be lower than those obtained for mammals, indicating a worse preservation of avian remains. As would be expected, bones are better preserved in non-ingested remains than in pellets and in nocturnal raptors than in diurnal raptors' assemblages. The highest values of relative abundance are displayed by the eagle owl. Among the skeletal remains recovered, sterna and pelvis are the best represented elements in nest and in non-ingested assemblages from most diurnal raptors. Regarding long bones, coracoids and humeri are usually the most frequent elements in non-ingested remains of diurnal raptors while tarsometatarsii are abundant in owl pellets (Laroulandie 2002; Bochenski 2005; Montalvo *et al.* 2011; Lloveras *et al.* 2014a; 2018a; Sanchis Serra *et al.* 2014). Results from the wing/leg ratio indicated that most diurnal raptors showed

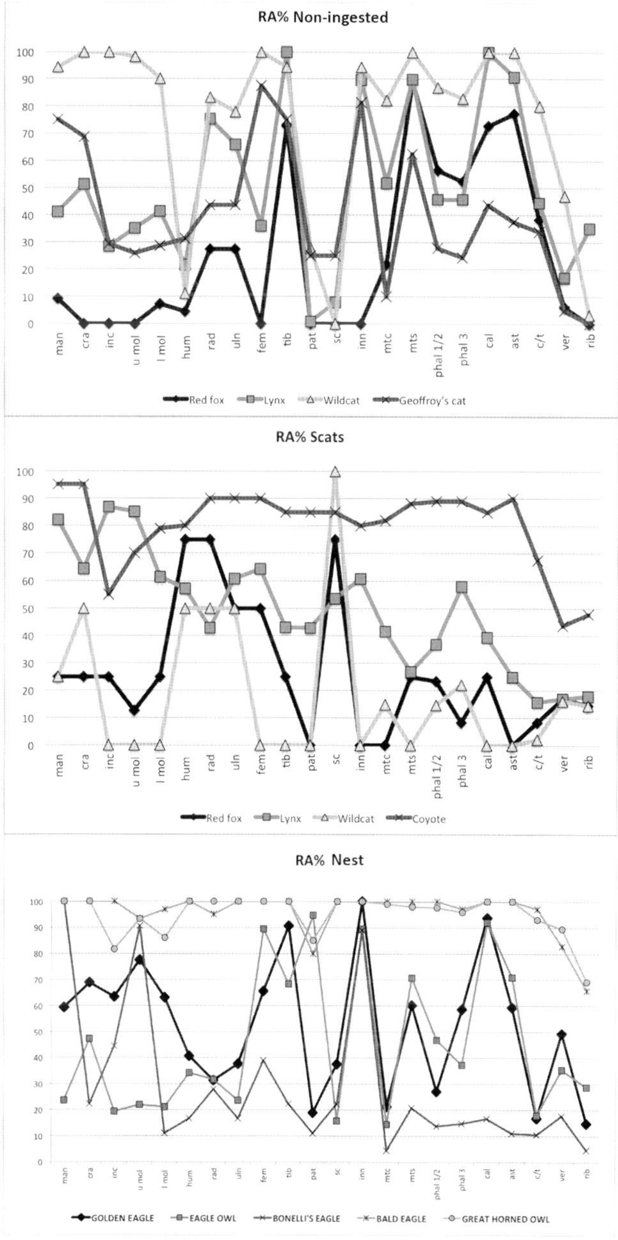

Figure 2.2. Comparisons of relative abundance (RA%) of different parts of the skeleton in rabbit samples originated by terrestrial carnivores and raptors (non-ingested, scats and nests). Abbreviations: man: mandible; cra: cranium; inc: incisors; u mol: upper molars; l mol: lower molars; hum: humerus; rad: radius; uln: ulna; fem: femur; tib: tibia; pat: patella; sc: scapula; inn: innominate; mtc: metacarpals; mts: metatarsals; phal 1/2: phalanges 1/2; phal 3: phalanges 3; cal: calcaneum; ast: astragalus; c/t: carpal/tarsal; ver: vertebrae, rib: rib.

a higher proportion of wing elements. In most owl pellets and Bonelli's eagle nest samples both parts are equilibrated. The preponderance of leg bones has only been recorded for imperial eagles in the pellets sample and in non-ingested

Table 2.2. Results of anatomical representation, breakage, digestion and teeth/beak/talon marks obtained on bird assemblages accumulated by different types of predators (raptors and terrestrial carnivores).

Bird assemblages	*Iberian lynx* Lynx pardinus	*Eagle owl* Bubo bubo	*Tawny owl* Strix aluco	*Long-eared owl* Asio otus	*Imperial eagle* Aquila heliaca	
References	Rodríguez-Hidalgo *et al.* 2016	Laroulandie 2002	Bochensky *et al.* 1993	Bochensky 2005	Bochensky *et al.* 1997	
Origin	Non-ingested	Pellets (only long bones)	Pellets (only long bones)	Pellets (only long bones)	Pellets	Non-ingested
N	128	321	3024	100	591	520
Mean RA%	36.7	63.9	–	–	–	–
>RA	rib-cr-pel	–	–	–	–	–
>RA long bones	sc-cmc	tmt-tbt-cmc	hu-ul	hu-ul	co-tmt	hu-co
Wing/Leg	L	=	=	=	L	W
Proximal/Distal	=	P	P	P	=	P
Core/Limb	L	L	L	L	L	L
Complete elements %	85	–	–	–	–	–
Complete long bones %	30.4	28	44.8	54.2	15.6	71
Average length (in mm)	16.6	–	–	–	–	–
%<10 mm	67	–	–	–	–	–
% Digested long bones	–	62.9	–	–	–	–
Pits & punctures	4	19.6	–	–	–	–
Bird assemblages	*Peregrine falcon* Falco peregrinus	*Egyptian vulture* Neophron percnopterus	*Bonelli's eagle* Aquila fasciata	*Golden eagle* Aquila chrysaetos	*Crested caracara* Caracara plancus	*White-tailed eagle* Haliaetus albicilla
References	Laroulandie 2002	Lloveras *et al.* 2014b	Lloveras *et al.* 2014a	Lloveras *et al.* 2018a	Montalvo *et al.* 2011	Bochenski *et al.* 2009
Origin	Non-ingested (only long bones)	Nest	Nest	Nest	Non-ingested	Non-ingested
N	114	46	385	94	504	2136
Mean RA%	30.3	12.8	49.9	13.9	17.3	_
>RA		str-pel-hu-co	str-pel-cr	str-pel-hu-co	str-hu-co-pel	str-pel-co
>RA long bones	hu-ul	hu-co	sc-co-hu	hu-co	hu-co	co-hu-sc
w/L	W	W	=	=	W	W
P/D	P	P	=	P	P	P
C/L	–	C	L	C	C	L
Complete elements %	–	58.7	49.9	45.7	–	71.6
Complete long bones %	64	73.3	14.4	63	63.8	92.3
Average length (in mm)	–	38.1	17.6	39.2	–	–
%<10 mm	–	–	35.4	41	–	–
% Digested remains	–	23.9	60.4	19.1	–	–

(Continued)

Table 2.2. (Continued)

Bird assemblages	*Peregrine falcon* Falco peregrinus	*Egyptian vulture* Neophron percnopterus	*Bonelli's eagle* Aquila fasciata	*Golden eagle* Aquila chrysaetos	*Crested caracara* Caracara plancus	*White-tailed eagle* Haliaetus albicilla
Null	–	0	28.6	80.9	–	–
Light	–	36.4	3.9	4.1	–	–
Moderate	–	63.6	17.5	18.6	–	–
Heavy	–	0	36.4	38.7	–	–
Extreme	–	0	13.6	14.5	–	–
Pits & punctures	36.8	28.3	6.2 (35% in NI)	14.6	27.6	9.1

remains accumulated by the Iberian lynx. Despite most predators accumulating more proximal elements; the ratio of proximal to distal elements indicates that in golden eagle accumulations proximal elements outnumbered distal bones to a much larger extent (>80%), which distinguishes this predator from other raptors. In nocturnal raptor created assemblages leg bones always outnumber core elements, but in diurnal raptors more variability is observed (Bochenski *et al.* 1997; 1999; 2009; Laroulandie 2002; Bochenski 2005; Montalvo *et al.* 2011; Lloveras *et al.* 2014a; 2018a).

Apart from rabbits and birds, different patterns of anatomical representation have also been established for other prey. Particularly, small rodents have been extensively studied and a large amount of data is available (see for instance Andrews 1990; Denys 2017; Denys and Cochard 2017). In general, these small mammals showed a different taphonomic pattern, which is related to their small size. Whole carcasses are transported to the nest or den, however a great loss of autopodium and axial skeleton elements is observed.

Breakage

The degree of breakage of small prey bones is another important variable studied in taphonomic analysis. It is evident that breakage produced by terrestrial carnivores will be heavier in remains accumulated from scats than in samples of non-ingested parts. Breakage increases as remains from scats are more abundant in an assemblage. Similarly, in samples originated by raptors, bones will be much more fragmented when coming from pellets. Considering this, three types of assemblages with different degrees of breakage can be found: (1) accumulations of non-ingested parts of carcasses; (2) accumulations of ingested remains from pellets or scats; and (3) a mixture of both, which is normally found, for example, in nest assemblages, where uneaten remains mix with regurgitated pellets. Besides, not all predators destroy the bones of their prey with the same intensity. Damage may vary depending on the predator/prey size relationship, the type predator (*e.g.* nocturnal raptors do not break bones as much as diurnal raptors or terrestrial carnivores do), the age, etc.

Different strategies can be used to assess the degree of breakage. It is usually described calculating the percentages of complete elements, isolated teeth (for mammals) and articulated elements (Andrews 1990). Likewise, the average maximum length of all identified skeletal elements and the percentage of remains under 10 mm are indicative of the breakage pattern in a bone assemblage (Lloveras *et al.* 2008a). Fragmentation of bones is also analysed using different bone fragment categories depending on bone type (for mammals and birds) in order to assess differences in the presence of certain bone portions (Andrews 1990; Bochenski *et al.* 1993; Lloveras *et al.* 2008a).

Comparisons reveal differences among most predators. In rabbit samples (Table 2.1, Fig. 2.3) the percentage of complete elements shows how the degree of breakage decreases in the following order: scats, pellets, nests, raptors non-ingested and terrestrial carnivores non-ingested assemblages. Some exceptions to this are the non-ingested accumulations originated by the Iberian lynx, which display even fewer complete bones than the scats assemblages do. This is probably because lynx scats contain large numbers of very small remains that are preserved complete. The lack of complete long bones in coyote non-ingested samples also stands out. That permits us to establish specific taphonomic signatures for each predator. The percentage of elements under 10 mm in a sample is also a good indicator that shows how breakage increase from felids scats to diurnal raptors pellets, to canid scats followed by nest accumulations and finally non-ingested remains. Also, exceptions to this tendency can be found, for instance, in the low percentage of small remains in the nest assemblages created by the Egyptian vulture (Fig. 2.3).

In bird samples (Table 2.2, Fig. 2.3), the percentage of complete elements also classifies pellets accumulations as the most fragmented followed by nest and non-ingested assemblages. In this case, two interesting exceptions are the avian remains from Egyptian vulture nests that are classified between non-ingested remains, and the non-ingested sample of bird bones created by the Iberian lynx that is situated among pellets samples. The latter is an interesting case that needs to be further explored, as up until now it is the only data about birds consumed by terrestrial carnivores.

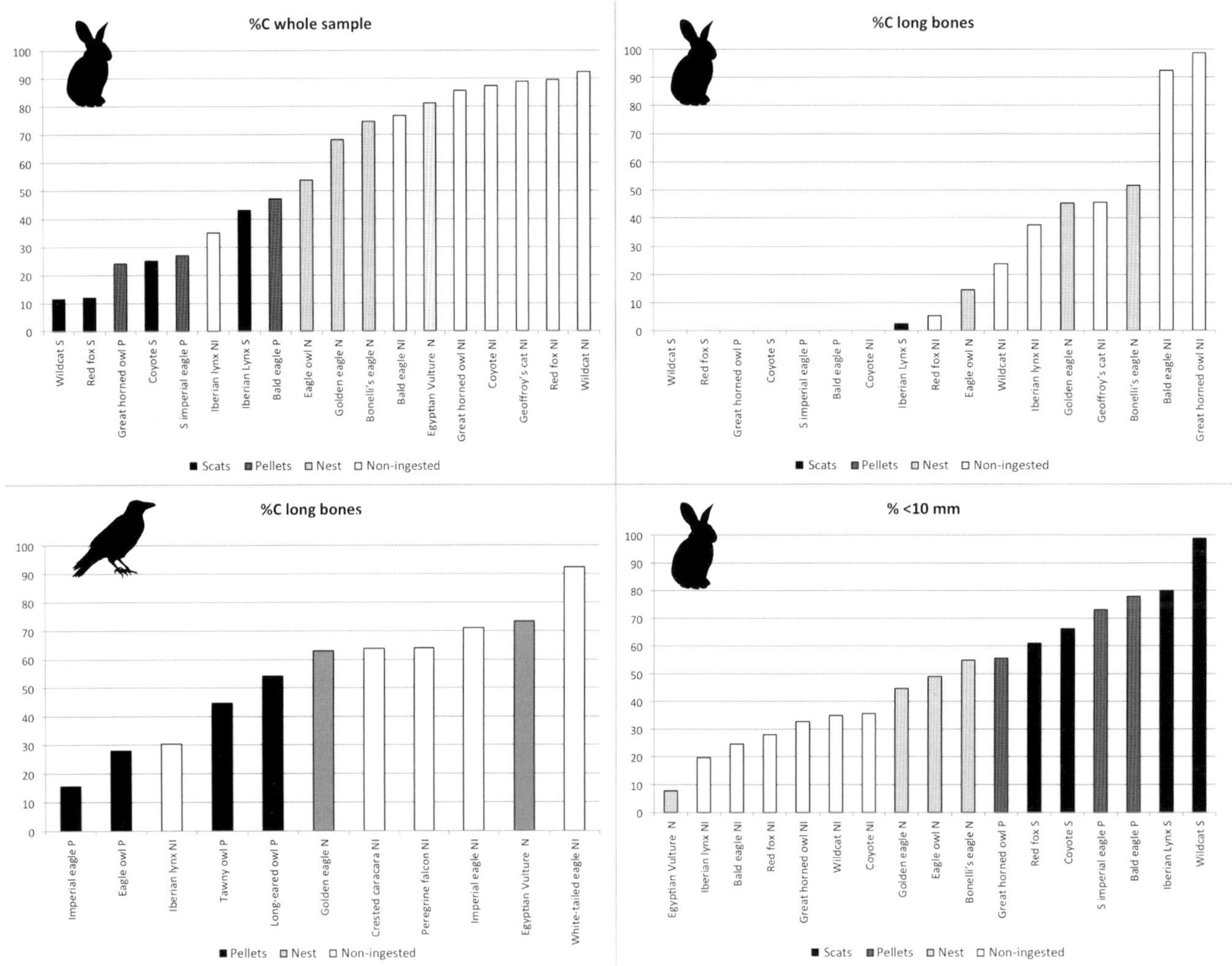

Figure 2.3. Comparisons of percentages of complete elements (%C) and percentages of elements with a maximum length less to 10 mm (%<10 mm) in different types of rabbit and bird assemblages (non-ingested, scats, pellets and nests) created by terrestrial carnivores and raptors.

Considering the grade of bone breakage in bird assemblages created by raptors, Bochenski (2005) divided the birds of prey into three groups. The first group included all diurnal raptor considering the bones found in their pellets. In this category, a high percentage of broken bones were registered. In contrast, the category of less modification also included diurnal raptor when the uneaten prey remains were considered. The results obtained in actualistic studies also show how specific bones are more fragmented than others or differences in the bone fragments present depending on the type of prey. Generally, crania, mandibles and long bones are more fragmented than metapodials, phalanges or vertebrae in most types of prey. However, in samples of small animals (such as small rodents or reptiles), large bones are less fragmented than in other larger prey samples such as rabbits (Andrews 1990; Denys 2017; Denys and Cochard 2017; Lloveras *et al.* 2018a). The size of the carnivores also plays a role in the breakage pattern, for instance, among nocturnal raptors, the barn owl (*Tyto alba*) and the eagle owl (*Bubo spp.)* display a larger proportion of complete (or slightly broken) small rodents skulls and mandibles compared to smaller owls (Denys 2017).

Bone surface modifications

The analysis of modifications caused by carnivores on the surface of small prey bones is very important to establish differences among predators. Alterations to the bone surface need to be systematically observed under a light microscope. Bone surface modifications mostly consist of two types of damage: digestion corrosion and marks caused by teeth or beaks and talons.

Digestion corrosion damage

Bones and teeth ingested by carnivores during feeding may be strongly damaged during digestion due to the action of the hydrochloric acid in the stomach of predators. Depending on the bone robusticity they may undergo only partial digestion damage or be completely destroyed by the action of the gastric juices. The partially digested bones and teeth are incorporated to carnivore's small prey assemblages into the animal's scats and pellets. The damage caused by digestion mostly consists of corrosion attrition that appears as pitting, cracking, polishing and rounding of fracture surfaces. This attrition is more conspicuous on the jagged ends of bones, which were broken during the chewing and ingestion process, and also on the intact articular ends (Andrews 1990; Fernández-Jalvo and Andrews 1992).

Andrews and Evans (1983) noted species differences in the degree of damage to digested skeletal remains, which are mainly due to the enzymes and acidity (pH) of the digestive tract of the predator. For this reason, different kinds of predators produce damage of variable intensity, and provide a character that can be used to distinguish between accumulating agents. In order to assess the importance of digestion corrosion in small prey assemblages, the percentage of digested bones and teeth are calculated. Also, the degree of damage is estimated applying different categories of digestion damage to bones and teeth. Specifically, five categories of digestion are distinguished: null (0); light (1); moderate (2); heavy (3); and extreme (4) (Fernández-Jalvo and Andrews 1992; Lloveras *et al.* 2008b).

From actualistic studies, Andrews (1990) established five categories of predators in relation to the damage they produce to skeletal remains after digestion. Category 1 was for the least amount of digestion damage recorded whilst category 5 was for the greatest. According to the results obtained by this author with micromammals, owls were normally placed in categories 1–3 while diurnal raptors and terrestrial carnivores in categories 4–5. Studies conducted with rabbit remains are consistent with these results. In Table 2.1 and Fig. 2.4 it can be observed how the highest values of digestion are obtained for terrestrial carnivores closely followed by diurnal raptors. The wildcat is the predator with the highest record of extreme degree digested remains. In some cases, terrestrial carnivores and diurnal raptors produce the same amount of damage, as in the case of the bald eagle and coyote. Nocturnal raptors samples, such as eagle owls, also display large numbers of digested remains but mostly in a light degree of damage. With bird remains reference data is scarcer (Table 2.2), however similar results have been obtained. Comparisons between rabbit and bird assemblages show that the same predator produce more damage by digestion on bird than on rabbit remains, probably due to the major fragility of bird bones. This is the case of eagle nest assemblages, where the percentage of heavy digested remains is clearly higher in bird samples (Fig. 2.4).

Tooth/beak/talon marks

During prey capture, transport and consumption predators create bone surface modifications that typically consist of punctures, pits, scores, notches and fractured and crenulated edges made with beaks, talons and teeth. Punctures are characterised as deep indentations that penetrate the bone cortical surface; they may breach the bone or only the cortical surface where crushed cortical and cancellous bone is visible around the base and margins of the puncture. Pits are characterised as shallow indentations that do not penetrate the entire bone cortical surface. Scores are linear marks that vary in length and have U-shaped cross-sections, often with smooth bottoms (Binford 1981; Andrews 1990).

The number and location of tooth/beak/talon marks vary depending on the predator involved. For this reason, the recognition and quantification of these alterations is also important to distinguish among different predators. In taphonomic studies damage to bone surfaces caused by beaks or talons is noted and counted and marks are classified according to the typology mentioned above. Punctures and pits are normally also classified by their number (isolated or multiple) and distribution (unilateral – *i.e.* located on one surface – or bilateral).

Actualistic studies show how differences are a consequence of diverse factors such as the predator involved (species, age, etc.), the type of prey or the origin of the assemblage. In ingested accumulations pits and punctures are rare because these samples are highly fragmented and besides marks become distorted or eliminated when bones pass through the digestive systems of the predator. Among terrestrial carnivores the Iberian lynx and wild cats produce similar percentages of damaged bones in non-ingested remains (0.9 and 1.2%), while the percentage of tooth damage is much higher in red foxes (9.5%) and Geoffroy's cat (19.8%) rabbit accumulations (Table 2.1). Within rabbit samples originated by the Iberian lynx it has been observed how the number of tooth marks increases (from 0.9% to 1.8%) with the presence of small marks (scores and pits) in samples created by a female lynx and her cubs during the denning season (Rodríguez-Hidalgo *et al.* 2016). The percentage of beak/talon marks found in nest assemblages of rabbits consumed by raptors is similar in diurnal raptors (close to 1%), but lower than in eagle owls that reach values around 2%. The highest numbers are those obtained for the Egyptian vulture, which reached values of 7.5–10.4% (Lloveras *et al.* 2014b; Sanchis Serra *et al.* 2014).

In bird assemblages the number of remains affected by beak/talon marks is clearly higher than in rabbits (Fig. 2.5). The greater percentages of marked bones are obtained for diurnal raptors, followed by nocturnal raptors, and terrestrial carnivores display the lower numbers (Table 2.2). Most of the studies

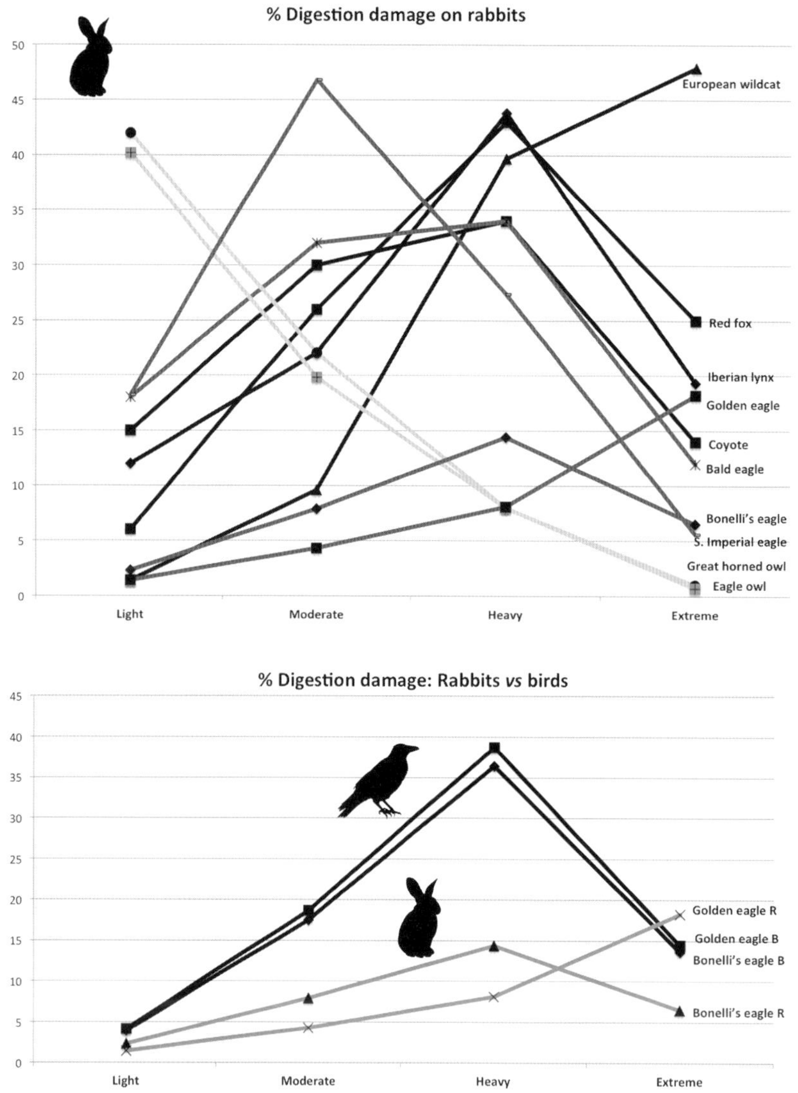

Figure 2.4. Comparisons of different degrees of digestion damage caused by terrestrial carnivores and raptors to rabbit and bird skeletal remains.

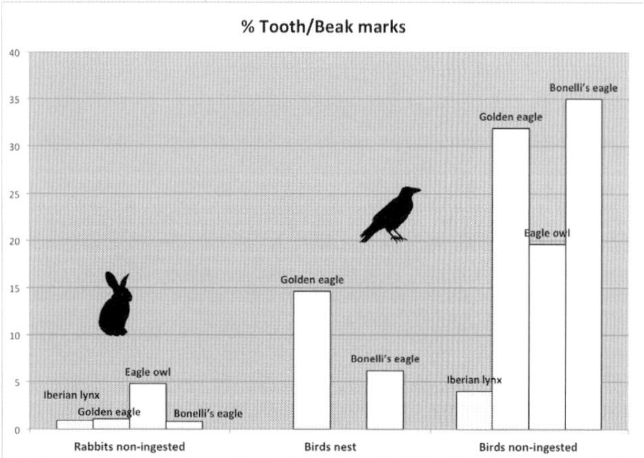

Figure 2.5. Comparisons of percentages of tooth/beak-marked bones on rabbit and bird assemblages originated by different type of predators.

show that the location of most beak/talon marks is on sterna and pelvis: Bochenski *et al.* (2009) found punctures on 70% of sterna and 38% of pelvis in white-tailed eagle (*Haliaaetus albicilla*) bird assemblages; punctures were observed on 39% of sterna and 51% of pelvis in golden eagles; Lloveras *et al.* (2014a) recoded punctures on 75% of sterna and 71.4% of pelvis in Bonelli's eagles nests. Coracoids, scapulae and humeri are usually the most affected long bones (Bochenski *et al.* 2009; Lloveras *et al.* 2018a). In assemblages of smaller prey (small rodents and reptiles) tooth or beak marks are absent or very rare (Andrews 1990; Lloveras *et al.* 2018a).

Conclusions

Being able to discriminate among taphonomic agents of bone accumulation is critically important to zooarchaeologists. The results commented in the above sections reveal that different predators are likely to differ in prey selection and in the extent and style of damage that they inflict on the skeletal remains of their prey. These differences are the key to establish the taphonomic pattern of each carnivore predator.

In the first chapter of the book *The Archaeology of Animals* (1987), Simon Davis commented on different methods and problems of the zooarchaeology. Among others, he raised the problem of the origin of archaeological bone assemblages and pointed to the importance of taphonomic studies to detect carnivores' contributions. More or less at the same time several pioneer studies appeared (Mayhew 1977; Dodson and Wexlar 1979; Andrews and Evans 1983; Andrews 1990). These studies started to characterise small prey assemblages, mostly micromammals, accumulated by different types of predators. Three decades later, in order to contribute to the solution of the problem raised by Davis (1987), the number of actualistic studies focused in establishing the taphonomic signature of different small prey predators have multiplied all over the world. Many thousands of different species of small prey skeletal remains have been analysed and the taphonomic pattern of many predators have been or are being created. At present, researchers working with archaeological faunal remains can find a large amount of reference data from these actualistic studies to apply to the analysis of small prey remains. These data are essential for many reasons, the identification of the predators involved in the origin of fossil accumulations permit us to obtain a more accurate image of local palaeoenvironments and to identify the fauna composition biases associated with predation. However, it is mainly important because it disproves or confirms human participation on the origin of small vertebrates accumulations in archaeological contexts. Data from actualistic studies are an essential tool to properly establish human subsistence strategies in the past.

The existence of an important quantity of taphonomic studies does not mean that all work is done. The number of carnivore predators is large and small prey taxa are still larger. Besides, the intra-specific variability existing (variability in small prey assemblages is introduced by both predator and prey type) makes the scenario more complicated and requires more studies in order to deepen our understanding each predator. Some carnivores have never been analysed and there are many predators and areas that have not been the subject of in-depth taphonomic analysis. For example, studies on bird remains consumed by terrestrial carnivores almost do not exist. This is an area that clearly needs to be further explored. Besides, in many archaeological assemblages, especially caves and shelters, it is possible that several mammalian and raptor carnivores take turns over time to use the space for nesting or denning creating palimpsests of different carnivores and anthropogenic contributions. All of this points to the necessity of increasing the neotaphonomic reference background to advance our knowledge of the role played by different predators on the formation of small prey assemblages.

Acknowledgements

I would like to thank Simon Davis for his teaching and friendship since I started to collaborate with the IPA many years ago. I would also like to thank the organisers of the meeting 'From the Mediterranean to the Atlantic: Themes in Old World Archaeology' and editors of this book for inviting me to participate on the celebration of Simon's book anniversary. I am very grateful to Jane Clarke for her valuable help with English edition. Financial support from research projects HAR2017-86509 from the Ministerio de Economía y Competitividad and SGR2017-00011 from the Generalitat de Catalunya, and from postdoctoral grant (BP-B00140-2014) from the Generalitat de Catalunya and the COFUND programme (Marie Curie Actions) are gratefully acknowledged.

References

Adrian, M.I. and Delibes, M. (1987) Food habits of the otter (*Lutra lutra*) in two habitats of the Doñana National Park, SW Spain. *Journal of Zoology* 212(3), 399–406.

Alvarez, M.C., Kaufmann, C.A., Massigoge, A., Gutiérrez, M.A., Rafuse, D.J., Scheifler, N.A. and González, M.E. (2012) Bone modification and destruction patterns of leporid carcasses by Geoffroy's cat (*Leopardus geoffroyi*): an experimental study. *Quaternary International* 278, 71–80.

Andrews, P. (1990) *Owls, Caves and Fossils*. London, Natural History Museum Publications.

Andrews, P. and Evans, E.M.N. (1983) Small mammal bone accumulations produced by mammalian carnivores. *Paleobiology* 9(3), 289–307.

Armstrong, A. (2016) Eagles, owls, and coyotes (oh my!): taphonomic analysis of rabbits and guinea pigs fed to captive raptors and coyotes. *Journal of Archaeological Science: Reports* 5, 135–155.

Aura, J.E., Villaverde, V., Pérez Ripoll, M., Martínez Valle, R. and Guillem, P. (2002) Big game and small prey: Paleolithic and Epipaleolithic economy from Valencia (Spain). *Journal of Archaeological Method and Theory* 9, 215–267.

Binford, L.R. (1981) *Bones: Ancient Men and Modern Myths*. New York, Academic Press.

Blasco, R. (2008) Human consumption of tortoises at Level IV of Bolomor Cave (Valencia, Spain). *Journal of Archaeological Science* 35(10), 2839–2848.

Blasco, R., Rosell, J., Rufà, A., Sánchez Marco, A. and Finlayson, C. (2016) Pigeons and choughs, a usual resource for the Neanderthals in Gibraltar. *Quaternary International* 421, 62–77.

Bochenski, Z.M. (2005) Owls, diurnal raptors and humans: signatures on avian bones. In T. O'Connor (ed.) *Biosphere to Lithosphere. New Studies in Vertebrate Taphonomy*, 31–45. Oxford, Oxbow Books.

Bochenski, Z.M., Boev, Z., Mitev, I. and Tomek, T. (1993) Patterns of bird bone fragmentation in pellets of the Tawny Owl (*Strix aluco*) and the Eagle Owl (*Bubo bubo*) and their taphonomical implications. *Acta Zoologica Cracoviensia* 36(2), 313–328.

Bochenski, Z.M., Huhtala, K., Sulkava, S. and Tornberg, R. (1999) Fragmentation and preservation of bird bones in food remains of the golden eagle *Aquila chrysaetos*. *Archaeofauna* 8, 31–39.

3

Fish bone studies in Iberia: an overview of 40 years of research from the LAZ-UAM (Madrid)

Arturo Morales Muñiz, Laura Llorente Rodríguez and Eufrasia Roselló Izquierdo

This paper reviews the development of archaeological fish analyses in the Iberian Peninsula. From almost total neglect until the early 1970s to the research now being done in terms of historical and ecological problems, fish remains studies have endured a slow and often not too straightforward development involving local and foreign scholars alike. Secular problems persist to this day and we argue that only through truly interdisciplinary collaboration will it be possible to address the many challenges, both cultural and biological, that fish remains can help explain.

Keywords: Archaeology, fish, fishing methods, trade, biogeography, conservation

Origins and early development of Iberian fish bone studies: a historical overview

Fish remains have been collected on Iberian sites since the end of the 19th century when archaeological excavations became routine within the discipline. As has been so often the case in Iberian archaeology, the Cantabrian region and the Portuguese shell middens of the Tagus and Sado valleys paved the way in terms of developments. For such reason, sites such as La Paloma, Altamira, Bolinkoba, Lumentxa, Santimamiñe and Samouqueira feature among the first to report fishes (Hernández Pacheco 1919; Aranzadi and Barandarián 1935; Breuil and Obermaier 1935; Marqués de Loriana 1941; but see also Ferreira 1956). In most cases, the scarce collections that hand collecting allowed were reported at best as 'pisces indet.' but already in the 1930s Telesforo de Aranzadi and José Miguel Barandiarán identified substantial numbers of Hake (*Merluccius merluccius*), Ballan wrasse (*Labrus berggylta*) and Needlefish (*Belone belone*) in Lumentxa cave (Barandarián, 1965).

The development of the New Archaeology and the flow of foreign archaeologists to Iberia that started in the mid-1960s did not change matters substantially. The interests of these researchers lay mostly with large mammals or, occasionally, molluscs (*i.e.* the most frequent groups in archaeological deposits). The fact that retrieval by hand kept on being the rule did not contribute to enhanced recovery of fish remains. Also, as was elsewhere the case until some three decades ago, fishes were not deemed important for Middle and Upper Paleolithic 'economies' thus their study never constituted a priority on the research agendas.

This scenario changed in the late 1960s when the veterinarians from the *Institut für Paläoanatomie, Domestikationsforschung und Geschichte der Tiermedizin* of Munich University led by Joachim Boessneck and Angela von den Driesch arrived in Spain. These scholars were called to study the faunal remains excavated by the archaeologists of the German Archaeological Institutes from Madrid and Lisbon. The Institute of Munich did not have reference collections of Iberian fishes at that time yet both scholars had done extensive work on Mediterranean sites, thus were fully acquainted with the archaeological relevance of fishes at places such as the Nile River. Since fish remains constituted substantial fractions on some of the Iberian coastal sites they analysed, they soon realised the need to address the study of these materials. Their initial move was to have Johannes Lepiksaar, a leading expert on the field, study their Iberian samples. In a matter of a mere three years out came three seminal papers on the Iron Age sites of Toscanos and Calle del Puerto and the Chalcolithic site of Zambujal that mark the onset of Iberian ichthyoarchaeological research proper (Lepiksaar 1973a; 1973b; 1976). These papers exemplify the virtues and drawbacks of the Central European School of doing archaeozoology. Among the former, one may mention a rigorous anatomical and taxonomic identification, the amount of fragmentation notwithstanding, the use of standardised measurements, and inferences about biological traits such as age, along with the analysis of pathologies. Among the drawbacks, these inductivists exhibited a certain reluctance to proceed with interpretations transcending bones. Obviously, this was far from a monolithic stand as Angela von den Driesch or Hans Peter Uerpmann conceded ample room for addressing cultural matters. But the senior author (AMM) recalls well Joachim Boessneck often repeating: 'let archaeologists make theories, that is none of our business!' In retrospect, one must admit that, in the case of fishes, such a theory-building exercise was a far-fetched task in the 1970s when the scarcity of studies did not allow one to proceed beyond singularities. Still, the reluctance to proceed beyond bone data, along with a tendency to publish in German, explains to some extent the surprisingly slight influence these scholars left on Iberian ichthyoarchaeologists. This despite works such as those on the fishes from Cueva de Nerja, Cerro del Mar and Munigua constituting seminal references of Paleolithic and Roman fish trade to this day (Boessneck and Driesch 1980a 1980b; Driesch 1980; Driesch and Boessneck 1985).

The early 1970s was the time when the senior author (AMM) became interested in fishes as part of his PhD research (Morales 1976). A substantial portion of that work took place at the Zoological Museum of Copenhagen where he was fortunate to join an extraordinary group of archaeozoologists, amongst which Knud Rosenlund was one of the few working on fishes. It was Knud who taught AMM the essentials of fish analysis and introduced him to Lepiksaar, his student Leif Jonsson and most of the remaining 'fish people' (*i.e.* Dirk Heinrich, Jean Desse and Dick Brinkhuizen). Knud also adviced AMM to visit Munich where a long-lasting friendship developed with Driesch and Boessneck (Fig. 3.1). As a member of the Scandinavian school of archaeology, Knud was keen on methods but also interpretations for which he stressed that context, before, during and after excavation, was critical. The two sentences Knud kept on repeating ('remember, an archaeozoologist must simply know about everything' and (talking about interpretation) 'if we can't do it, no one can') left a lasting impression on AMM. The senior author and, by extension, Iberian ichthyoarchaeology owes Knud Rosenlund a great debt of gratitude (Fig. 3.2).

Figure 3.1. Joachim Boessneck and Angela von den Driesch with (a very young!) Arturo Morales at the Deutsches Archäologisches Institut *in Madrid, 1979 (Photo DAI).*

Figure 3.2. Knud Rosenlund (Photo Arturo Morales).

At the start of the 1980s, a paper meant also as a call for action reviewed the state of fish bone studies after that first decade of effort (Morales 1983). There it was stressed the need for building up reference collections and revealed the critical role PhDs played in the development of the field serving for the most part as updated catalogues of finds (Uerpmann 1971; Altuna 1972; Driesch 1972; Davidson 1972). The year 1980 was a landmark for Iberian Archaeoichthyology. For one thing, it witnessed the publication of the three seminal papers by Boessneck and von den Driesch and also the study of the first systematically fine-sieved Iberian fish assemblage (Morales 1980). In July that year, AMM became a co-founder of the Fish Remains Working Group (FRWG) that became instrumental for fostering archaeoichthyology throughout the world.

Starting in the 1980s, Iberian archaeologists took an interest in faunas and started to routinely sieve and flotate sediments. Not surprisingly, fish remains started to emerge throughout the Peninsula, often in substantial numbers. Partly as a result of this, partly as the need to integrate faunal analysts in archaeological teams grew (a need that in Spain gained impetus once Atapuerca became the flagship of Iberian biological anthropology), the decade also witnessed the emergence of the first generation of Spanish fish analysts. Most of them had a strong regional focus and became best known for their PhDs. This was the case of Nuria Juan-Muns i Plans (Catalonia), Maria José Rodrigo (Valencia), Carmen Gloria Rodríguez Santana (Canary Islands) and Maria del Carmen Ferré Álvarez (Galicia) (Juan-Muns i Plans 1992; Rodríguez Santana 1994; Ferré Álvarez 2003). In Portugal, An Lentacker (University of Gent) completed a PhD on the fishes from the shell middens of the Tagus and Sado rivers (Lentacker 1986). These valuable contributions, paying substantial attention to archaeological detail, focused on restricted cultural periods (*i.e.* Paleolithic (Catalonia), Mesolithic (Portugal), Neolithic (Valencia), Iron Age (Galicia and Canary Islands)). Although their methodological proposals did not deviate from traditional protocols, the database they provided forever changed the scope of archaeoichthyological research in the Iberian Peninsula.

The Masters' and PhD theses of Eufrasia Roselló-Izquierdo (ERI) departed somewhat from that setup. Her Masters' was the first contribution of a long-term project addressing the comparative osteology of Iberian fishes for archaeoichthyological purposes (Roselló Izquierdo 1988a). Such work was important as a field tool and a means to neutralise the lack of reference collections. ERI's PhD and two additional ones she later supervised followed in the footsteps of that Masters' and designed dichotomic keys for the identification of isolated bones elaborating on the never-published proposals put forward by Johannes Lepiksaar (Roselló Izquierdo 1988b; 1989a; Cañas Díaz 1992; Patón Domínguez 1993).

Amongst the variety of methodological issues addressed in ERI's PhD, one may mention osteometry, taphonomy, size estimation and multivariate analyses. These combined with inferences about fishing, fish trade, biogeography, etc. to yield a proposal that exemplified the ways archaeoichthyology was carried out at the Laboratorio de Arqueozoologia of the Universidad Autónoma de Madrid (LAZ-UAM). The 11 fish collections that made up the bulk of the analyses derived from all parts of Iberia and ranged from the Paleolithic to the Middle Ages. The PhD also featured an updated catalogue of finds that, already by the late 1980s, allowed one to envision a time when Iberian archaeoichthyology could proceed from strict data gathering, description and singularities into the far more appealing realms of pattern seeking and data integration. Such fact notwithstanding, the Conclusions chapter on that PhD made it clear that the time for that kind of research had not arrived yet. In the late 1980s proceeding beyond singularities was still a rather far-fetched exercise for Iberian faunas and warning papers were written to pinpoint the dangers of doing so (*e.g.* Morales 1990a; 1990b). This was to no avail. Many Spanish archaeologists had shifted by that time from the almost total neglect of animal bones to a routine use of faunal data obtained by other people to back up theoretical proposals. One major trigger for this shift was an epistemological revolution taking place within archaeology we need to mention at this point.

Iberian ichthyoarchaeology: an epistemological digression

The initial steps of Iberian ichthyoarchaeology in the late 1960s and 1970s took place in the wake of a processualist current known as the New Archaeology. This aimed at focusing research on processes that one could analyse from a hypothetico-deductive perspective in which hypotheses only become testable under specific frames of reference.

The fish studies mentioned in the previous section conformed well with the processualist proposal. Fish remains were considered data proper, rather than sources of data by scholars whose ultimate goal centered upon obtaining as many and 'reliable' (*i.e.* falsifiable, refutable) data as possible. Of great concern in that context, for example, was the fact that diagnostic traits of fish bones could change with size (fish grow through life!) or geographical provenience. Failure to acknowledge those limitations could result in misidentifications with dangerous knock-down effects at higher levels of the interpretive ladder.

The one main drawback of the processualist approach was the suggestion to abandon normativist schemes affecting culture and focus instead on presumably 'invariable' phenomena, in particular diet. This implied restricting research to the most basic of the cultural spheres, namely that of subsistence. Reluctance to address issues revolving around the more fascinating social and symbolic realms (see Joachim Boessneck's comment above) allowed others less concerned with strict scientific procedures do the work.

From this perspective, it was unfortunate that the 1980s irruption of Iberian archaeologists into archaeozoology coincided with this shift taking place in the discipline from processualism to post-processualism. This meant a drastic change of scope and method for faunal analyses. The main problem is that post-processualism is not empirical science. Processualists use actualistic data, obtained through observation or experiment, to build models along the testable tenet of 'if…then….so'; later, the model is confronted with the archaeological record and validated, refined or rejected. Post-processualism, on the other hand, is a philosophical current whose intellectual roots lie in post-modern thinking, itself socially and linguistically conditioned by so-called 'production contexts'. Not surprisingly, post-processualism centers upon those epiphenomenal realms of culture (*i.e.* social and symbolic) that are by their nature unpredictable, as well as on historic processes that are singular by definition, thus incomparable. With such provisos in mind, the past is often created (*i.e.*, 'the Palaeolithic diet', 'the secondary products revolution', 'the Roman fishery', etc.) based on analogies and isolated bits of contextual information rather than on empirical data proper. For all practical purposes, the 'past' becomes an untested projection of a homogenised view of what often turns out to be a most diverse universe. For post-processualists one may say that truth lies in the eye of the beholder. A model, post-processualists contend, is always better than a bunch of disconnected data. The problem is that such models, being untestable by their very nature, are impossible to falsify. If the proposition 'all models are false, some are useful' is true, then the post-processualist tenets should be a matter of concern for all since usefulness, as truth, is often utterly idiosyncratic.

Post-processualist currents have had a far more lasting influence on Iberian archaeologists because, in contrast with countries such as the US or the UK, Iberian archaeology has been always deeply rooted within the humanities where the scientific method is not often a major issue. The shift from processualism to post-processualism, where models take priority over data, reached Iberian ichthyoarchaeology. Two examples will suffice to illustrate this.

The first case involves the so-called Roman 'fisheries' and Roman 'fish productions'. For Spanish archaeologists in the 1980s the paradigm here centered upon the 'Roman-tuna' binomium. A wealth of physical evidence in terms of *Tituli picti*, artwork, documents, etc. exist testifying to the relevance of tunas (presumably the Bluefin, *Thunnus thynnus*, L.) for the Roman fishing activities and trade industries (Marzano 2013). Yet equally evident is that many of those tuna fishing 'evidences', including amphorae, salting vats or fishing gear, served equally well to capture, process and trade many other kinds of fishes. More problematic for us at the time was that, with the exceptions of the aforementioned Calle del Puerto and Cerro del Mar, no tuna assemblages were available for study. This, in itself, was striking since the large tuna bones are unlikely to be missed even when retrieved by hand. The issue then was not how important tunas were for the Romans but rather whether ichthyoarchaeological evidence could confirm the data from non-faunal records. Most of those we 'talked' to (there were strong words against one of our presentations at a meeting in Ceuta, back in 1987) were convinced that fish bones could never add anything of relevance to Ponsich and Tarradell's seminal proposals of 1965. Our experience at the fish factory of Santa Pola best illustrates this clash of data against ideas. The salting vats of this Roman installation featured a thick crust of tiny fish, mostly sardines, but neither there nor anywhere else were tuna remains found. While our paper remarked upon such facts, the excavators failed to mention the abundance of small fishes yet devoted several pages to describe the fishing of tunas at Santa Pola (Roselló 1989; Sánchez *et al.* 1989). Data

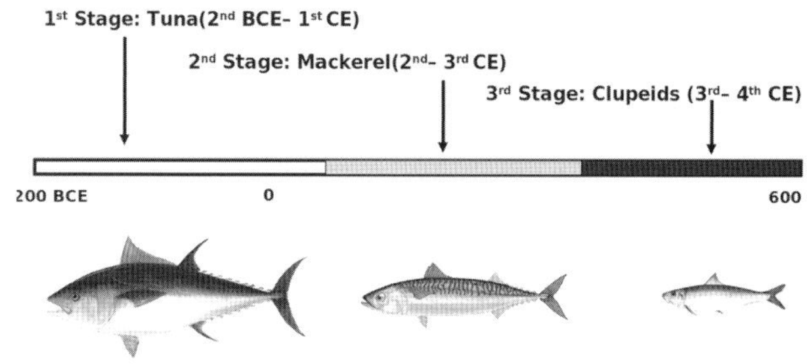

Figure 3.3. Temporal evolution of Roman fish productions. Time intervals define the dominant items in each stage given that all groups are documented throughout the Roman period in non-industrial Mediterranean fish assemblages (modified from Morales-Muñiz and Roselló-Izquierdo 2016). All drawings are public domain from the National Oceanic and Atmospheric Administration (tuna), Robert Cada (mackerel) and H. Gervais (1877) (sardine).

do not fare well when confronted with ideas and 'established wisdom'. In the early 1990s we even wrote a paper warning about the dangers of making unwarranted assumptions in the absence of data (Morales 1993). As predicted there, for the past 20 years Iberian sites have been yielding tuna remains and the story they unveil is far more interesting than the traditional paradigm made us believe. In this way, the greatest tuna accumulations systematically appear outside the factories (sensible: tuna bones are smelly, not edible and take a lot of space) and many other fish species abounded, often to the exclusion of tunas, on the processing sites. Diachronic studies also revealed that the 'industrial' fishing of tunas started centuries before with the Phoenicians, whose trade networks possibly allowed the marketing of surplus fishes. Likewise, and unsuspected by previous proposals, Roman factories evidenced shifts of migratory species through time for which neither satisfactory biological nor cultural explanations exist to this day (Morales and Roselló 2008; 2016; Fig. 3.3). As is so often the case, these insights added new lines of enquiry to the debate and proved instrumental for an intensification of the collaboration among scholars. Although matters are far from settled, most now concede that tuna was probably not even the main target of the Roman fish industries and that the fishing of non-migratory (*i.e.* local) taxa went on undisturbed at all times (Morales *et al.* 2017b; Garcia *et al.* 2018).

The power of ideas over data one also evidences in papers dealing with the nature of prehistoric fishing in the Cantabrian region. Again, the model put forward here was one of fishing focusing on a single 'species', in this case the Atlantic Salmon (*Salmo salar*, L.). The debate exemplified the old motto 'absence of evidence is not evidence of absence' but also the risk involved in taking data at face value, disregarding context. One noteworthy aspect most authors dealing with this issue failed to consider was that of sea level fluctuations (Galparsoro *et al.* 2010). Given that marine fishing is likely to leave most evidence near the shore, the fact that prior to the Neolithic the sea level of the Cantabrian coast was set below present-day levels, implies that most fishing evidence now lie underwater. For all practical purposes, sites we now call coastal were inland sites prior to the Neolithic. This seems as an issue of serious concern for anyone attempting to interpret the fish record. Another troublesome issue involves the agents of the fish accumulations. In archaeological sites, the tendency has been to consider fish remains evidence of fishing by humans, but this assumption may not always prove warranted. The problem is more pressing in the Cantabrian region where most prehistoric sites are located in caves that were occupied at times by other animals. To confirm that fishes are the sole result of human activity one needs to carry out taphonomical analyses yet very few are available for fishes in Cantabria as of this writing (Roselló and Brinkhuizen 1994; Roselló and Morales 2011b; 2014; Roselló *et al.* 2016). Interestingly, in all these sites, animals turned out to be always involved in at least part of the fish accumulations. Neither these taphonomical limitations nor the ludicrously small size of most collections or the Genus *Salmo* including three osteologically undistinguishable yet ecologically distinct species prevented reviews theorising extensively on the subject since the mid-1970s (Clark 1976; 1983; Strauss 1983; Meléndez *et al.* 1986; Strauss *et al.* 2002; Turrero *et al.* 2008; 2013; 2014; Adán *et al.* 2009; Álvarez Fernández 2011). Some of the most recent papers keep on characterising this fishing as an essentially freshwater activity while other papers restrict its practice to the Pleistocene-Holocene transition or postulate it as a (sic.) '*regular*' activity only from the Late Magdalenian onwards. No consideration is given here to time as a major agent of bone loss, inappropriate retrieval techniques, sea level changes or animal accumulators as biasing agents. Failure of these papers to mention those publications our group has produced over the past 40 years warning about these dangers may partly explain this inertia to interpret phenomena from the standpoint of questionable paradigms.

As was the case of the 'Roman tunas', our papers have been for years yielding bits of far more fascinating stories than the classical paradigms predicted. At the Santa Catalina rock shelter, for example, the 46 marine fish taxa (including no less than 33 species) question the salmon-focused fishing, revealing that presently non-Iberian species such as cod (*Gadus morhua*, L.) constituted one of the main targets of the 'fishery', on equal terms with salmon during the Upper Magdalenian (Roselló *et al.* 2016). Other papers not only evidence that marine fishes were taken far earlier, in the Aurignacian and Gravettian periods, but that even Neanderthals at the nearby site of Cueva Millán (northern Burgos) might have been fishing trout and carpfishes already by Mousterian times (Roselló and Morales 2005–2006; 2011b; 2016). Placing ideas before data has proved unfortunate – more so when data are wanting – also because established wisdom is difficult to eradicate. We still lack an encompassing theory to account for the fishing done during prehistory in the Cantabrian region. Until that moment arrives, however, one would do well to keep an open mind and remember that the golden midpoint standard between hyper-inductivism and model proposal lies not in post-processualism but with the hypothetical-deductive method of doing science. This notwithstanding the necessarily contingent nature of historical processes.

Iberian ichthyoarchaeology: developments from the 1990s until today

The developments in Iberian archaeoichthyology taking place from 1990 onwards exhibit a diversification at the methodological, geographical and temporal levels that makes it difficult to thread a coherent, unitary account. The number of collections studied after a careful retrieval of remains has increased steadily. This quantitatively and qualitatively expanded the databases, allowing for incipient/restricted inter-site comparisons. The collaboration between excavators and faunal analysts shifted from strict post-excavation feedback to discussing ideas and excavation strategies prior to and during excavation. As a result, both archaeologists and fish bone analysts are now more fully aware of each other's aims and methods, the synergies emerging from such awareness often shared with historians, molecular biologists, geologists, etc.

Obviously, not all excavators have been equally sensitive to these developments, the Palaeolithic prehistorians often leading the way, but matters do seem to have been improving on all fronts since the 1990s. At the risk of oversimplifying we will attempt to provide a cursory review of developments grouping these according to five operational, though not necessarily excluding, categories namely: (1) site reports; (2) methods; (3) reviews; (4) specific issues plus interpretation; and (5) integration of fish studies into broader scope analyses. In this way:

1. Iberian archaeoichthyology started with site reports and these have kept on constituting the bulk of the effort to this day. A secular trend with an emphasis on prehistoric sites in the early years and a gradual move into more recent times with medieval sites and, above all, Modern Age sites clearly lagging behind. A move started in recent years to push fish studies back to times before the appearance of Anatomically Modern Humans (Roselló and Morales 2005–2006; Cortés *et al.* 2011).
2. Iberian archaeoichthyology has traditionally refrained from methodological proposals. The majority of these deal with taphonomical issues in particular the detection of signatures that set apart fish assemblages produced by humans from those produced by other agents (Morales

1984; Van Neer and Morales 1992; Morales and Van Neer 1994; Morales and Roselló 1998a; Morales *et al.* 1999; Russ 2010; Roselló *et al.* 2015; Guillaud *et al.* 2018).

3. Despite the comparatively small number of site reports, reviews have not been scarce in Iberian archaeoichthyology. Most of the aforementioned PhDs were conceived as reviews of a particular area but our group at LAZ-UAM has been producing reviews of various kinds since the 1983 paper of AMM. Some reviews are regional in scope (Morales and Roselló 2006; 2008; 2009; 2012; 2016) whereas others focus on a particular species (Morales *et al.* 1991; Ludwig *et al.* 2011). A substantial number of reviews were meant to inform lay readers (basically archaeologists) about the possibilities fish studies offer and their state of development in our two countries (Morales and Roselló 1988b; 2007), being often difficult to set apart these from reviews focusing on more specific issues (Morales *et al.* 2016).
4. Few papers deal only with data interpretation although interpretation makes part of site reports and a substantial portion of PhDs. A few papers are restricted to interpretations deriving from the application of methods such as ancient DNA or trophic level analysis (Morales and Roselló 2004; Ludwig *et al.* 2009) and comparative analyses of specific issues may also qualify within this category (Morales *et al.* 1994; Morales *et al.* 2007; Morales 2010; García *et al.* 2018). Additional subjects include paleo-environmental inferences (Rodrigo 1994; Kettle *et al.* 2006), choice (Roselló *et al.* 1994), artwork (Morales and Roselló 1984/85) and the shaping of fish assemblages by shifting sea levels (Roselló and Morales 1994b; 2014; Sternberg 2007).
5. The integration of fishes within a general context, be this local (site) or regional, is the most recent development of archaeoichthyological research in Iberia. The pioneer works at the sites of Cueva de Nerja and Castillo de Doña Blanca became possible because AMM and ERI were both members of the archaeological teams (Roselló and Morales 1994a; Pellicer and Morales 1995). In fact, it took almost another decade and a half before fishes became routinely integrated into holistic site analyses (Cortés *et al.* 2008; see also Roselló and Morales 2011a; 2015; Cortés *et al.* 2012; Roselló *et al.* 2015; 2016). At this time, analysts from the historical and biological sciences have converged on a project to study a species of commercial interest, as is the hake (Morales *et al.* 2017a).

With the exception of the 1983 review and a methodological paper from 1984, all of the aforementioned papers are post-1990 exemplifying a phenomenon we have insisted all along this paper, namely that until a certain threshold is reached in terms of the amount of data available, no meaningful elaboration of data can be reliably attempted.

Future developments: a wealth of possibilities, both old and new

The Iberian Peninsula is a privileged place for a fish analyst to work. The geographical location, straddling the Atlantic and the Mediterranean, but also bridging Africa with Europe, offers multiple possibilities to address phenomena dealing with fishes and fishing. The history and 'co-evolution' of Iberians and fish is as old as it is complex and moves beyond Iberia after Spain and Portugal became maritime powers in the 15th century. It is the duty and the challenge of the faunal analyst to turn this richness into fruitful work.

Challenges do not stop at this level. The pace of scientific developments is now so fast that almost every day new methods arise opening possibilities we could barely imagine a few years ago. Very often, however, new methods also allow one to frame old debates from a new perspective. Take the aforementioned issue on a preferential fishing of 'salmon' in Cantabrian prehistory. Techniques such as ancient DNA and ZooMS (Zooarchaeology by Mass Spectrometry) should allow one to not only determine the species of a great majority of the bones that today remain identified only at the level of the Genus, but also the thousands of splinters that, at best, qualify only as 'pisces indet.'. Doing so would allow one to determine what salmonids are present at a given site and roughly in what proportions. These data, in turn, allow the analyst to better calibrate the importance of each taxon at any given assemblage, correlate specific abundances with all sorts of contextual data, and allow for a more reliable comparison of assemblages.

The same reasoning applies to stable isotope analyses. Their use in archaeological fishes has been mostly restricted to determine the contribution of aquatic resources to human diet and even in these cases methodological problems remain (Richards *et al.* 2005). Some of these have to do with shifting values among species (sardines and tunas have very different trophic levels, right?), but also within a single species since juveniles and adults often feed on very different prey. Other problems have to do with variable isotopic values within the different amino acids of a given protein, etc. Be it as it may, the possibilities these studies offer are enormous and increasing every day. These include identification of fish species according to stock to determine whether the fishing at a given site is local or comes for other grounds, trade networks, etc. (Barrett *et al.* 2008; 2011; Sisma-Ventura *et al.* 2018). Stable isotopes harbour a huge potential to unveil the nature of former, though not necessarily 'pristine', aquatic environments, the structure of former food webs and how fish size changed through time. If the Compound Specific Isotopic Analysis technique (CSIA) of nitrogen stable isotopes (δ15N) proves feasible for archaeological fish remains, stable isotopes may even allow the evaluation of 'top-down' effects (*e.g.* overfishing) and 'bottom-up' effects (climate change and pollution levels). Although these biomolecular approaches are still in its infancy when it comes to Iberian fish assemblages, the work in progress as of this writing looks promising (Llorente Rodriguez in prep.).

While new methods develop, site reports must keep on. Spain and Portugal are still lagging way behind European countries and, as said, without adequate databases, comparative analysis and data elaboration may falter. The attractiveness of the new methodologies, also in terms of funding possibilities, should not lead us into thinking that the work thus far accomplished is enough. On the contrary, it may so happen that, in terms of impact factors, future fish analyses from pre-sapiens hominines may prove as relevant as biomolecular analyses of the most sophisticated kind.

To achieve good results on all fronts will also requires good reference collections – a secular drawback of Iberian Archaeoichthyology – and intensive collaboration with scholars other than archaeologists. The historical archives and humanities libraries constitute data treasures that only historians are in a position to tap. Equally crucial, now that archaeologists realise the possibilities of archaeological faunas for addressing wildlife management and conservation issues, would be to make ecologists, fishery biologists and wildlife managers coalesce into joint research groups to create the synergies necessary to design holistic research projects to protect our natural heritage.

With such hope in mind we conclude this contribution.

Acknowledgements

We want to thank the organisers of the conference 'From the Mediterranean to the Atlantic: Themes in Old World Zooarchaeology' for their invitation to such a productive meeting. Not only was it successful but equally important friendly. We are also grateful for participating in an event to honor Simon Davis and his immense contribution to archaeozoology in general and Iberia in particular.

This research benefited from Grants HAR 2014-55722-P (*Ictioarqueologia de la Prehistoria cantábrica: Modelos para la caracterización de las primeras pesquerías europeas*), HAR 2017-88325-P (*Historical archaeobiology of the European hake* (Merluccius merluccius, *L.1758) in the NE Atlantic: The Iberian evidence (AD 10th–18th*) from the Spanish Ministerio de Economía y Competitividad. The European Commission supported L. Llorente through the FISHARC-IF 658022 Marie-Curie-Sklodowska-If fellowship for Career development.

References

Adán, G.E., Álvarez-Lao, D., Turrero, P., Arbizu, M. and García-Vázquez, E. (2009) Fish as diet resource in North Spain during the Upper Paleolithic. *Journal of Archaeological Science* 36(3), 895–899.

Altuna, J. (1972) Fauna de los yacimientos de Guipúzcoa. *Munibe* XXIV (1–4), 1–464.

Álvarez-Fernández, E. (2011) Humans and marine resource interaction reappraised: Archaeofaunal remains during the late Pleistocene and Holocene in Cantabrian Spain. *Journal of Anthropological Archaeology* 30, 327–343.

Aranzadi, T. de and Barandiarán, J.M. de (1935) *Exploraciones en la Caverna de Santimamiñe (Basondo-Cortezubi).* In: *3ª Memoria-Yacimientos Azilienses y Paleolíticos. Exploraciones en la Caverna de Lumentxa (Lequeitio).* Bilbao, Excma. Diputación de Vizcaya.

Barandiarán, J.M. (1965) Excavaciones en Lumentxa (campaña de 1963). *Noticiario Arqueológico Hispánico* 7, 56–61.

Barrett, J., Johnstone, C., Harland, J., Neer, W.V., Ervynck, A., Makowiecki, D., Heinrich, D., Hufthammer, A.K., Enghoff, I.B., Amundsen, C., Christiansen, J.S., Jones, A.K.G., Locker, A., Hamilton-Dyer, S., Jonsson, L., Lõugas, L., Roberts, C. and Richards, M. (2008) Detecting the medieval cod trade: a new method and first results. *Journal of Archaeological Science* 35, 850–861.

Barrett, J.H., Orton, D., Johnstone, C., Harland, J., Van Neer, W., Ervynck, A., Roberts, C., Locker, A.M., Amundsen, C., Bødker Enghoff, I., Hamilton-Dyer, S., Heinrich, D., Hufthammer, A.K., Jones, A.K.G., Jonsson, L., Makowiecki, D., Pope, P., O'Connell, T.C., de Roo, T., and Richards, M. (2011) Interpreting the expansion of sea fishing in medieval Europe using stable isotope analysis of archaeological cod bones. *Journal of Archaeological Science* 38, 1516–1524.

Boessneck, J. and Driesch, A. von den (1980a) Tierknochenfunde aus vier südspanischen Höhlen. *Studien über frühe Tierknochenfunde von der Iberischen Halbinsel* 7, 1–83.

Boessneck, J. and Driesch, A. von den (1980b) Knochenfunde aus dem römischen Munigua (Mulva), Sierra Morena. *Studien über frühe Tierknochenfunde von der Iberischen Halbinsel* 7, 160–185.

Breuil, H. and Obermaier, H. (1935) *La cueva de Altamira.* Madrid, Tipografía de Archivos.

Cañas Díaz, J.M. (1992) *Contribución al Atlas Osteológico de los Teleósteos ibéricos II. Osteología Comparada de la Familia de los Labridae Ibéricos.* Unpublished PhD thesis, Universidad Autónoma de Madrid. Madrid.

Clark, G.A. (1976) *El Asturiense Cantábrico. Bibliotheca Prehistorica Hispana* XIII. Madrid, Consejo Superior de Investigaciones Científicas, Instituto Español de Prehistoria.

Clark, G.A. (1983) *The Asturian of Cantabria. Early Holocene Hunter-Gatherers in Northern Spain.* Anthropological Papers of the University of Arizona, Number 41. Tucson, The University of Arizona Press.

Cortés, M., Jiménez, F., Simón, M.D., Gibaja, J.F., Faustino, A., Martinez, F., Rodrigo, M., Flores, J.A., Paytan, A., López, J.A., Peña, L., Carrión, J.S., Morales, A., Roselló, E., Riquelme, J.A., Dean, R.M., Salgueiro, E., Martínez, R.M., De La Rubia, J.J., Lozano, M.C., Vera Peláez, J.L., Llorente, L., and Bicho, N.F. (2012) The Mesolithic-Neolithic transition in southern Iberia: A Maghrebian neolithisation wave coupled to the 7.4 ky cal. BPclimatic event? *Quaternary Research*, 77 (2): 221–234. doi:10.1016/j.yqres.2011.12.003.

Cortés, M., Morales, A., Simón, M.D., Bergadà, M.M., Delgado, A., López, P., López-Sáez, J.A., Lozano, M.C., Riquelme, J.A., Roselló, E., Sánchez, A. and Vera, J.L. (2008) Palaeoenvironmental and cultural dynamics of the coast of Malaga (Andalusia, Spain) during the Upper Pleistocene and Holocene. *Quaternary Science Reviews*, 27(23–24): 2176–2193. doi: 10.1016/j.quascierev.2008.03.010.

Cortés-Sánchez, M., Morales-Muñiz, A., Simón-Vallejo, M.D., Lozano-Francisco, M.C., Vera-Peláez, J.L., Finlayson, C., Rodríguez-Vidal, J., Delgado-Huertas, A., Jiménez-Espejo, F.J., Martínez-Ruiz, F., Martínez-Aguirre, M.A., Pascual-Granged, A.J., Bergadà-Zapata, M.M., Gibaja-Bao, J.F., Riquelme-Cantal, J.A., López-Sáez, J.A., Rodrigo-Gámiz, M., Sakai, S., Sugisaki, S., Finlayson, G., Fa, D.A. and Ferreira Bicho, N. (2011) Earliest known use of marine resources by Neanderthals. *PlosONE* 6(9): e24026. doi:10.1371/journal.pone.0024026.

Davidson, I. (1980) *Late Paleolithic Economy in Eastern Spain*. Unpublished PhD thesis, University of Cambridge, Cambridge.

Driesch, A. von den (1972) Osteoarchäologische Untersuchungen auf der Iberischen Halbinsel *Studien über frühe Tierknochenfunde von der Iberischen Halbinsel* 3, 1–212.

Driesch, A. von den (1980) Osteoarchäologische Auswertung von Garum-Resten des Cerro del Mar. *Madrider Mitteilungen* 21, 151–154.

Ferré Álvarez, M.C. (2003) *Contribución al estudio de la Arqueictiofauna Holocena en Galicia*. Unpublished PhD thesis, Departamento de Bioloxía Animal, Universidade de Santiago de Compostela.

Galparsoro, I., Borja, A., Legorburu, I., Hernández, C., Chust, G., Liria, P. and Uriarte, A. (2010) Morphological characteristics of the Basque continental shelf (Bay of Biscay, northern Spain), their implications for Integrated Coastal Zone Management. *Geomorphology* 118, 314–329.

Ferreira., O. da Veiga (1956) – Faune Malacologique; Crustaces et Poissons. Muge (Moita do Sebastião). *Congreso internacional de Ciencias Prehistóricas y Protohistóricas*, 339–353. Zaragoza.

García, E., Roselló, E., Bernal, D. and Morales, A. (2018) Salazones y salsas de pescado en la antigüedad. Un primer acercamiento a las evidencias de paleocontenidos y depósitos primarios en el ámbito euro-mediterráneo. In D. Bernaland A. Arévalo (eds) *Las factorías de salazones de Traducta. Primeros resultados de las excavaciones arqueológicas en la c/San Nicolás de Algeciras (2001–2006)*, 267–292. Junta de Andalucia. Sevilla.

Guillaud, E., Morales-Muñiz, A, Roselló-Izquierdo, E., Béarez, P. and Moncel, M.H. (2018) Taphonomy of Yellow-legged Gull (*Larus michahellis* Naumann, 1840) pellets from the Chafarinas Islands (Spain). *Canadian Journal of Zoology* 97(2), 100–111, https://doi.org/10.1139/cjz-2018-0139.

Hernández Pacheco, E. (1919) *La Caverna de La Peña de Candamo (Asturias)*. Madrid, Comisión de Investigaciones Paleontológicas y Prehistóricas 24, Museo Nacional de Ciencias Naturales.

Kettle, A.J., Roselló-Izquierdo, E., Morales-Muñiz, A., Heinrich, D. and Vollestad, A. (2010) Refugia of marine fish in the Northeast Atlantic during the Last Glacial Maximum: Concordant assessment from archaeozoology and palaeotemperature reconstructions. *Climate of the Past Discussions* 6, 1–39.

Juan-Muns i Plans, N. (1992): *La pesca com alternativa económica per als Yàmana, nòmadas canoers del Canal Beagle (Tierra del Fuego, Argentina)*. Unpublished PhD thesis, Universitat Autónoma de Barcelona, Barcelona.

Lentacker, A. (1991) Archeozoologisch Onderzoek van Laat-Prehistorische Vindplaatsen Uit Portugal. PhD thesis, RUG Faculteit Wetenschappen, Laboratorium voor Paleontologie, Gent.

Lepiksaar, J. (1973a) Fischreste aus einer tartessischen Siedlung in Huelva. *Studien über frühe Tierknochenfunde von der Iberischen Halbinsel* 4, 32–34.

Lepiksaar, J. (1973b) Fischknochenfunde aus der phönizischen Faktorei von Toscanos. *Studien über frühe Tierknochenfunde von der Iberischen Halbinsel* 4, 109–119.

Lepiksaar, J. (1976) Die Fischknochen aus Castro do Zambujal. *Studien über frühe Tierknochenfunde von der Iberischen Halbinsel* 5, 139–142.

Loriana, Marqués de (1941) La cueva de Bolinkoba. Un yacimiento vizcaíno inédito. *Archivo Español de Arqueología* 14, 194–507.

Ludwig, A., Arndt, U., Debus, L., Roselló, E. and Morales, A. (2009) Ancient DNA analyses of Iberian sturgeons. *Journal of Applied Ichthyology* 25, 5–9.

Ludwig, A., Morales, A. and Roselló, E. (2011) Sturgeon in Iberia from Past to present. In P. Williot *et al.* (eds) *Biology and Conservation of the European Sturgeon Acipenser sturio*, L., 1758, 131–146. Springer Verlag. Berlin.

Marzano, A. (2013) *Harvesting the Sea. The Exploitation of Marine Resources in the Roman Mediterranean*. Oxford, Oxford University Press.

Meléndez, M., Straus, L.G. and Clark, G.A. (1986) The ichthyology of La Riera Cave. In L.G. Straus and G.A. Clark (eds) *La Riera Cave, Stone Age Hunter-Gatherer Adaptations in Northern Spain*, 285–288, Tempe, Arizona State University (Anthropological Research Papers 36).

Morales, A. (1976) *Contribución al estudio de las faunas mastozoológicas asociadas a yacimientos prehistóricos españoles*. Unpublished PhD thesis, Universidad Complutense, Madrid.

Morales, A. (1979) Informe sobre los restos faunísticos de la cueva de Cuartamentero (Asturias). *Trabajos de Prehistoria* 36, 497–509.

Morales, A. (1980) Los peces fósiles del yacimiento achelense de Áridos-1 (Arganda, Madrid). In M. Santonja, N. López and A. Pérez (eds) *Ocupaciones achelenses en el Valle del Jarama*, 93–104, Madrid, Publicaciones de la Excma. Diputación Provincial de Madrid (Arqueología y Paleoecología 1).

Morales, A. (1983) Fish remains from Iberian sites: a general review. *Icthyo-Osteo-Archaeology News* 2, 6–1.

Morales, A. (1984) A study of the representativity and taxonomy of the fish faunas from two Mousterian sites on northern Spain, with special reference to the trout (*Salmo trutta* L., 1758). In N. Desse-Berset (ed.) *Second Fish Osteoarchaeological Meeting Editions du C.N.R.S.*, 41–59, París.

Morales, A. (1990a) Multiple hypothesis, unrefutable theories: a case sample from the policultive theory. In J. Schibler, J. Sedlmeier and H. Spycher (eds): *Beitrage zur Archäozoologie, Archäologie, Anthropologie, Geologie und Paläontologie*: 131–140. Basel, Helbing & Lichtenhahn.

Morales, A. (1990b) Arqueozoología teórica: usos y abusos reflejados en la interpretación de las asociaciones de fauna de yacimientos antrópicos. *Trabajos de Prehistoria* 47, 251–290.

Morales, A. (1993) Where are the tunas? Ancient Iberian fishing industries from an archaeozoological perspective. In A. Clason, S. Payne and H.P. Uerpmann (eds) *Skeletons in her Cupboard*, 135–141. Oxford, Oxbow Books.

Morales, A. (2010) Inferences about prehistoric fishing gear based on archaeological fish assemblages. In t. Bekker-Nielsenand D. Bernal Casasola (eds) *Ancient Nets and Fishing Gear*, 25–53. Aarhus and Cadiz,Aarhus University Press and Servicio de Publicaciones, Universidad de Cádiz.

Morales, A. and Roselló, E. (1984/85) Algunas consideraciones de índole zoológica en torno al pez representado en la cueva del Pindal. *Ars Praehistorica* III/IV, 247–251.

Morales, A. and Roselló, E. (1988) Ictioarqueología: nuevas técnicas al servicio de la reconstrucción prehistórica con algunos datos sobre el País Vasco. *Munibe* (suplemento 6), 97–104.

Morales, A. and Roselló, E. (1998a) Casual or intentional? Comments on fish skeletal representation from Iberian archaeological settlements. In P. Anreiter, L. Bartosiewicz; E. Jerem and W. Meid (eds) *Man and the Animal World*, 383–394. Budapest, Archaeolingua.

Morales, A. and Roselló, E. (1998b) Aportación de la Arqueoictiología al conocimiento de las culturas pre- y protohistóricas de la Península Ibérica. *Arbor* 635–636 CLXI, 235–267.

Morales, A. and Roselló, E. (2004) Fishing down the food web in Iberian Prehistory? A new look at the fishes from Cueva de Nerja (Málaga, Spain) In J.P. Brugal J.Desse (eds*) Petits Animaux et Societés Humaines. Du Complément Alimentaire aux Ressources Utilitaires*, 111–123. Antibes, Éditions APDCA.

Morales, A. and Roselló, E. (2006) La pesca en la Antigüedad Andaluza: Una perspectiva biológica. In *Historia de la Pesca en el Ámbito del Estrecho*, 37–79. Sevilla, Instituto de Investigación y Formación Agraria y Pesquera. Junta de Andalucia.

Morales, A. and Roselló, E. (2007) Pesca prehistórica. In Sociedad Española de Cetáceos (SEC) *Todos por la Mar*, 95–97. Madrid, Organismo Autónomo de Parques nacionales. Ministerio de Medio Ambiente.

Morales, A. and Roselló, E. (2008) 20,000 years of fishing in the Strait: archaeological fish and shellfish assemblages from southern Iberia. In J. Erlandsonand R. Torrey (eds): *Human Impacts on Ancient Marine Environments*, 243–278. Berkeley, University of California Press.

Morales, A. and Roselló, E. (2012) Especies pescadas, especies ingeridas: el consumo de pescado y moluscos marinos en las sociedades fenicio-púnicas. In B. Costaand J.H. Fernández (eds,) *Sal, pesca y salazones fenicios en Occidente. XXVI Jornadas de Arqueología Fenicio-Púnica (Eivissa, 2011)*, 123–156. Formentera, Museo Arqueológico de Ibiza y Formentera, Ibiza.

Morales-Muñiz, A. and Roselló-Izquierdo, E. (2016) Fishing in Mediterranean prehistory: an archaeo-ichthyological overview. In T. Bekker-Nielsen and R. Gertwagen (eds) *The Inland Seas: Towards an Ecohistory of the Mediterranean and the Black Sea*. (Geographica Historica 35), 23–56. Stuttgart, Franz Steiner.

Morales, A. and Rosenlund, K. (1979) *Fish Bone Measurements. An Attempt to Standardize the Measuring of Fish Bones from Archaeological Sites*. Copenhagen, Steenstrupia.

Morales, A. and Van Neer, W. (1994) Species abundance models as potential discriminators of natural and anthropogenic paleo-ichthyocenoses with a case sample from African sites. *Offa* 51, 386–394.

Morales, A., Antipina, Y., Antipina, A. and Roselló, E. (2007) An ichthyoarchaeological survey of the ancient fisheries from the northern Black Sea coast. *Archaeofauna* 16, 117–172.

Morales-Muñiz A., Gonzalez-Gomez de Agüero E., Fernández-Rodríguez C., Saborido Rey F., Llorente-Rodriguez L., Lopez-Arias, B. and Roselló-Izquierdo E., (2017a) Hindcasting to forecast. An archaeobiological approach to the Europeanhake (*Merluccius merluccius*, Linnaeus 1758) fishery: Iberia and beyond. *Regional Studies in Marine Science* (2017) 21, 2352–4855, https://doi.org/10.1016/j.rsma.2017.11.001.

Morales Muñiz, A., González Gómez de Agüero, E., Fernández Rodríguez, C., Ephrem, B., López-Arias, B., Llorente Rodriguez, L., Saborido Rey, F. and Roselló Izquierdo, E. (2017b) Fishes as indicators of seasonality in Roman non-industrial fisheries: an overview from the southern NE Atlantic. In R. González Villaescusa, K. Schörle, F. Gayet and F. Rechin (eds) L'exploitation des ressources maritimes de l'Antiquité. Activités productives et organisation des territoires. XXXVIIe Rencontres Internationales d'Archéologie et d'Histoire d'Antibes & XIIe Colloque de l'Association AGER, 172–192. Antibes, Éditions Apdca.

Morales Muñiz, A., Llorente Rodríguez, L., Jiménez Cano, N., López Arias B. and Roselló Izquierdo, E. (2016) La Ictiofauna. La identificación de los restos de peces de yacimientos arqueológicos. In L. Lloveras, C. Rissech, J. Nadal and J.M. Fullola (eds) *What Bones Tell Us. Lo Que Nos Dicen Los Huesos*. Monografies del SERP, 12, 77–86. Barcelona, Universitat de Barcelona.

Morales, A., Morales, D.C. and Roselló, E. (1991) Sobre la presencia del bacalao (*Gadus morhua*) en La Cartuja sevillana de Santa María de las Cuevas (Siglos XV–XVI). *Archivo Hispalense* T. LXXIV nº 226, 17–26.

Morales Muñiz, A., Moreno García, M., Roselló Izquierdo, E., Llorente Rodríguez, L. and Morales Muñiz, D.C. (2011) 711 Ad: ¿El Origen de una Disyunción Alimentaria? *Zona Arqueológica*, 15(2), 301–319.

Morales, A., Roselló, E. and Hernández, F. (1998) Late Upper Paleolithic subsistence strategies in Southern Iberia: Tardiglacial faunas from Cueva de Nerja (Málaga, Spain). *European Journal of Archaeology* 1(1), 9–50.

Morales, A., Roselló, E., Lentacker, A. and Morales, D.C. (1994) Archaeozoological research on medieval Iberia: Fishing and fish trade on Almohad sites. *Trabalhos de Antropologia e Etnologia* 34(1–2), 453–475.

Morales, A., Roselló, E., Rosenlund, K. and López, J.L. (1999) Iberian and Danish ichthyoarchaeological assemblages: Patterns of diversity from a paleocultural persepctive. In C. Becker, H. Manhart, J. Peters and J. Schibler (eds) *Historia Animalium ex Ossibus. Beiträge zur Paläoanatomie, Archäologie, Ägyptologie, Ethnologie und Geschichte der Tiermedizin. Festschrift für Angela von den Driesch*, 309–331. Rahden/Westf, Verlag Marie Leidorf GmbH.

Morales, A., Roselló, E. and Ruíz, M. (2001) A renewable fish exploitation model for application on ichthyoarchaeological assemblages. In H. Buitenhuisand W. Prummel (eds) *Animals and Man in the Past*, 44–52. Groningen, ARC-Publicatie 41.

Morales, D.C., Roselló, E. and Morales, A. (2009) Pesquerías medievales hispanas: Las evidencias arqueofaunísticas. In: Sociedad Española de Estudios Medievales (de.) *La Pesca en la Edad Media,* 145–165. Madrid, Servicio de Publicaciones Universidad de Murcia, CSIC y Xunta de Galicia.

Patón Dominguez, D. (1993) *Contribución al Atlas Osteológico de los Teleósteos ibéricos III. Osteología Comparada de los Mugilidae Ibéricos*. Unpublished PhD thesis, Universidad Autónoma de Madrid, Madrid.

Pellicer, M. and Morales, A. (eds) (1995) *Fauna de la Cueva de Nerja I. Salas de la Mina y de La Torca, Campañas 1980–1982*. Trabajos sobre la Cueva de Nerja, nº 5. Málaga, Patronato de la Cueva de Nerja.

Ponsich, M. and. Tarradell, M. (1965) *Garum et industries antiques de salaison dans la Meditérranée Occidentale*. Paris, Bibliothèque de L'Ecole des Hautes Etudes Hispaniques XXXVI.

Richards, M.P., Jacobi, R., Cook, J., Pettit, P.B. and Stringer, C.B. (2005) Isotope evidence for the intensive use of marine foods by late Upper Palaeolithic humans. *Journal of Human Evolution* 49, 390–394.

Rodrigo, M.J. (1994) Remains of *Melanogrammus aeglefinus* (Linnaeus, 1758) in the Pleistocene-Holocene passage in the cave of Nerja, Málaga/Spain. *Offa* 51, 348–351.

Rodríguez Santana, C.G. (1994) *Las Ictiofaunas Arqueológicas del Archipiélago Canario: Una aproximación a la pesca entre los canarios, guanches y auaritas*. Unpublished PhD thesis, Universidad de La Laguna, Santa Cruz de Tenerife.

Roselló, E. (1988a) *Contribución al Atlas Osteológico de los Teleósteos Ibéricos. I. Dentario y Articular.* Madrid, Colección de Estudios nº 14, Ediciones de la Universidad Autónoma de Madrid.

Roselló, E. (1988b) Animal osteological files in archaeology: proposal of a standardized identification system with a case sample from the mullets (G. *Mullus*). *Ichthyo-osteo-archaeology News* 5, 7–16.

Roselló, E. (1989) *Arqueoictiofaunas ibéricas. Aproximación metodológica y bio-cultural*. Madrid, Ediciones de la Universidad Autónoma de Madrid.

Roselló, E. (1989) Informe preliminar de la ictiofauna de Santa Pola (prov. Alicante). *Saguntum* 22, 439–445.

Roselló, E. (1991) Segundo informe sobre la ictiofauna magdaleniense de la Cueva de Tito Bustillo (provincia de Asturias). *Boletín de Ciencias de la Naturaleza I.D.E.A.* 41, 67–76.

Roselló, E. and Brinkhuizen, D. (1994) Laminak II/Spain: Alternative taxonomies as approaches to the interpretation of a fish fauna. *Offa* 51, 401–409.

Roselló, E. and Morales, A. (1992) Grouping patterns in Iberian ichthyoarchaeological assemblages from coastal sites. *Archaeofauna* 1, 11–22.

Roselló, E. and Morales, A. (eds) (1994a) *Castillo de Doña Blanca. Archaeo-Environmental Investigations in the Bay of Cádiz, Spain (750–500 BC)*. Oxford, B.A.R. (International Series) 593.

Roselló, E. and Morales, A. (1994b) The fishes. In E. Roselló and A. Morales 1994a, 91–142.

Roselló, E. and Morales, A. (2005–2006) Ictiofaunas Musterienses de la Península Ibérica: ¿Evidencias de pesca Neanderthal? *Munibe (Antropologia-Arkeologia)* 57(1), 183–195.

Roselló-Izquierdo, E. and Morales-Muñiz, A. (2011a) Evidencias de pesca en las ocupaciones de Santimamiñe. *Kobie, Serie Excavaciones Arqueológicas en Bizkaia* 1, 239–246.

Roselló, E. and Morales, A. (2011b) Estudio de los peces del yacimiento de Aitzbitarte III (zona de entrada). In J. Altuna, K. Mariezkurrena and J. Ríos, J. (eds) *Ocupaciones Humanas en Aitzbitarte III (País Vasco) 33.600–18.400 BP (Zona de entrada de la cueva),* 507–516. Vitoria, EKOB, Colección de Patrimonio Cultural Vasco 5.

Roselló, E. and Morales, A. (2014a) La ictiofauna de época Argárica de Cabezo Pardo (San Isidro, Alicante). In J.A. López Padilla (ed.) *Cabezo Pardo. Excavaciones arqueológicas en el yacimiento de la Edad del Bronce*, 306–319. Alicante, Memorias excavaciones arqueológicas MARQ, 6.

Roselló Izquierdo, E. and Morales Muñiz, A. (2014b) Las ictiofaunas de Santa Catalina (Lequeitio, Vizcaya): un registro singular para la prehistoria cantábrica. *Kobie Serie Bizkaiko Arkeologi Indusketak – Excavaciones Arqueologicas en Bizkaia* 4, 161–262.

Roselló Izquierdo, E. and Morales Muñiz, A. (2015) Los peces del pozo-depósito de Tabacalera (unidad estratigráfica 26) In C. Fernández Ochoa, Orejas Saco del Valle, P. García Díazand F. Gil Sendino (Eds) *La Fábrica de Tabacos de Gijón. Arqueología e Historia de un espacio milenario*, 252–259. Gijón, Ayuntamiento de Gijón.

Roselló Izquierdo, E. and Morales Muñiz, A. (2016) Los peces Gravetienses de la zona interior de la Cueva de Aitzbitarte III. In J. Altuna, K. Mariezkurrenaand J. Ríos (eds) *Ocupaciones Humanas en Aitzbitarte III (País Vasco) 33.600–18.400 BP (Zona profunda de la cueva)*, 332–338. Vitoria, EKOB, Colección de Patrimonio Cultural Vasco, 5, Vitoria.

Roselló-Izquierdo, E., Berganza-Gochi, E., Nores-Quesada, C. and Morales-Muñiz, A. (2016) Santa Catalina (Lequeitio, Basque Country): An ecological and cultural insight into the nature of prehistoric fishing in Cantabrian Spain, *Journal of Archaeological Science: Reports* 6, 645–653. http://dx.doi.org/10.1016/j.jasrep.2015.06.002.

Roselló Izquierdo, E., Llorente Rodríguez, L. and Morales Muñiz, A. (2015) The fishes from Cova Fosca (Castellón, Spain): Lost signatures of a hunter gatherer tradition? *CuPAUAM,* 41, 87–96.

Roselló, E., Morales, A. and Morales, D.C. (1994) La Cartuja/Spain: Anthropogenic ichthyocenosis of culinary nature in a paleocultural context. *Offa* 51, 323–331.

Russ, H. (2010) A *Taphonomic Approach to Reconstructing Upper Palaeolithic Hunter-Gatherer Fishing Strategies*. Unpublished PhD thesis, University of Bradford, Bradford.

Sánchez, M.J., Blasco, E., Guardiola, A. (1989) Descubrimiento de una factoría bajoimperial de salazón de pescado en Santa Pola (prov. Alicante). *Saguntum* 22, 413–438.

Sisma-Ventura, G., Tütken, T., Zohar, I., Pack, A., Sivan, D., Omri, L., Giboa, A. and Bar-Oz, G. (2018) Tooth oxygen isotopes reveal Late Bronze Age origin of Mediterranean fish aquaculture and trade. *Scientific Reports* 8(1), 14086.

Sternberg, M. (2007) Pêche et exploitation des milieux aquatiques du VIIIe s. au VI e s. av. n.-è. à la Rabita de Guardamar. In P. Rouillard, E. Gailledrat and F. Sala Sellés (eds) *Fouilles de la Rabita de Guardamar II. L'établissement protohistorique de la Fonteta (fin VIIIe-fin VIe s. av. J.-C.)*. Collection de la Casa de Velasquez 96, 372–398. Madrid, Casa de Velázquez.

Straus, L.G. (1983) From Mousterian to Magdalenian: Cultural evolution viewed from Vasco-Cantabrian Spain and Pyrenean France. In E. Trinkaus (ed.) *The Mousterian Legacy,* 73–111. Oxford, B.A.R. (International Series) 164.

Straus, L.G., González Morales, M., Martínez, M.Á.F. and García-Gelabert, M.P. (2002) Last Glacial human settlement in eastern Cantabria (northern Spain). *Journal of Archaeological Science* 29(12), 1403–1414.

Turrero, P., Ardura, A., García-Vázquez, E. (2014) Paleolithic vs. Epipaleolithic fisheries in northern Iberia. *Quaternary Research* 82(1), 51–55.

Turrero, P., García Vázquez, E., Álvarez Laó, D.J., Arbizu, M. and Adán, G. (2008) Los peces como recurso en el Norte de la Península Ibérica durante el Paleolítico Superior. In J.I. Ruíz Omeñaca, L. Piñuela and J.C. García Ramos (eds) *XXIV Jornadas de la Sociedad Española de Paleontología,* 15–18. Colunga, Museo del Jurásico de Asturias (Muja).

Turrero, P., Horreo, J. L, López, B., Pola, I.G., Arbizu, M., García Vázquez, E. (2013) Chronological changes in Upper Paleolithic fisheries revealed by Museum Archival Material. *Palaios* 28, 228–232.

Uerpmann, H.P. (1971) Die Tierknochenfunde aus der Talayot-Siedlung von S'Illot (San Lorenzo/Mallorca). *Studien über frühe Tierknochenfunde von der Iberischen Halbinsel* 2, 1–95.

Van Neer, W. and Morales, A. (1992) Fish middens: Anthropogenic accumulations of fish remains and their bearing on archaeoichthyological assemblages. *Journal of Archaeological Science* 19, 683–695.

4

On the use of micromammals for paleoenvironmental reconstruction: Qesem Cave as a case study

Orr Comay and Tamar Dayan

Micromammals have been used for paleoenvironmental reconstructions for decades, mostly by relying on current species habitat requirements. Recently, two new approaches have been proposed: the Coexistence Approach, which relies on species lists and quantitatively assesses the range of environmental variables in each species' modern distribution, and the Weighted Averaging Partial Least Squares (WAPLS), which relies on relative abundances and requires a training set (reference sample) of modern assemblages from known environments. WAPLS reconstructs a single predefined environmental variable, which is known a-priori to impact the fauna in question. In the Mediterranean Levant, previous studies showed that vegetation density (from grasslands to maquis) is the dominant factor shaping the micromammal fauna. Here we apply WAPLS on Qesem Cave, a Middle Paleolithic Levantine Acheulo-Yabrudian site, which was already reconstructed using both species habitat requirements and the Coexistence Approach, thus enabling us to compare results of these three methods. Modern habitat requirements of species provided evidence of a variety of paleohabitats, including grasslands and forests, with no indication of their relative dominance. The Coexistence Approach concluded that the paleoclimate was colder and drier than today's, a conclusion that relies on specimens identified tentatively either as a very little extant known species (Myomimus setzeri), or as its extinct congener (Myomimus judaicus). WAPLS suggested that grasslands were the dominant habitat, but this method did not account for species missing from its training set that may have provided valuable information. Overall, a combination of species habitat requirements and WAPLS provided the most detailed yet reliable reconstruction. Both the Coexistence Approach and WAPLS should be used with caution, based on the ecological knowledge on species distributions and their driving factors.

Keywords: Micromammals, paleoecology, modelling, Middle Paleolithic, Levant

Introduction

Micromammals have been used as paleoenvironmental indicators for decades (*e.g.* Chaline 1977; Avery 1982; Andrews 1990; Tchernov 1996; Reed 2011; Maul *et al.* 2016; Comay and Dayan 2018a). The most common approach is to describe the habitat requirements of each species, and conclude that the paleoenvironment met all these requirements, at least during some part of the assemblage accumulation. While this approach is empirically sound, the resulting reconstruction is qualitative and not always detailed, and sometimes suggests the existence of considerably different habitats such as grasslands and forests.

Recently, two quantitative methods, originally designed for other paleoenvironmental proxy, were used for the first time on micromammals assemblages: the Coexistence Approach (reviewed in Utescher *et al.* 2014; micromammal application in Maul *et al.* 2016), which was developed for fossil plants, and Weighted Averaging Partial Least Squares (ter Braak and Juggins 1993; micromammal application in Popov *et al.* 2014; Comay and Dayan 2018a), which was developed for paleolimnology.

Paleoenvironmental reconstruction

Comay and Dayan (2018a, 2163) proposed a Weighted Averaging Partial Least Squares (WAPLS; ter Braak and Juggins 1993) model that reconstructs a Levantine Mediterranean site's vegetation structure from its micromammal fauna. Comay and Dayan (2018a, 2165) calculated a model for barn owl-derived microvertebrate assemblages. Since *Tyto alba* was the likely bone collector at Qesem Cave (Smith *et al.* 2013, 3337), the model can be applied to Qesem Cave's data.

WAPLS assumes that the reconstructed variable has a causal relationship with species abundance, and each species has an optimum value (ter Braak and Juggins 1993, 486; Juggins and Birks 2012, 445). In our case, the WAPLS model assumes that vegetation structure has a direct causal link to micromammal faunal composition, and different species are found in varying habitats – some in open grasslands, others in maquis – along a vegetation gradient (see below). The optimum values are calculated directly from the training data, rather than being predetermined, and thus are empirically based. In our case, the training set comprised numerous *Tyto alba* prey assemblages from Mediterranean Israel (Comay and Dayan 2018a, 2163). In addition, WAPLS attempts to find data components (axes) that have the largest covariation with the reconstructed environmental variable (the Partial Least Squares regression).

Vegetation structure is evaluated from the Vegetation Index for Mediterranean climates, ranked from 0 to 3 on the following scale: 0 – urban, 1 – grassland, 2 – shrubland, and 3 – maquis or forest (Comay and Dayan 2018b, 5). Mixed environments merit non-integer values calculated as weighted means. For example, a habitat whose environment is 50% shrubland and 50% maquis would have a Vegetation Index of $50\%\cdot3+50\%\cdot2 = 2.5$. We applied Qesem Cave's calculated MNI data (Table 4.1) to reconstruct the Vegetation Index for both its microvertebrate concentrations. Table 4.1 details the coefficient of each taxon to the calculated Vegetation Index, with higher coefficients indicating higher abundance in denser vegetation structures. See Comay and Dayan (2018a) for further discussion of the model.

Results and discussion

Species habitat requirements

All Qesem Cave's fossil microfauna species (Smith *et al.* 2016, 236) occur today under Mediterranean climate in the Middle East (Mendelssohn and Yom-Tov 1999), save for the extinct rat *Rattus haasi* and possibly also the tentatively identified dormouse *Myomimus judaicus/setzeri* (*M. judaicus* is extinct; *M. setzeri* occurs in alpine habitats in west Iran; Kryštufek *et al.* 2008, 3). This suggests a Mediterranean climate. Habitat requirements of the species found are varied, including riverbanks, rocks, shrublands and woodlands (Maul *et al.* 2016, 223).

The Coexistence Approach

Maul and his colleagues (2016, 225) concluded that during the deposition of Qesem Cave's Concentration 2, the climate was colder than today's, and their quantitative analysis resulted in winter minimum temperature lower by at least 10 °C (−2.3 °C compared to a modern mean of 8.3 °C). Precipitation was also suggested to be lower, especially during winter (which is the rainy season in the present climate), but the seasonal distribution of precipitation might have been more even than today (*e.g.* summer rains). These deviations from modern climate rely solely on the presence of *M. judaicus/setzeri*, whose nearest living relative is *M. setzeri*. Nevertheless, the Coexistence Approach found no indication that the climate was considerably different from today's during the deposition of Qesem Cave's Concentration 1.

WAPLS

The WAPLS model based on *Tyto alba* prey predicted a Vegetation Index of 1.210 for Qesem Cave's microvertebrate Concentration 1 and 1.161 for Concentration 2. The model's coefficients by taxonomic relative abundance (percentage of NISP of all taxa that were included in the model) are graphed in Fig. 4.1. These values are relatively low, indicating that grasslands dominated the landscape.

While not contradicting *per se*, the two quantitative approaches reconstruct Qesem Cave's paleoenvironment quite differently. The Coexistence Approach provides a detailed climate description yet provides little insight on the vegetation structure, while WAPLS makes no attempt at climate reconstruction yet stresses vegetation structure.

Our model did not account for the occurrence of *M. setzeri*, a species that had a very large impact on Qesem Cave's previous paleoenvironmental reconstruction (Maul *et al.* 2016, 223), leading to the conclusion that winters were considerably colder and drier than today. This omission reflects our reliance only on observations of *Tyto alba* prey taxa from pellets collected in Mediterranean Israel. Nevertheless, *M. setzeri* is extremely rare (both recent and archaeozoological specimens are scarce) and very little known (Kryštufek and Vohralík 2005, 119; Kryštufek *et al.* 2008, 2). Obuch (2001, 147) surveyed over 42,000 owl prey specimens from Turkey, Syria, Iran and Israel, and found *M. setzeri* only in Iran and Turkey and its relative abundance reaches a maximum of 3.2% of owl prey individuals in any given assemblage. Moreover, this species' IUCN assessors (Kryštufek *et al.* 2008, 4), who provided the distribution used for extracting its environmental tolerance via the Coexistence Approach (Maul *et al.* 2016, 222), write that 'The range is very poorly known. [...] Very poorly known species, virtually nothing known about biology or ecology. [...] majority of information is available from owl pellets.' This suggests that an attempt to achieve sufficient samples of its occurrence to relate between its presence and a given

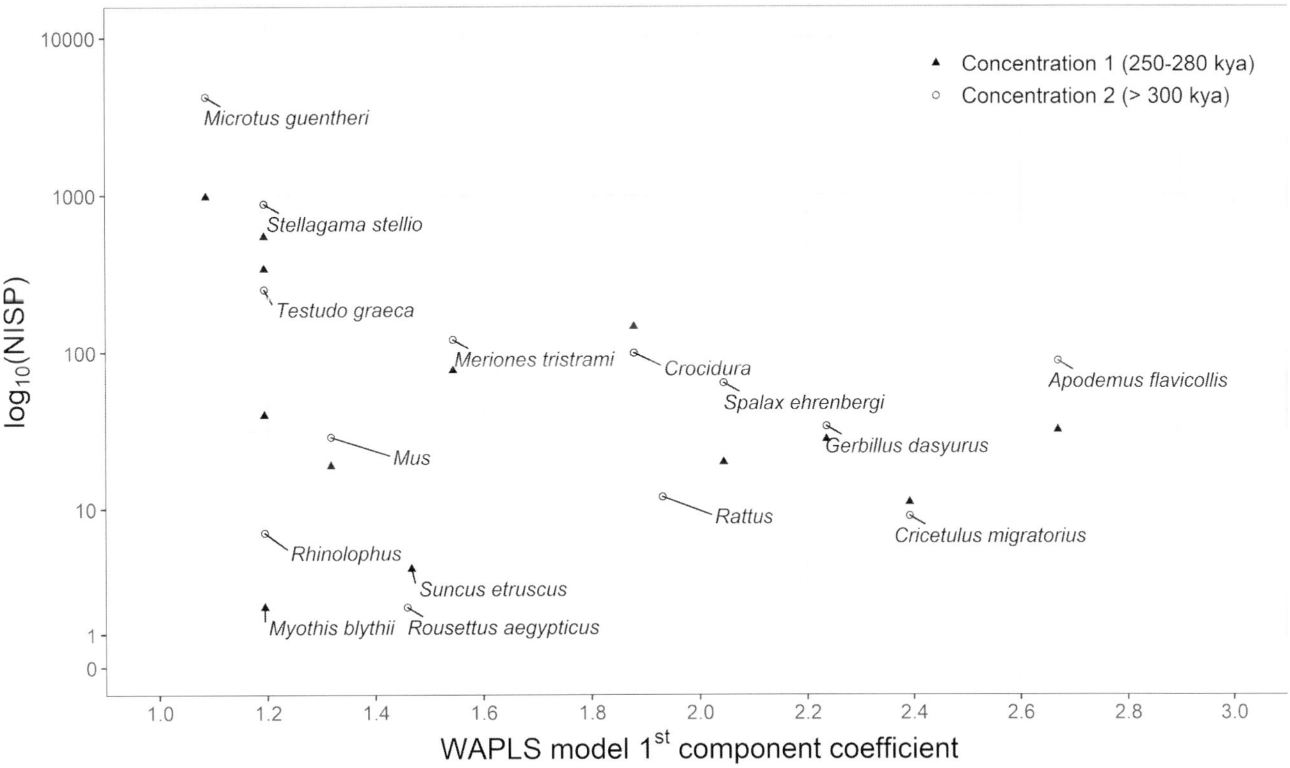

Figure 4.1. Qesem Cave's microvertebrate taxa that were included in the Weighted Averaging Partial Least Squares (WAPLS) model by their model coefficient. Higher coefficients indicate that species are found in denser vegetation in the recent training set and vice versa. See text for further details.

environmental variable is likely to be prohibitively labor intensive. In addition, as the Qesem Cave specimens are not identified to the species level, it is possible that they belong to the extinct *M. judaicus*, whose biology and habitat requirement could differ from *M. setzeri*. In Qesem Cave's Concentration 2, this species is represented by three molars, accounting for about 0.03% of the NISP identified below the order level, while it does not occur at Concentration 1 (Smith *et al.* 2016, 236).

Our approach to this limitation is parsimonious: what was not demonstrated and supported by empirical evidence would have to be assumed unknown. By choosing to use only observations from recent Israeli owl pellets, we have decided not to draw inferences from the presences of the rare, mountain-dwelling *Myomimus judaicus/setzeri* (Table 4.1), whose presence had a great impact on the results of the Coexistence Approach in Qesem Cave (Maul *et al.* 2016, 225). In our case, the omission could be justified by the facts that this species (*M. setzeri*; *M. judaicus* is extinct) is very little known (fewer than ten specimens in collections worldwide with most information originating in owl pellets; Kryštufek *et al.* 2008, 4), and that the specimens from Qesem Cave were not even attributed to *M. setzeri* unambiguously.

Focusing on species taken into account in our *Tyto alba*-based WAPLS model, taxa with low coefficients (indicating that they tend to be more abundant in barn owl prey assemblages from habitats with open vegetation) dominate both fossil concentrations (Table 4.1; Fig. 4.1). Namely, the great dominance of *Microtus guentheri* in both concentrations points to a landscape of grasslands. The rare presence of woodland or even arboreal taxa (*Apodemus flavicollis*, *Sciurus anomalus*) suggests some forest habitat was also present, but at a considerably smaller scale than the grasslands. Combining the insights from all three approaches, we conclude that paleoenvironment of Qesem Cave was a grassland with Mediterranean climate, although some forest or maquis had also been present but to a lesser degree. Despite the emphasis on grasslands rather than shrublands or maquis, we reject the interpretation that climate was drier during the deposition of Concentration 2, based on independent evidence detailed below.

Vegetation cover is connected with temperature and precipitation. Some data on speleothems could provide indications of a generally higher precipitation, and those of pollen could indicate a general forest expansion during Qesem Cave's occupation. In Israel, speleothems form only in relatively rainy climates, where annual precipitation exceeds 200 mm (Frumkin *et al.* 2011, 443). Vaks *et al.* (2010, 2650) defined periods of intensive and limited speleothem deposition in the central and southern Negev (southern Israel), where no speleothems are formed today. An intensive speleothem deposition period occurred 350–290 ka and was followed by a period of limited deposition 290–285 ka.

Similarly, Lisker *et al.* (2010, 1209) described speleothems that were formed ~350–260 ka at the Dead Sea caves (southern Israel). These periods are penecontemporaneous with Qesem Cave's occupation period (420–200 ka; Gopher and Barkai 2016, 1) and also with its microvertebrate concentrations (>300 ka for Concentration 2 and 250–280 ka for Concentration 1; Smith *et al.* 2016, 234). A recent study of fossil pollen records from north-eastern Greece dated to 240–312 ka (during the glacial period Marine Isotope Stage [MIS] 8 and the onset of the interglacial MIS 7e) correlated between relatively warm periods (in glacial terms) and forest expansion (Fletcher *et al.* 2013, 118). The latter occurred during ~312–277 ka, while during 277–246 ka the climate was relatively cold and steppe-like in north-eastern Greece, although the implications to the contemporary paleoenvironment in the southern Levant are unclear. Overall, this evidence suggests that the climate during the deposition of Qesem Cave's Concentration 2 was at least as rainy as today's. This contests the results of the Coexistence Approach (Maul *et al.* 2016, 230).

In light of our grassland dominant reconstruction, the high percentage of Persian fallow deer (*Dama mesopotamica*; 73–76% of NISP in various layers) in Qesem Cave's macrofauna (Stiner *et al.* 2011, 218) might seem intriguing, as one might expect greater dominance of grassland species. The Persian fallow deer occupies a range of woodlands, such as tamarisk, oak and pistachio woodlands (Werner *et al.* 2015, 5). However, Persian fallow deer browse in low vegetation habitats at night (Bar-David 2002, 96), so a grassland habitat for Qesem Cave may not contradict their natural history. The second most abundant macrofaunal species is the extinct wild aurochs (*Bos primigenius*; 4–20%), likely a grassland species. According to Stiner *et al.* (2011, 218), the absence of mountain gazelles (*Gazella gazella*) in Qesem Cave is likely not because of the lack of suitable habitats, but due to biogeographic barriers (*i.e.* the Sahara Desert) that blocked this species' migration northward into the Levant until ~200 ka. These authors suggest that 'savannah conditions dominated the Levant roughly 700 ka'. Stiner *et al.* (2011, 222) already suggested that the hominins that inhabited Qesem Cave selected larger (prime age, not juveniles) individuals as game and thus the taphocoenosis probably does not reflect a normal body size distribution of the prey population. Therefore, the macrofaunal assemblage probably reflects the (hominin) predator bias more than the paleo-community composition, and thus does not contradict our reconstruction.

Conclusions

In sum, we suggest using caution when reconstructing the paleoenvironment based on micromammal assemblages. The natural history of species used for the Coexistence Approach should be well known, including potentially non-environmental causes for limited distributions such as competition or anthropogenic influence. The training set of the WAPLS model should include all dominant species in the assemblage used for reconstruction. Moreover, only those environmental variables that were demonstrated to limit species distributions or abundances should be reconstructed. Lastly, independent paleoenvironmental proxies should be assessed as well, to either corroborate or contradict the conclusions obtained from analyzing the micromammal fauna. We hope this approach would assist in creating detailed and well based paleoenvironmental reconstructions.

Acknowledgements

OC wishes to thank The Planning and Budgeting Committee of The Council for Higher Education in Israel for generously supporting this research through a postdoctoral fellowship for natural history collections-based research.

References

Andrews, P. (1990) *Owls, Caves and Fossils*. Chicago, The University of Chicago Press.

Avery, M. (1982) Micromammals as palaeoenvironmental indicators and as interpretation of the late Quaternary in the Southern Cape province, South Africa. *Annals of the South African Museum* 85(2), 183–374.

Bar-David, S. (2002) *Modeling Population Growth in Time and Space – The Reintroduced Persian Fallow Deer*. Unpublished thesis, University of Tel Aviv.

Barkai, R., Gopher, A., Lauritzen, S.E. and Frumkin, A. (2003) Uranium series dates from Qesem Cave, Israel, and the end of the Lower Palaeolithic. *Nature* 423, 977–979.

Barkai, R., Lemorini, C., Shimelmitz, R., Lev, Z., Stiner, M.C. and Gopher, A. (2009) A blade for all seasons? Making and using Amudian blades at Qesem Cave, Israel. *Human Evolution* 24, 57–75.

Ben-Dor, M., Gopher, A., Hershkovitz, I. and Barkai, R. (2011) Man the fat hunter: the demise of Homo erectus and the emergence of a new hominin lineage in the Middle Pleistocene (ca. 400 kyr) Levant. *PLoS ONE* 6, 1–12.

Chaline, J. (1977) Rodents, evolution and prehistory. *Endeavor* 1(2), 44–51.

Comay, O. and Dayan, T. (2018a) From micromammals to paleoenvironments. *Archaeological and Anthropological Sciences* 10, 2159–2171. doi: https://doi.org/10.1007/s12520-018-0608-8.

Comay, O. and Dayan, T. (2018b) What determines prey selection in owls? Roles of prey traits, prey class, environmental variables and taxonomic specialization. *Ecology and Evolution* 8 (6), 3382–3392. doi: 10.1002/ece3.3899.

Comay, O. and Dayan, T. (2018c) Taphonomic signatures of owls: New insights into micromammal assemblages. *Palaeogeography., Palaeoclimatology., Palaeoecology* 492, 81–91. doi: 10.1016/j.palaeo.2017.12.014.

Fletcher, W.J., Müller, U.C., Koutsodendris, A., Christanis, K. and Pross, J. (2013) A centennial-scale record of vegetation and climate variability from 312 to 240 ka (Marine Isotope Stages 9cea, 8 and 7e) from Tenaghi Philippon, NE Greece. *Quaternary Science Reviews* 78, 108–125.

Frumkin, A., Bar-Yosef, O. and Schwarcz, H.P. (2011) Possible paleohydrologic and paleoclimatic effects on hominin migration and occupation of the Levantine Middle Paleolithic. *Journal of Human Evolution* 60, 437–451.

Gopher, A., Ayalon, A., Bar-Matthews, M., Barkai, R., Frumkin, A., Karkanas, P. and Shahack-Gross, R. (2010) The chronology of the late Lower Paleolithic in the Levant based on U-Th ages of speleothems from Qesem Cave, Israel. *Quaternary Geochronology* 5, 644–656.

Gopher, A. and Barkai, R. (2016) State of the art of the multidisciplinary research at Middle Pleistocene Qesem Cave, Israel, 2015 – an introduction. *Quaternary International* 398, 1–5.

Hershkovitz, I., Smith, P., Sarig, R., Quam, R., Rodriguez, L., Garcia, R., Luis Arsuaga, J., Barkai, R. and Gopher, A. (2011) Middle Pleistocene dental remains from Qesem Cave (Israel). *American Journal. Physical Anthropology* 144, 575–592.

Hoffman, R. (1988) The contribution of raptorial birds to patterning in small mammal assemblages. *Paleobiology* 14(1), 81–930.

Juggins, S. and Birks, J.B. (2012) Quantitative environmental reconstructions from biological data. In J.B. Birks, A.F. Lotter, S. Juggins and J.P. Smol (eds) *Tracking Environmental Change using Lake Sediments Data Handling and Numerical Techniques*, 431–494. London, Dordrecht, Heidelberg and New York, Springer.

Kryštufek, B. and Vohralík, V. (2005) *Mammals of Turkey and Cyprus. Rodentia I: Sciuridae, Dipodidae, Gliridae, Arvicolinae*. Univerza na Primorskem, Znanstveno-raziskovalno središče, Koper.

Kryštufek, B., Amori, G. and Mitsain, G. (2008) *Myomimus setzeri*, Setzer's Mouse-tailed dormouse. The IUCN Red List of Threatened Species 2008: e.T14089A4389600. http://dx.doi.org/10.2305/IUCN.UK.2008.RLTS.T14089A4389600.en.

Lisker, S., Vaks, A., Bar-Matthews, M., Porat, R. and Frumkin, A. (2010) Late Pleistocene palaeoclimatic and palaeoenvironmental reconstruction of the Dead Sea area (Israel), based on speleothems and cave stromatolites. *Quaternary Science Reviews* 29, 1201–1211.

Maul, L.C., Bruch, A.A., Smith, K.T., Shenbrot, G., Barkai, R. and Gopher, A. (2016) Palaeoecological and biostratigraphical implications of the microvertebrates of Qesem Cave in Israel. *Quaternary International* 398, 219–232.

Maul, L.C., Smith, K.T., Barkai, R., Barash, A., Karkanas, P., Shahack-Gross, R. and Gopher, A. (2011) Microfaunal remains at Middle Pleistocene Qesem Cave, Israel: Preliminary results on small vertebrates, environment and biostratigraphy. *Journal of Human Evolution* 60, 464–480.

Mendelssohn, H. and Yom-Tov, Y. (1999) *Fauna Palaestina. Mammalia of Israel*. Jerusalem, The Israel Academy of Sciences and Humanities.

Obuch, J. (2001) Dormice in the diet of owls in The Middle East. *Trakya University Journal of Scientific Research Series B* 2, 145–150.

Popov, V.V., Di Canzio, E. and Giaccio, B. (2014) Late Quaternary small mammals and paleotemperatures in Bulgaria and Italy. *Acta Zoologica. Bulgarica* 66(1), 89–108.

Reed, D.N. (2011) Serengeti micromammal communities and the paleoecology of Laetoli, Tanzania. In T. Harrison (ed.) *Paleontology and Geology of Laetoli: Human Evolution in Context. Volume 1: Geology, Geochronology, Paleoecology and Paleoenvironment, Vertebrate Paleobiology and Paleoanthropology*. doi: 10.1007/978-90-481-9956-3_12. Springer Science+Business Media B.V. 2011.

Shimelmitz, R., Barkai, R. and Gopher, A. (2011) Systematic blade production at late Lower Paleolithic (400–200 kyr) Qesem Cave, Israel. *Journal of Human Evolution* 61, 458–479.

Smith, K.T., Maul, L.C., Barkai, R. and Gopher, A. (2013) To catch a chameleon, or actualism vs. natural history in the taphonomy of the microvertebrate fraction at Qesem Cave, Israel. *Journal of Archaeological Science* 40, 3326–3339.

Smith, K.T., Maul, L.C., Flemming, F., Barkai, R. and Gopher, A. (2016) The microvertebrates of Qesem Cave: A comparison of the two concentrations. *Quaternary International* 398, 233–245.

Stiner, M.C., Barkai, R. and Gopher, A. (2009) Cooperative hunting and meat sharing 400–200 kya at Qesem Cave, Israel. *Proceedings of the National Academy of Sciences. U.S.A.* 106, 13207–13212.

Stiner, M.C., Gopher, A. and Barkai, R. (2011) Hearth-side socioeconomics, hunting and paleoecology during the late Lower Paleolithic at Qesem Cave, Israel. *Journal of Human Evolution* 60(2), 213–233.

Tchernov, E., (1996) Rodent faunas, chronostratigraphy and paleobiogeography of the southern Levant during the Quaternary. *Acta Zoologica Cracoviensa* 39, 513–530.

ter Braak, C.J. and Juggins, S. (1993) Weighted averaging partial least squares regression (WA-PLS): an improved method for reconstructing environmental variables from species assemblages. *Hydrobiologia* 269, 485–502.

Utescher, T., Bruch, A., Erdei, B., François, L., Ivanov, D., Jacques, F.M.B., Kern, A.K., Liu, Y.-S. (C.), Mosberger, V. and Spicer, R.A. (2014) The Coexistence Approach – theoretical background and practical considerations of using plant fossils for climate quantification. *Palaeogeography, Palaeclimatology, Palaeoecology* 410, 58–73.

Vaks, A., Bar-Matthews, M., Matthews, A., Ayalon, A. and Frumkin, A. (2010) Middle–Late Quaternary paleoclimate of northern margins of the Saharan-Arabian Desert: reconstruction from speleothems of Negev Desert, Israel. *Quaternary Science Reviews* 29, 2647–2662.

van der Voet, H. (1994) Comparing the predictive accuracy of models using a simple randomization test. *Chemometrics and Intelligent Laboratory Systems* 25, 313–323.

Verri, G., Barkai, R., Bordeanu, C., Gopher, A., Hass, M., Kaufman, A., Kubik, P., Montanari, E., Paul, M., Ronen, A., Weiner, S. and Boaretto, E. (2004) Flint mining in prehistory recorded by in situ-produced cosmogenic Be-10. *Proceedings of the National Academy of Sciences U.S.A.* 101, 7880–7884.

Verri, G., Barkai, R., Gopher, A., Hass, M., Kubik, P.W., Paul, M., Ronen, A., Weiner, S. and Boaretto, E. (2005) Flint procurement strategies in the Late Lower Palaeolithic recorded by in situ produced cosmogenic Be-10 in Tabun and Qesem Caves (Israel). *Journal of Archaeological Science* 32, 207–213.

Werner, N.Y., Rabiei, A., Saltz, D., Daujat, J. and Baker, K. (2015) *Dama mesopotamica*, Persian Fallow Deer. The IUCN Red List of Threatened Species.

5

Traditional sheep and goat husbandry in Cyprus: the effects of scale and its identification in archaeological assemblages

Angelos Hadjikoumis

Khirokitia village, Cyprus, 2013
Herder: *...so, are you a friend of Σίμος* (Simon's Hellenised name in Khirokitia village)?
Me: *Yes, I've known him for a few years and he is a good colleague of mine.*
Herder: *I thought so!*
Me: *What made you think that?*
Herder: *...because he also likes talking about sheep and goat, like you!*

Zooarchaeological, as any archaeological, interpretation relies on analogical reasoning. In many cases, the analogies employed as interpretative aids derive from ethnographic studies. Zooarchaeology in particular has been heavily reliant on ethnography to inform its interpretative framework, especially on animal husbandry practices. The majority of relevant studies, however, concern specialised pastoralists in large-scale systems. Medium- and small-scale animal husbandry practices, especially in mixed agro-pastoral systems, are less documented ethnographically. The scarcity of such ethnographic analogies is detrimental to zooarchaeological interpretation, as scale is a fundamental characteristic of any animal husbandry system and permeates most of its aspects. This study constitutes an effort to address this gap through an ethnozooarchaeological study of, now extinct, medium-scale traditional caprine husbandry practices in 20th-century Cyprus. Data collected through semi-structured interviews with sheep and goat herders are presented to illustrate how scale shapes important aspects of caprine management. Besides providing information of practical use to zooarchaeologists working in Mediterranean environments, this study also serves as a source of analogy for the detection and interpretation of ancient medium-scale systems.

Keywords: Ethnozooarchaeology, zooarchaeology, Cyprus, Mediterranean, caprine husbandry, archaeological interpretation

Introduction

As is the case for all archaeological interpretations (*e.g.* Wylie 1982), evidence of pastoral activities in the past is interpreted through analogical reasoning. In this endeavour, the most relevant source of analogy for archaeologists is ethnography. For the zooarchaeologist, an even more apt source is ethnozooarchaeology (*sensu* Albarella 2011), which differentiates from ethnography in that it is purpose-focused on collecting data suitable to address specifically zooarchaeological questions. Despite the abundance of ethnographic studies of zooarchaeological relevance in the literature, there is a bias in favour of large-scale specialised pastoralism (*e.g.* Digard 1981). The scarcity of ethnographic analogies for small- and medium-scale animal husbandry practices, currently limits the interpretative framework available to the zooarchaeologist. The archaeological record suggests that large-scale and highly specialised pastoralism was a chronologically later, and generally rarer, occurrence (Halstead 1996) than small-scale and medium-scale animal husbandry. Yet, the available interpretative framework for the latter is less developed and lacks in diversity.

The vast majority of caprine herds in this study can be considered as examples of *medium-scale* husbandry. Referring to caprine husbandry practices as *small-*, *medium-* and *large-scale*, however, is a subjective characterisation

Table 5.1. Herd composition and numbers, divided into 'predominantly sheep', 'predominantly goat' and 'mixed herds'.

Predominantly sheep herds				*Predominantly goat herds*			
Herder	*Practices*	*Sheep*	*Goat*	*Herder*	*Practices*	*Sheep*	*Goat*
1	Traditional	150	6	3	Traditional	0	50
2	Traditional	105	3	8	Traditional	0	200
3	Traditional	50	0	9	Traditional	0	160
4	Traditional	140	0	18	Traditional	10	150
5	Traditional	135	0	21	Traditional	0	180
7	Traditional	120	0	*6*	*Traditional*	*0*	*25*
10	Traditional	120	14	2	Recent	0	105
12	Traditional	100	0	6	Recent	0	100
13	Traditional	75	0	13	Recent	0	265
14	Traditional	70	0	16	Recent	0	200
15	Traditional	115	0		Recent	100	800
17	Traditional	125	0	*Mixed herds*			
22	Traditional	80	0	*Herder*	*Practices*	*Sheep*	*Goat*
23	Traditional	135	0	19	Traditional	80	30
4	Recent	245	0	14	Recent	150	65
12	Recent	300	20	20	Recent	200	100
20	Recent	300	2	22	Recent	150	200
23	Recent	400	0				
11	*Recent*	*31*	*1*				

Herds are considered 'mixed' if they contain a minimum of 25% sheep in a goat herd or vice versa. Within each category, traditional practices are presented first, followed by the more recent practices. All herders were occupied full-time in sheep/goat herding with the exception of two part-time herders (6 and 11), indicated in italics.

his main income from agriculture. Moreover, managing a small herd was an unpopular economic strategy, as it required a similar level of commitment as larger herds do, but was insufficient to sustain a family. Herder 10 added that only a few affluent families managed up to 200–300 animals, while poorer families would have to make ends meet with as few as 50–80 animals. A few herders with experience in the 1950s and 1960s mentioned that even smaller herds were viable. Herder 3 for example, was managing about 50 animals in the 1950s and 1960s. Several others with long family tradition also mentioned that in the late 19th/early 20th century families could survive on herds of up to 50 sheep/goat.

Sheep and goat herd sizes did not differ greatly in numbers, but a trend towards larger goat than sheep herds is evident (Table 5.2). Some of the largest caprine herds pre-1974 consisted of 160 (herder 9), 180 (herder 21) and 200 (herder 8) goats, while most sheep herds tended to fall within the 100–150 range. The reason for this is hinted at by the environmental setting of the areas where those larger goat herds were foraging. The aforementioned herders (8, 9 and 21) managed their herds in areas with few or no cultivations. The largest of these goat herds (herder 8) was

Table 5.2. Mean, minimum (Min) and maximum (Max) of 'predominantly sheep' and 'predominantly goat' herds (also see Table 5.1), under traditional and recent practices.

Species	*Mean*	*Min*	*Max*
Predominantly sheep, pre-1974 (N=14)	110	50	156
Predominantly goat, pre-1974 (N=5)	150	50	200
Predominantly sheep, post-1974 (N=4)	317	245	400
Predominantly goat, post-1974 (N=5)	314	100	800

foraging in an abrupt high-altitude landscape (100–1000 m asl) far from any cultivations (Fig. 5.1). The same herder also exploited a population of about 1000 feral goats through occasional hunting or capture of animals, but this was not included here as that population was not managed.

After 1974, caprine herds gradually increased in size as traditional husbandry practices were being eroded away (Table 5.2), as it has also been noted for other Mediterranean areas (*e.g.* Seguí 1999). From 1974 to about the late 1990s, herds tended to fall in the upper end of the 100–200 range, but closer to the present day, there were several examples of herds with more than 300 animals (herders 12, 16, 20, 22 and 23). An exception to this trend was herder 11, who kept small numbers of animals as he focused on fattening lambs for sale, so as to complement his main income. Commenting on the reasons behind the increase in herd numbers, the herders attributed these changes to mechanisation, cheap immigrant labour, as well as legislative and market pressures. Additional reasons mentioned included the availability of imported animal feed and an increase in available foraging areas due to further abandonment of agricultural land and the countryside in general. Echoing this general trend, several herders mentioned that nowadays, one would have to scale up and manage herds in excess of 500 animals to make a living. This view also illustrates the point that the absolute number of animals necessary to support a family is variable.

Herd size constraints

In 1935, during the colonial period in Cyprus (1878–1960), the maximum number of animals managed by a shepherd was restricted by law to 80 (Harris 2007). Some herders mentioned that in the summer months the law was relaxed and 120 animals were allowed per herder, but the author is yet to encounter documented evidence of this aspect in the legislation. It is possible that the herders were referring to local variations and circumstances. Most importantly, however, the herders confirmed that the legislation did not pose serious constraints, as it could be circumvented by obtaining multiple licences within a network of family and friends.

The most important restriction to herd size was labour requirements. In the absence of mechanisation, most herders could manage at most a few dozen animals, especially when focused on labour-intensive dairying activities. The exact number of animals each herder could manage depended on the availability of additional labour input from the family (usually young boys for herding and spouses for milking/dairying). Another major constraint in herd numbers was the density of cultivation in the foraging area and the consequent responsibility of the herders to avoid damaging the crops. In densely cultivated areas, herders had to be particularly vigilant to avoid conflicts with farmers. This rationale is also reflected in a practice followed by sheep herders in areas with extensive cereal cultivations that involved the fusion of two herds (*i.e.* reaching 200–300 sheep) after summer harvest and prior to autumn sowing, as the need for close supervision of animals was relaxed. The presence of cultivations also heavily affected decisions on whether to manage sheep or goats, with sheep being preferred in cereal-/cash crop-cultivated areas and goats in areas of wild vegetation or a patchwork of wild and cultivated vegetation. The traditional pattern of land use was characterised, for most of the agricultural year, by 100–150 strong sheep herds managed in areas of cereal cultivation (central plain and coastal lowlands) and 100–200 strong goat herds managed in the foothills of the two main mountain ranges (Fig. 5.1), away from densely cultivated areas.

Age-at-death and male:female ratios

Age-at-death data obtained through the interviews were the focus of a different paper (Hadjikoumis 2017) and, hence, findings are only briefly summarised here, focusing on their relevance to the scale of herding. The interviewees mentioned species (*i.e.* sheep or goat) and economic strategies (*i.e.* focus on milk and/or meat), rather than scales of operation, as the most important factors determining age-at-death. Nevertheless, the trend towards an increase in herd size in the last few decades was accompanied by a significant increase in the average age-at-death of lambs and kids (Hadjikoumis 2017, tables 15.2, 130). The herders attributed this, almost exclusively, to technological advances, adoption of modern husbandry knowledge and, most importantly, market forces favouring heavier lambs/kids. The opposite trend was observed in adult sheep and goat, both males and females (Hadjikoumis 2017, tables 15.3–4, 131–32). In this case, besides recent technological and economic developments, several herders attributed this trend also to an increase in the scale of operations in the last 10–20 years. The rationale behind the younger age-at-death among grown-up animals in larger herds is that there is more scope for faster replacement of animals once their productivity starts dropping in terms of milk and reproduction. Put simply, the herders claimed that the larger the herd, the more feasible it is to approach optimisation in productivity, especially when combined with modern technical knowledge and means for its implementation in more recent practices. Another notable concomitant change with the increase in herd numbers is the weakening of emotional bonds between herder and animal. In several cases, traditional herders mentioned that they kept specific animals even to senile age (>15 years) merely due to affection. To sum up, the age-at-death information from traditional Cypriot caprine husbandry shows a tendency towards younger age-at-death among adults, associated with an increase in herd numbers.

Male:female ratios do not exhibit a strong directionality in their fluctuations (Table 5.3). Overall, traditional caprine herding in Cyprus favoured male:female ratios of 1:23–1:40 for sheep and 1:20–1:45 for goat, with the mean at 1:30 and 1:28 respectively. A few herders commented that at the beginning of their careers they started with herds consisting of 10–20 females and a male (*e.g.* herder 15: 17 ewes and 1 ram), thus suggesting that small herds require proportionately more males than medium and large herds. It is important, however, to note that this is not necessarily the case if communal access to male animals for reproduction is practiced or provided as a service in exchange for goods (Xioutas 2001).

Moreover, a few herders hinted at the need for more males in goat than sheep herds with the rationale that female goats exhibit a more restricted ovulation period compared to ewes. More males are therefore required to achieve conception within one ovulation cycle. The fact that such a

Table 5.3. Male: female ratios among adult animals, provided by the interviewed sheep/goat herders. The size of the herd for which information was provided is indicated in parenthesis.

Herder	*Male: female ratio*
1	Sheep 1:37 (150)
2	Sheep/goat 1:26 (105)
3	Sheep/goat 1:25 (50)
4	Sheep 1:35 (140)
5	Sheep 1:23 (135)
6	Goat 1:20 (100)
7	Sheep 1:33 (120)
8	Goat 1:23 (200)
9	Goat 1:25 (160)
10	Sheep 1:40 (120)
11	Sheep 1:25 (31)
12	Sheep 1:28 (100)
13	Sheep 1:30 (75)
14	Sheep 1:30 (70) Goat 1:22 (65)
15	Sheep 1:35 (115)
16	Sheep/goat 1:28 (100/200)
17	N/A
18	Goat 1:30 (150)
19	Sheep 1:23 (80) Goat 1:30 (30)
20	Sheep 1:25
21	Goat 1:45 (180)
22	Sheep 1:27 (80)
23	Sheep 1:33 (135)
Mean	Sheep 1:30 (112) Goat: 1:28 (119)

trend is only very slightly reflected in the information they provided (Table 5.3) might be attributed to two factors. The firstly could be the lack of motivation (and/or knowledge) in traditional caprine husbandry to focus on a short birth season (Hadjikoumis *et al.* 2019). Secondly, in their accounts of male:female ratios, herders only considered subadult/adult bucks without including younger bucks (usually >1 year) used in reproduction and for promoting the 'male effect' (*e.g.* Murata *et al.* 2014). In fact, there is a strong belief among traditional Cypriot herders that in bucks, semen quality deteriorates with age, while in rams it improves or at least does not deteriorate (Xioutas 2001, 15).

Mobility

The type of mobility documented by the interviewed herders is of particular relevance to zooarchaeologists studying faunal assemblages from insular or coastal Mediterranean environments, especially contexts where animal husbandry is geared towards generic milk-meat exploitation, tightly knitted with agriculture. All interviewed herders commented on the mobility of their herds but also provided additional useful information on other related activities, such as sheep washing in the sea. Their herds foraged in lands owned by their or adjacent communities and involved a moderate level of mobility with distances of 3–12 km covered (Table 5.4), without any form of transhumance exercised. This level of mobility involved daily round trips from the herd's base to different foraging areas. Moreover, there was little difference between sheep and goat herds in the distances covered. Traditional caprine husbandry in 20th-century Cyprus was defined by this level of mobility but, in the more distant past, longer distance mobility was also an option. One of the herders mentioned that when he was a child, his family travelled 30 km on foot with their goats and sheep to stay for 3–4 months with relatives, taking advantage of access to stubble and agricultural waste after cereal harvest. It is plausible, thus, that in prehistoric (*e.g.* Neolithic) small communities for example, family ties such as those mentioned by the herder may have been activated seasonally to improve labour distribution and productivity in both farming and pastoral tasks.

Another aspect of traditional Cypriot husbandry related to mobility was the now-extinct practice of washing sheep, mainly in the sea but also in freshwater pools wherever an accessible coast was unavailable or too far (Table 5.4). Washing at sea was almost ubiquitously practised by sheep herders when distances from the sea were no more than 10–11 km; otherwise they opted for freshwater pools, natural and artificial. As for the reasons cited for this practice, the herders listed a series of beneficial effects for their animals such as the healing of wounds (especially after shearing) due to the disinfectant quality of salt, the elimination of parasites, and even the overall well-being of the animals. Goat herds were usually not subjected to this practice but, in cases of mixed herds, they were also driven with the sheep in the sea. Besides contributing to the health of caprine herds, such locations in the landscape played the additional role of meeting points that enabled herders to exchange/sell animals, strike other deals of economic nature and exchange information about the landscape and other communities.

Another mobility-related, and also now-extinct, practice in traditional Cypriot caprine husbandry was nocturnal grazing, mainly of sheep, from May through the summer, especially in the more arid eastern Cyprus. After the harvest of cereals, sheep and goat herds were allowed to feed on agricultural waste, stubble and fallen seed. During this period, usually two herders combined herds and took turns in sleeping outdoors to tend the animals. Consequently, mobility increased in the summer and was reduced in winter, also due to weather conditions. Moreover, between years, herders changed their itineraries in accordance to

Table 5.4. Information on the mobility (foraging range) of caprine herds managed traditionally, combined with distance from and interaction with the sea.

Herder	*Composition*	*Foraging range*	*Distance from sea & interaction*
1	Mainly sheep	4–5 km	2–3 km, sheep washed every 15 days in late spring and summer
2	Mainly sheep	6 km	2 km, sheep (and goat) washed every 20–30 days in summer
3	Sheep & goat	6–7 km	8–9 km, no interaction
4	Sheep	5 km	4–5 km, sheep washed
5	Sheep	Usually 3–4, occasionally up to 11–12 km	12 km, sheep washed a few times each summer
6	Goat	Little mobility (small-scale)	3 km. Most herders, although not him, washed sheep/goat in summer
7	Sheep	5 km	2 km, sheep were washed once a week in the sea after shearing (late March–early April)
8	Goat	Mostly free-range goats	N/A
9	Goat	2–3 km	Goats forage near the sea
10	Mainly sheep	4–5 km	25 km, no interaction (sheep washed in water pools for irrigation)
11	Sheep	<2 km	8–9 km, no interaction
12	Sheep	3–4 km	1 km, regularly washed sheep after shearing (April–August)
13	Sheep & goat	7 km	10–11 km, no interaction (sheep washed in nearby stream)
14	Sheep & goat	5 km	17 km, no interaction (washed sheep in pools of water or streams)
15	Sheep	3–4 km	2–3 km, sheep washed in sea every 8–10 days, June–October
16	Sheep & goat	Sheep 1 km, goat 2–3 km	4–5 km, sheep (not goat) washed in the past after shearing
17	Sheep	3–5 km	3 km, sheep washed in summer
18	Mainly goat	4–5 km	16–20 km, no interaction generally in the area (mainly goat herds)
19	Sheep & goat	5 km	10–11 km, sheep and goat washed in sea monthly (May–September)
20	Mainly sheep	3 km	3–4 km, sheep washed twice a year
21	Goat	3 km	20 km, no interaction
22	Sheep	usually 4–5 km, rarely 7 km	7 km, sheep washed twice a year (June–July)
23	Sheep	1.5 km	25–30 km, no interaction. Sheep washed in natural/artificial pools (summer)

community-wide agreements designating agricultural and pastoral use of the land. Most herders also owned land and cultivated crops such as cereals and pulses for their animals to feed on.

Discussion

The different aspects of traditional Cypriot caprine husbandry, presented here and elsewhere (Hadjikoumis 2017; 2018; Hadjikoumis *et al.* 2019), compose the picture of a pastoral system well-integrated with agriculture and adapted to a semi-arid Mediterranean insular environment. Though unique in its specific aspects, this system's overall form is almost identical to the one defined by Halstead (1996, 23) as 'sedentary pastoralism'. Agropastoral systems involving medium-scale caprine husbandry can vary widely in terms of their agricultural and pastoral components, as well as management techniques and economic strategies. There are, however, commonalities stemming from the practicalities of caprine husbandry on the medium scale, which can serve as valuable analogies in archaeological contexts. As a norm, however, the remains of individual herds are not deposited separately and hence, are not amenable to study in the same way as in ethnoarchaeology. Nevertheless, the scale of ancient pastoral systems often leaves recognisable patterns in different categories of data, which helps zooarchaeologists identify it through analogy to herds documented ethnoarchaeologically.

Herd size and composition

One of the defining elements of traditional caprine husbandry in 20th-century Cyprus was a herd size composed of 50 to 200 animals, in predominantly (or exclusively) sheep or goat herds (Table 5.1). The factors that affected the reported herd sizes are relevant to zooarchaeological interpretation. The herd sizes documented (Table 5.1) were deemed adequate by the herders to cover most of their family needs. The exact size of a herd that adequately supports a family, however, can vary significantly depending on economic context, as well as variability between families and the concept of family in different periods. While 100–120 sheep/goat were sufficient to make a living in 20th-century Cyprus, this was not necessarily the case in earlier periods. The trend within the 20th century in Cyprus was towards smaller herds in the first half of the century, with herds as small as 50–60 animals adequately supporting families of five children or more (Given 2000; Xioutas 2001, 113). Similar information has also been documented in Greece (Halstead 2014, 293; 1998, 213) and Spain (Seguí 1999, 69–70). Extrapolation of herd sizes from 20th-century Cyprus to ancient periods requires caution due to several unknowns on the economies and family/social structures involved, but it is more reasonable to assume that medium-scale caprine husbandry in the past would have involved numbers towards the lower end of the range reported by Cypriot herders.

Besides family needs, herd sizes were also heavily affected by the herder's responsibility to protect plant cultivations, at least in agricultural contexts such as those present in 20th-century Cyprus (wheat/barley, viticulture, and arboriculture). Some of these agricultural options were not available in prehistory, but it can be safely assumed that ancient herders would have also had to protect cultivations from their animals. All herders interviewed in this study made it abundantly clear that density of cultivations in their herd's forage area dictated how many animals they could control without causing damage. The trend for larger average herd size in goat (Table 5.1), compared to sheep herds, is partly attributable to the fact that goats traditionally were led to forage in areas of wild vegetation, further away from cultivations. In archaeological contexts, the integration of other lines of evidence (especially archaeobotanical, field systems and land use evidence), can help in approximating the level of care required by caprine herders to avoid agricultural damage.

Several herders highlighted the importance of good herding dogs in facilitating animal control, while others mentioned the use of a few goats to push sheep herds away from cultivations faster. The identification of the remains of these animals in zooarchaeological assemblages in contexts of predominantly sheep-based agropastoral economies may be pointing towards their use in controlling sheep herds and averting agricultural damage. Moreover, plant cultivation by the herders so that their animals can feed on directly before ripening, constituted yet another measure to divert pressure from cultivations destined for human consumption. Consequently, cases where unripe seeds and other elements of plants (*e.g.* legume pods or cereal ears) are identified in ancient animal dung, could be also interpreted in terms of land ownership by herders, besides as solely the result of agricultural damage. The distinction between these two scenarios is particularly difficult but archaeological interpretations are only further enriched by the consideration of both, as well as providing the incentive for refined archaeobotanical analyses that could potentially render them separately identifiable.

Age-at-death and sex ratios

The zooarchaeological potential of the age-at-death data collected in this study has been more extensively discussed elsewhere (Hadjikoumis 2017), although the focus was on management rather than scale. While economic strategy is the dominant factor in deciding the age-at-death of lambs and kids, scale affects the age-at-death of older animals. Even medium-scale systems such as those described here are prone to maintaining most reproductive animals beyond their optimal age range. Indeed, in traditional Cypriot caprine husbandry it was common to keep female sheep and goat in declining productivity beyond 10 years and even much older. This is attributable to the more restricted capacity for herd replacement and even the stronger emotional bonds, both emanating from the relatively restricted herd size of medium-scale husbandry. On the other hand, large-scale transhumant pastoralists such as the Baxtyari, free from such constraints, can afford to replace ewes at 6–10 and does at 7–8 years (*e.g.* Digard 1981, 36), and similar ages are reported from large communal sheep herds in Turkey (Payne 1973, 301). It is, therefore, conceivable that herders involved in medium- or small-scale caprine husbandry in the past were also prone to keeping animals well beyond their productively optimal age. In 20th-century Cyprus at least, maintaining female caprines beyond their most productive age range occurred under a strategy primarily aiming at milk and tender meat production, rather than wool.

Scale had a smaller effect on male:female ratios. Beyond the direct interpretative value of the 1:28–30 average ratio (Table 5.3) for medium-scale caprine herds, a few herders suggested that larger herds require fewer males. This approach is supported by ethnographic information on large-scale caprine herders such as the Baxtyari of Iran involving ratios of 1:30–50 for sheep and goat (Digard 1981, 36) and 1:50–60 ratio reported for a large communal sheep herd at Aşvan in Turkey (Payne 1973, 301). This ratio is, however, dependent on a variety of other factors, such as the seasonality of birth and knowledge of animal reproduction. Overall, it constitutes an important trend to be taken into account in zooarchaeological interpretations, especially involving the integration of sex ratios with age-at-death in determining scale (*i.e.* higher male:female ratio combined with younger age-at-death among adults in large-scale vs lower ratios combined with older age-at-death in medium-/small-scale husbandry). Such combinations of multiple lines of evidence can also potentially help in telling apart deliberate economic strategy from the effects of scale (*e.g.* keeping caprines to old age due to lack of capacity for earlier replacement rather than a focus on wool production).

Mobility

Mobility for most caprine herds in 20th-century Cyprus was a short distance affair, involving daily round-trips to distances of 3–12 km, with distances of around 5 km being the norm for most of the year (Table 5.4). Among other factors, the level of mobility is affected by scale, with large-scale pastoralism being characterised by long-distance mobility (Digard 1981; Halstead 1996, 22) and small-scale caprine husbandry being almost stationary. It can be logically assumed that scale had similar effects in ancient husbandry systems as well, due to the practicalities involved in each scale of operations. Leading a small herd to forage farther away than the outskirts of a village would represent a bad investment of an experienced herder's time, resources and skill. This is also illustrated by the traditional practice of delegating the herding of small numbers of caprines (especially goats) to young members of a family. At the other end of the scale, the logistical complexity and scope for agricultural devastation associated with moving caprine herds on the large scale, dictates that such movements are well-organised in advance and the herds involved are kept at safe distance from growing cultivations. The level of mobility of caprine herds in Cyprus (Table 5.4) falls between the two approaches laid out above, and is characteristic of medium-scale caprine husbandry integrated in an agropastoral system.

The most general attributes of the system such as the geographic and environmental context (*i.e.* large Mediterranean island with semi-arid climate in this case) act as the backdrop for this type of mobility, and thus require consideration before applying information to archaeological data. The low frequency of climatic extremes, the dependence on water wells with strict access rights in some areas, the relatively small scale of the insular environment and the integration of the pastoral and agricultural land-uses, all contributed to the evolution of moderate levels of mobility associated with medium-scale husbandry in Cyprus and possibly similar archaeological contexts. As far as the animal species managed are concerned (sheep, goat, or mixed), it is intriguing that all herders noted that goats can forage faster and farther than sheep, but this difference is not reflected in the distances covered by their herds (Table 5.4). This discrepancy suggests that this level of mobility does not positively 'select' goat herds for their advantages cited by the herders. Nevertheless, the practice of including a few goats in predominantly sheep herds (Table 5.1), in this specific context, was based on the logic that they are leading sheep away from cultivations faster, thus reducing the risk of agricultural damage. Although it does not affect mobility *per se*, it provides yet another factor contributing positively to the integration of caprine husbandry and agriculture.

Other factors exerting an effect on mobility levels, as well as relating to scale, include the practices of combining two herds (mainly sheep) over the summer months and washing animals. The former practice had the purpose of alleviating pastoral labour requirements to allow much needed rest or diversion of labour to other activities (cheese making, processing of agricultural harvest, etc.). Moreover, it illustrates once more the effect that the need to protect growing crops had on the scale of caprine husbandry, as well as its intra-annual adaptations in response to constraints (*i.e.* plant growing phase) and opportunities (stubble and fallen seed after cereal harvest). The latter practice, of washing sheep mainly in the sea, is rather unrelated to scale as it can be potentially adopted by all scales of husbandry, but it does provide an incentive to herders for covering longer distances. Its documentation in the ethnographic record (*e.g.* Ryder 1984; Santillo Frizell 2004), however, is also relevant to archaeology. The often centuries-long use of the same locations for washing animals led to modifications, potentially identifiable in the archaeological record, such as the constructed steps or ramps and the levelling of rock surfaces from animal trampling. On the interpretative level, this practice (more detail in Hadjikoumis 2018) also provides an additional mechanism for the transmission of cultural information and carrying out economic transactions.

The approximation of distances involved, the exact landscapes exploited and the identification of finer details on mobility-related practices in the archaeological record, require the application of multiple lines of evidence. Some of the most promising proxies in studying mobility in the past include the identification of relevant structures (*e.g.* field systems with 'corridors' to allow safe animal access, modified washing spots, etc.), archaeobotanical data addressing questions on foraging environments and animal diet, as well as the taxonomic and 'demographic' composition of the herds. Concerning the fusion of herds after harvest and prior to sowing (*i.e.* summer), any category of data (taxonomic, age-at-death, stable isotopes, etc.) that could help detect a seasonal absence of herds from the settlement would constitute potential evidence for this practice.

Concluding remarks on scale: archaeological opportunities and challenges

Through its study of traditional animal husbandry practices, ethnoarchaeology contributes plausible management variants. These enrich zooarchaeology interpretative framework and promote the development of analytical tools to detect them in the archaeological record. This study contributes new ethnozooarchaeological data on caprine husbandry, relevant to zooarchaeological analysis and interpretation. The husbandry system presented in this study and elsewhere (Hadjikoumis 2017; 2018; 2019), constitutes a valuable source of analogies for medium-scale

6

Among hyenas: Nery Delgado, Albert Gaudry, Edouard Harlé and the hyenas of Furninha cave (Portugal)

João Luís Cardoso

The excavation of the Furninha cave, located on the coast of the Peniche Peninsula (Central Portugal), was directed by Nery Delgado in 1879/1880, and provided the largest European assemblage of Hyaena prisca/Hyaena hyaena prisca. *This is only comparable to the collection from the cave of Lunel-Viel, in the French Midi, from which the taxon was defined in 1828 by Marcel de Serres and collaborators. Delgado (1884) considered the existence of two distinct species:* Hyaena vulgaris *(=* H. hyaena*), the striped hyena, a species presently living in North and East Africa, as well as West and South Asia; and the extinct* Hyaena prisca, *which is of larger size. Albert Gaudry, who was consulted by Delgado, suggested the presence of a single species, known from Late Middle Pleistocene sites, which would anatomically differ only in size and robustness from the current striped hyena. This same collection was later studied by Edouard Harlé, who followed Gaudry's opinion, which has generally been accepted in the course of the 20th century. However, more recent studies carried out by the author, as well as J.-P. Brugal and collaborators (2012) have revealed that, besides its larger size and robustness, there are also morphometric differences between the current striped hyena and the Furninha hyena. These differences are supported by statistical tests, including PCA (principal component analysis). Although only one species appears to be present at Furninha, the difference in size observed by Nery Delgado, possibly due to geographic variation, suggests some morphological and size difference between* H. prisca *and the current striped hyena (*Hyaena hyaena*).*

Keywords: Furninha; *Hyaena prisca;* Nery Delgado; Albert Gaudry; Edouard Harlé

Joaquim Filipe Nery Delgado: a brief biography

Nery Delgado (1835–1908) was a Portuguese army engineer, geologist and archaeologist (Fig. 6.1). From 1857, he was remarkably active as assistant of the Second Geological Commission of Portugal. He assumed the direction of this institution in 1882, immediately after the death of Carlos Ribeiro. The value of his performance is attested by the successive promotions in his military career, reaching the rank of Lt. General. After 51 years at that institution, his sudden death in 1908 ended this glorious career when he contracted double pneumonia during his then to be last fieldwork season.

He was also was notable for his geological survey of the country, which resulted in the 1:500,000 scale edition of the Geological Map of Portugal published in 1876. This was almost immediately followed by a second edition two years later. It should be noted that a first, hand-coloured, sketch of this map, was awarded a prize at the Paris Universal Exhibition in 1867. In addition, he was actively involved in the stratigraphic and paleontological characterisation of the main geological formations of Portugal, especially those of Palaeozoic age. In this context, the Bilobites of the Beira quartzites and those related to the Palaeozoic of the Alentejo are particularly noteworthy (Choffat 1908).

As a geologist and archaeologist, he was internationally renowned, as proved by the correspondence he maintained with some of the most notable geologists and archaeologists of his time (Cardoso and Melo 2001). Nery Delgado, with Carlos Ribeiro and Francisco Pereira da Costa, embodies the emergence in Portugal of a scientifically-based research

Figure 6.1. Nery Delgado (1835–1908). Archive and photo by the author.

methodology for prehistory, closely linked to that of geological studies, whose methodological approach he was prepared to borrow (Cardoso 2020).

Furninha cave (Peniche)

'Furninha da descida do mar', or simply 'Furninha' (small cave) is a karstic cavity located on the escarpment of the southern coast of the Peniche Peninsula, in a Jurassic limestone formation, whose opening was caused by marine erosion (Fig. 6.2). During the last interglacial the sea level was 15 m above the current level, which is demonstrated by a sharp platform in front of the cave entrance. It was during this period that the opening of the cave formed, along with a vertical well in its interior. The base of the well was filled by a conglomerate that can be correlated with the more recent sea level of 5–8 m, observed on the exterior of the cave.

In the report of the Geological Works for the economic year 1879/1880, details of the excavation were reported (Relatório 1881, 19–20): 'The deputy [adjunto] of the section, Major Nery Delgado [...] In the Peniche cave, which was completely empty, besides the objects in the upper deposit of Neolithic age, [...], a partially extinct fauna [...] was obtained in the lower deposit and at different successive levels [...], that evidently was contemporary of man, as evidenced by the numerous bones, broken longitudinally, accompanying flint flakes and some instruments of this same substance, in which is evident an intentional action or intervention of an intelligent being [...]. This officer made some short-term excursions to the sites where the caves' explorations were carried out, which he immediately had to examine' (Translated from Portuguese by the author; English version revised by Roz Gillis).

This description allows us to conclude that Nery Delgado was not permanently at the excavations of Furninha cave. Labels preserved today on some of the pieces of the LNEG Geological Museum also allow us to situate the first excavations taking place in 1865. The archaeological excavations carried out in the cave in 1879 and in the early 1880s by Nery Delgado were essentially due to the need to present new archaeological and paleontological findings at the IX Session of the 1880 International Congress, held in Lisbon. The previous Geological Commission had been deprived of almost all the collections, due to their move to the Escola Politécnica, (Polytechnic School) following the 1868 extinction of the Commission decreed by the Government (Cardoso 2013).

The lower deposit consisted of alternating sandy and faunal beds that accumulated above the basal conglomerate level of the 5 to 8 m beach. These were only observed in the vertical well identified inside the cave. The sedimentary sequence accumulated above this basal conglomerate level appears to be contemporaneous with the early Würm period. This is suggested by a uranium/thorium dating, though with large uncertainty limits (80 886 +42 423, –31 265 BP), since it was not possible to date any bone by radiocarbon method due to the lack of collagen (Cardoso 1993).

Some Middle and Upper Palaeolithic materials without known stratigraphy do occur, though these were not mentioned by Delgado (1884). The Upper Palaeolithic is represented by several techno-complexes, some of which have long been identified (Breuil and Zbyszewski 1945). They were recently reviewed (Bicho and Cardoso 2010), as well as the Neolithic remains (Cardoso and Carvalho 2010/2011), associated with two Neolithic necropolises.

Pleistocene faunas originate exclusively from the lower deposit explored in the well that opened vertically within the cave, corresponding approximately to a sequence of about 8 m in depth. The stratigraphic sequence was composed of alternating beds of sterile yellowish sands, as a result of wind transport from the vast exposed emerged area to the west, with seven bone levels. The bones were in an excellent state of preservation with abundant manganese staining on the surface. The fauna content appears to be similar in all

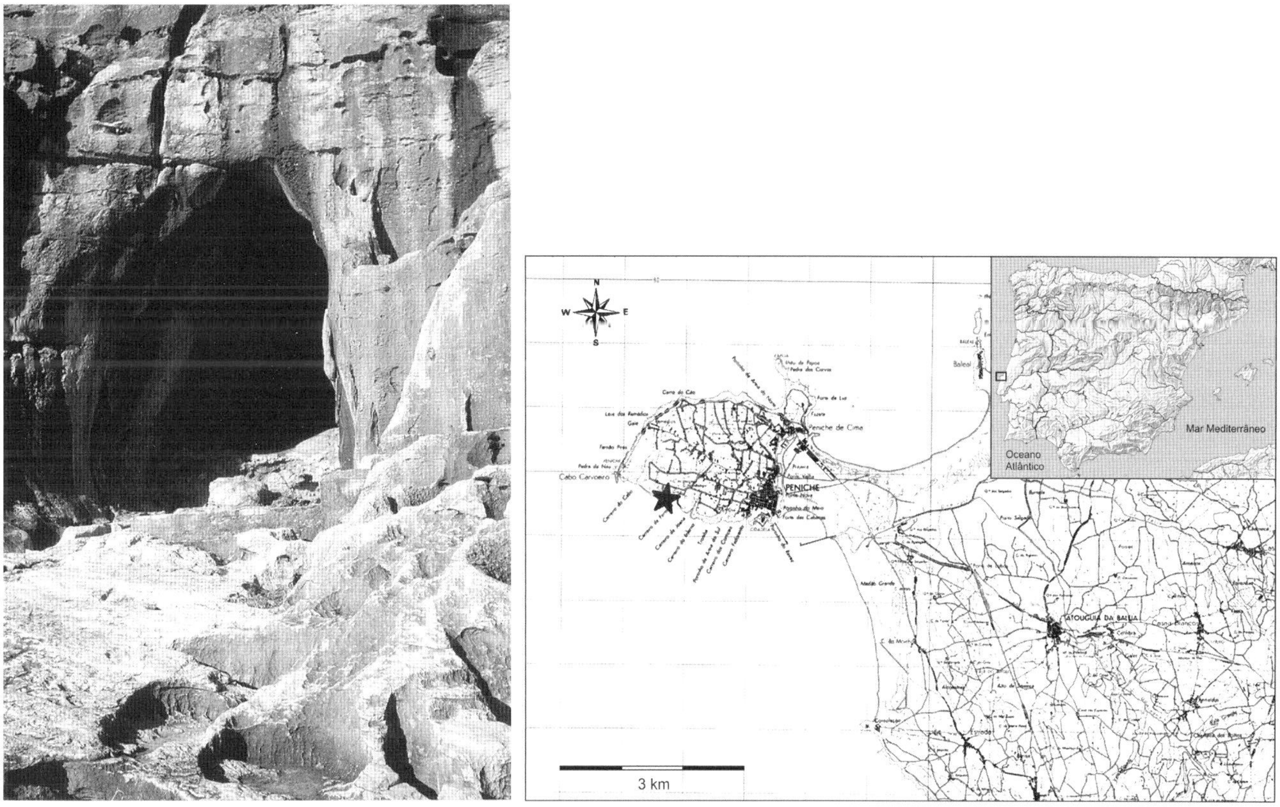

Figure 6.2. Location of the Furninha cave and view of its entrance. Photo and map by the author.

levels, which has led various authors to consider it as a single assemblage (Harlé 1910/1911; Cardoso 1993).

The identified seven bone levels reveal the occupation of the cavity essentially by hyenas, whose remains accumulated in great quantity along with their prey (horse, deer and aurochs) and coprolites show a recurring presence. The cave also served as a temporary refuge for other carnivores, such as bear, wolf, lynx and fox (Cardoso 1993). Humans used the cave at the same time as a shelter. This is evidenced by Middle Palaeolithic material, coeval to the Pleistocene occupation, along with a child's jaw, presumably belonging to a Neanderthal (Delgado 1884), collected from the third and richest bone level.

The importance conferred by Nery Delgado to the description of the Pleistocene deposits is demonstrated by his careful recording. The details of find location, in terms of level and depth, were marked individually on hundreds of the collected pieces. This care and attention to detail, as well as the unusual ability to report on the collected scientific data, confers to Nery Delgado the status of a pioneer of modern archaeological methodology (Zilhão 1993; Cardoso 2008).

The Furninha hyenas and their story

Among the remains of extinct faunas collected in the Furninha cave, the presence of *Hyaena hyaena prisca* deserves to be highlighted. Some authors prefer to call it simply *Hyaena prisca*, considering it as a separate species from the extant *Hyaena hyaena* (also known as *Hyaena striata*).

Marcel de Serres in 1828 was the first to propose the existence in Europe of a hyena species, which he called *Hyaena prisca,* which was different form the European cave hyena (*Crocuta crocuta spelaea*). With Dubreuil and Jeanjean, he characterised the first remains attributed to this species, from their explorations of the Lunel–Viel caves (SE of France) (de Serres *et al.* 1828). In the same year, another article was dedicated to the same faunal remains (Christol and Bravard 1828) designating them as belonging to *Hyaena monspessulana*, but without illustrating or describing the taxon. For this reason the original designation by de Serres and his collaborators has prevailed, despite the fact that Christol and Bravard's article was published first.

Currently the Furninha collection constitutes the most important *Hyaena prisca* fossil record on the European continent, even more so than those from the Lunel–Viel assemblage, which are stored at the Laboratoire de Géologie du Quaternaire de Marseille–Luminy (CNRS), and were examined personally by the author in 1989.

Following de Serres *et al.'s* (1828) publication, *Hyaena prisca* was considered to be a form of *Hyaena hyaena* in some Pyrenean deposits of the Middle Pleistocene:

Montsaunès (Harlé 1894) and Es–Taliens (Harlé 1895). Conversely, the bone remains from Montmaurin were classified as *Hyaena brunnea* (Boule 1902).

We can, however, be confident that the Furninha's material does not belong to *H. brunnea*. A modern jaw mould from Southern Africa observed at the Laboratoire d'Anatomie Comparée (MNHN de Paris) has a clearly distinct morphology that separates this piece from the Furninha remains (Cardoso 1993, 391). The Furninha remains were attributed to *H. hyaena* by Harlé (1910/1911), except one or two fragments later mentioned by Bonifay (1971), although she did not examine them personally. This conclusion put an end to a long period of uncertainty about the specific attribution of the remains of hyenas collected from Furninha. This uncertainty dates back to the time when Nery Delgado (1884) studied them. The eminent Portuguese geologist considered the occurrence of two forms (Delgado 1884, 239–240): 'One [of them] more frequent, intimately related to *H. vulgaris* [*i.e. H. hyaena* in modern terminology], and the other much rarer and of a larger size, which we suppose to be *H. prisca*' (original text in French).

The uncertainty of the taxonomic status of these two forms justified the consultation by Nery Delgado of the French palaeontologist Albert Gaudry (1827–1908) (Fig. 6.3), to whom he wrote two letters in 1885. These were accompanied by photographs of a skull and two hemimandibles, one of them attributed to the photographed skull. The official justification for such an initiative was presented in the following terms:

> Having the wise professor of Palaeontology at the Paris museum, Gaudry, presented to the Institute a note, in which he noted with surprise that the species of fossil hyena found in France's quaternary deposits is not the same as that found in Algeria, but rather that which is seen in Southern Africa, and today is not found north of the 17th degree latitude; the head of the section, who had previously expressed a different view from Portugal, arguing that the two species coexisted here in the fourth epoch, felt that he should address himself to this expert and to receive his opinion on the matter; and by the correspondence exchanged with him, he was pleased to see his appraisals confirmed (Relatório 1886, 18).

The drafts of these missives written by Nery Delgado are preserved at the LNEG Historical Archive. Unfortunately, Gaudry's response to these letters has been not found. Similarly, the original letters received by Gaudry are not included in the archive of his correspondence, kept at the MNHN (Paris), as was personally verified by the author in 2015.

Following Edouard Harlé's (1850–1922) interest in studying the Pleistocene mammals and birds of the Portuguese territory, Nery Delgado refers to the prior studies undertaken by himself and Albert Gaudry in two letters sent to the French scholar in 1897 and 1899. These two documents can be considered as the genesis of the paper presented by Harlé on the hyena of Furninha (Harlé 1909) and his remarkable synthesis of the Pleistocene mammals and birds known from Portugal (Harlé 1910/1911, 30).

Figure 6.3. Albert Gaudry (1827–1908). Marble portrait in the Muséum National d'Histoire Naturelle, Paris. Photo by the author.

More recent work (Kurtén 1968) has supported Albert Gaudry and Edouard Harlé's opinion that the hyena fossil represented by the Furninha remains, belong to a fossil form of the modern-day striped hyena, which could be referred to as *Hyaena hyaena prisca.* The only distinction between the modern and fossil forms is the greater size and robustness of the latter.

The correspondence between Nery Delgado and Albert Gaudry

Two letters written by Nery Delgado in French addressed to Albert Gaudry, are translated below in full, keeping the original form. The letters are in a draft form as the final version received by Gaudry could not be found.

1st letter. Pack 91 – Folder 1 (Fig. 6.4)

Autograph draft letter sent by Nery Delgado to Albert Gaudry.

Lisbon, April 8, 1885

Dear Sir and most honoured master.

I hope you excuse me for the inconvenience I am about

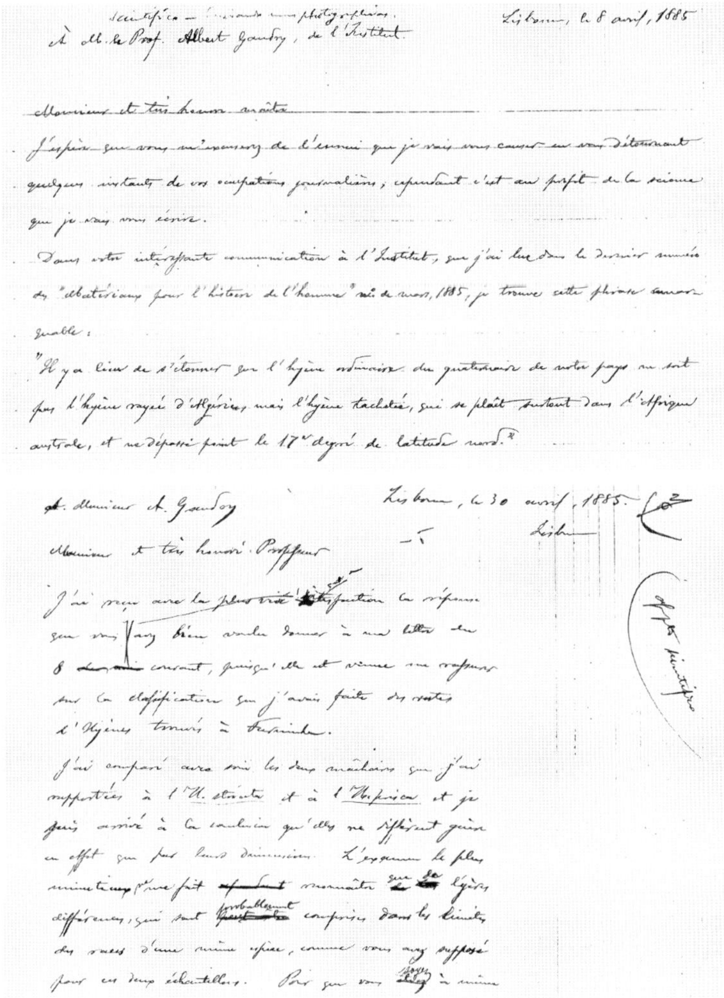

À M. le Prof. Albert Gaudry, de l'Institut Lisbonne, le 8 avril, 1885

Monsieur et très honoré maître

J'espère que vous m'excuserez de l'ennui que je vais vous causer en vous détournant quelques instants de vos occupations journalières; cependant c'est au profit de la science que je vais vous écrire.

Dans votre intéressante communication à l'Institut, que j'ai lue dans le dernier numéro des "Matériaux pour l'histoire de l'homme" nº de mars, 1885, je trouve cette phrase remarquable:

"Il y a lieu de s'étonner que l'hyène ordinaire du quaternaire de notre pays ne soit pas l'hyène rayée d'Algérie, mais l'hyène tachetée, qui se plaît surtout dans l'Afrique australe, et ne dépasse point le 17e degré de latitude nord."

A Monsieur A. Gaudry Lisbonne, le 30 avril, 1885

Monsieur et très honoré Professeur

J'ai reçu avec la satisfaction la réponse que vous avez bien voulu donner à ma lettre du 8 courant, puisqu'elle est venue me confirmer sur la classification que j'avais faite des restes d'Hyènes trouvés à Furninha.

J'ai comparé avec soin les deux maxillaires que j'ai rapportées à l'H. striata et à l'H. prisca et je suis arrivé à la conclusion qu'elles ne diffèrent guère en effet que par leurs dimensions. L'examen le plus minutieux n'a fait reconnaître que de légères différences, qui sont probablement comprises dans les limites des races d'une même espèce, comme vous avez supposé pour ces deux échantillons. Pour que vous ... à même

Figure 6.4. Partial reproductions of the two drafts of the letters sent by Nery Delgado to Albert Gaudry in 1885. LNEG (Lisbon). Photo by the author.

to cause you by diverting you for a few moments from your daily occupations; however, it is for the benefit of science that I am going to write to you. In your interesting communication to the Institute, which I read in the last issue of 'Matériaux pour l'histoire de l'homme' (Materials for the History of Man), issued in March, 1885, I found this remarkable sentence:

> 'It is astonishing that the ordinary hyena of the Quaternary of our country is not the striped hyena of Algeria, but the spotted hyena, which occurs particularly in southern Africa, and does not go beyond the 17th degree of north latitude.'

This sentence contains an important correction to what I wrote in my description of the Furninha cave (Report of the Prehistoric Congress of Lisbon, page 240), where, speaking of the remains of hyena found in the Quaternary deposit of this cave, I argued that the two species of hyena were only found in northern and southern Africa. (See Blainville: genus Hyaena*, page 74). Moreover, as the species of hyena which was more abundantly in Furninha's cave does not appear to differ from the* H. crocuta*, it occurred to me the idea to let you know the main pieces on which I based my classification.*

I wrote in my description of Furninha (pp. 239–240):

> 'Among the remains of hyena, one can easily distinguish two species; one more frequent, closely related to *H. vulgaris* [*i.e. H. hyaena*], which currently lives in northern Africa; and the other, much rarer and larger, which we supposed to be *H. prisca* (M. de Serres)'.

The presence of the first species in the cave of Furninha is very interesting, because in another cave, near Cercal, recently explored by my illustrious superior, Mr. Carlos Ribeiro, a cave that is only a short distance from that of Furninha (about 37 kilometres), this species seems to be missing, and in its place H. spelaea *has been discovered, which has, as we know, the most intimate relation of affinity with* H. crocuta*, which currently lives in southern Africa.*

I should further add, at least for the foreign reader, to make easier the understanding of this passage, that the Furninha cave is located on the coast, while the cave of Cercal is placed on the slope of a hill in the interior, where the water of the Tagus and the Atlantic mingle, at an altitude of over 300 meters, where the climate is now very different from that of Peniche.

Since I do not have at my disposal a better way than photography to let you know the main pieces on which I based myself to write this passage, I made reproductions of two skulls and some jaws, which I am pleased to send you, giving you more details to complement my description above.

If these pieces are valid for the distinction I have established, and if you approve of it, my classification would acquire the authority it now lacks; if, on the contrary, they are worthless for that purpose, I must declare it so that no one is misled by the false interpretation I have given to these documents.

In the affirmative case or in the negative case, if this letter deserves an answer, I hope you will allow me to [???], either as confirmation of the study I have done, or to put the question in its true light.

However, I must say that when I wrote the description of Furninha I did not have any suitable pieces for comparison, nor did I have the time to send my copies to a competent verifier to determine their reliably; I had no other choice but use some descriptions and plates, having then had the opportunity to benefit especially from the magnificent work of Blainville.

That there is a difference between the hyena species found in Portugal, it is a point which seems to me beyond doubt; that these species represent spotted hyena, striped hyena, and H. prisca, *that is what I was not in a position to assert peremptorily.*

The larger skull and the corresponding jaw (I), which I consider to belong to H. crocuta, *were accompanied by other pieces of the skeleton of the same individual, and were found together, wrapped in quaternary red silt. The difference in size and thickness of the bones of this skull compared with those of the other skulls from Furninha is found in other pieces of the skeleton. The other, smaller, skull and jaw (II) that I reported as belonging to* H. vulgaris, *belonged to the 3rd fossiliferous level of the Furninha cave and were mixed with many pieces of the skeletons of different individuals (see Report, p. 253). From the same level comes the large jaw (III), which seems to me to have close relations with* H. prisca *(Mem. of the Museum, tome 7, plate XXIV, Fig. 1).*

I take this opportunity to thank you, Sir, for sending the note on the new palaeontology gallery which has just been installed at the Museum under your wise guidance, and I congratulate you for this great service you have just given to science and especially to your capital.

Accept, dear Sir, the expression of my most distinguished and devoted sentiments.

This interesting letter shows the quality of Nery Delgado's analysis regarding the hyena species represented in Furninha. Despite the limited literature available, as he himself admitted, he found that the hyena remains from Furninha, especially the skulls, were morphologically different from the hyena skull collected from the Cercal cave. This is now known as Fontainhas cave in the Montejunto mountain range, where he correctly attributed the skull to the European cave hyena (*Crocuta crocuta spelaea*).

Regarding the Furninha set, the differences pointed out were biometric only, as indicated in the difference in size between two hemi-mandibles. Such differences in Nery Delgado's opinion supported the attribution of the larger specimens to *H. prisca,* while the smaller should be attributed to *H. vulgaris*, today's North African species and presently designated as *Hyaena hyaena* (= *H. vulgaris*, *H. striata*) (Delgado 1884).

The only aspect that Nery Delgado did not consider was the intraspecific biometric variation inherent to the hyena species present in Furninha, a conclusion only possible if he had made a statistical study of the collected biometric data. This practice was at the time uncommon. But the essential conclusion, namely the confirmation of a single genus, *Hyaena,* on the basis of the sparse scientific literature available to him (largely confined to the beautiful de Bougainville atlas), shows the quality of the work done.

In the second letter sent in reply to Albert Gaudry, transcribed below, the Portuguese geologist comments on the conclusion that Gaudry had previously sent to him. This letter was later mentioned to Edouard Harlé.

2nd letter. Pack 91 – Folder 1

Autograph draft letter send by Nery Delgado to Albert Gaudry.

Lisbon, April 30, 1885

Dear Sir and very honoured Professor

I received with the liveliest satisfaction the answer you kindly gave to my letter of the 8th of this month, since it came to reassure me as to the classification I had made of the remains of hyenas found at Furninha.

I carefully compared the two jaws that I have associated to H. striata *and* H. prisca *and I have come to the conclusion that they differ only in their size. The most minute examination only allowed me to recognise some slight differences, which are probably within the limits of races of the same species, as you have supposed for these two specimens. In order for you to be able to recognise these differences, I am sending you today four new photographs, which I am happy to offer you as well as the others, assuming they are of interest to you.*

The differences that I have noticed are the following. The disposition and the shape of the teeth are nearly the same in both specimens; however, in the largest jaw (III) the anterior roots of the two premolar teeth and especially of the main tooth are much more developed than in the jaw

(II); on the other hand, the point of the posterior root of the main tooth of the larger jaw is less acute than in the same tooth of the smaller jaw. I still find small differences in the size and shape of the ascending branch, in the size of the condyle and its inclination in relation to the plane of the jaw, and in the form of the angular apophysis. Finally, I will add that the masseterian pit is proportionally wider and deeper in jaw III than in jaw II.

But are these characters sufficient to distinguish two different species, or will they barely be sufficient to distinguish two races belonging to the same species? It seems to me that you are leaning towards this latter opinion; in any case, I am very satisfied to have learned that you accept the distinction of the two species of Furninha and Cercal, which is for me the most important point.

Please, Sir and honoured Master, accept the expression of my respectful sentiments and my entire devotion.

Based on the photographs sent to him by Nery Delgado, Albert Gaudry considered the differences identified, between the skulls and mandibles of the Cercal cave (= the Fontainhas cave) and the Furninha specimens, attributing them respectively to *Crocuta* and *Hyaena*.

As for the presence of two distinct species in Furninha based on differences in size and morphological details, Gaudry considered such differences to be explainable within the range of variation of the same species, in this case *Hyaena hyaena*, as mentioned above.

Gaudry's view is also known through a passage of his letter sent to Nery Delgado, which was reproduced in a letter sent to Edouard Harlé years later:

> Your beautiful research in the caves of Portugal proves the existence of two very well characterised species of hyena: *Hyaena crocuta* (race *spelaea*) of which you sent me three photographs labelled as n.I, and *Hyaena striata* (*vulgaris*) of which you sent me three photographs labelled as No.II. Concerning the mandible, whose photograph is labelled as number III, it is larger than your *Hyaena striata*, but, as far as I can judge from the photograph, it is very similar in shape. It does not represent a *Hyaena arvernensis*, because in this species the carnassial is proportionately more elongated. Concerning the *Hyaena prisca* of Montpellier, I confess that I do not know of any distinction from *Hyaena striata*. Perhaps as there were larger races in the Quaternary than today, to which we give the names of *Bos taurus* (race *primigenius*), *Bison europaeus* (race *priscus*), *Cervus elaphus* (race *canadensis*), *Arctomys marmotta* (race *primigenia*), *Felis leo* (race *primigenia*), *Hyaena crocuta* (race *spelaea*), your jaw no. III indicates a larger race of *Hyaena striata* that could be called *Hyaena striata* (race *prisca*).

Gaudry's statement is clear. He was unable to distinguish the remains of the fossil form, *H. prisca* from the current remains of *H. striata* (*H. hyaena*), except by size, as is the case with many other large mammal species. He suggests that the hyena remains of Furninha could be designated as *Hyaena striata* race *prisca* or, following modern terminology, *Hyaena hyaena prisca*. This conclusion was later reaffirmed in a paper he wrote with his then disciple Marcellin Boule (Gaudry and Boule 1892, 120), stating that the hyena of Furninha 'only differs from the striped hyena because of its considerably larger size', a conclusion that was adopted ever since by several authors, including the author of this article (Cardoso 1993; 1996).

The correspondence between Nery Delgado and Edouard Harlé

About 12 years after the correspondence between Nery Delgado and Albert Gaudry about the specific determination of the hyena remains from Furninha, Edouard Harlé (1850–1922), residing in Toulouse, expressed his interest in studying the Furninha collection.

The research rigor and perfectionism of Nery Delgado meant that he was pleased by Harlé's interest, as that would help to clarify the *status* of the Furninha hyena. Harlé's interest was justified by the fact that, unlike the Quaternary faunas of France, those of the Iberian Peninsula were largely understudied (Harlé 1910/1911: 23). It was therefore important to obtain the opinion of such researcher who, at the time, specialised in the study of hyenas. The first letter by Nery Delgado to Edouard Harlé is reproduced below.

3rd letter. Pack 91 – Folder 5

Draft letter from Nery Delgado to Edouard Harlé, copied by an officer.

Lisbon, June 11, 1897

Dear Sir,

In reply to your letter of May 26th, I have the honour to send to you by this mail several photographs which represent some samples of hyenas (the most characteristic specimens) found at Furninha (Peniche). [...]

Albert Gaudry has accepted the distinction which I made of the remains of Hyenas into three different forms, but he considers Hyaena prisca *as a variety of* Hyaena striata *and not as a distinct species.*

As far as the remains of Es-Taliens are especially concerned, I have not found in my collection any piece which would resemble the one represented in your Fig. 3. Each tubercle is only as large as this one; besides, in my specimens, this tooth always shows a direction almost perpendicular to that of the carnassial. Perhaps the hyena of Es-Taliens, which you consider to be a striped hyena, corresponds to the larger race of this species (Hyaena striata*) that Mr. Gaudry named* race prisca.

I hope I have satisfied your desire; however, I will gladly give you further explanations, if you will need them.

I beg you, Sir, to accept the expression of my highest consideration.

The morphological differences between the Furninha and the Es–Taliens specimens are carefully described by Nery Delgado, demonstrating the thoroughness he adopted in the analysis of this question. The differences he mentioned are consistent with the classification of the Es-Taliens specimens as *Hyaena brunnea* (Gaudry and Boule 1892) rather than *H. striata* (Harlé 1895). Nery Delgado insisted on his view that two different species of hyenas existed at Furninha, regardless of Gaudry's opinion. In the second letter sent to E. Harlé, Nery Delgado discusses the morphological features of the hyenas of Monsaunés and Es–Taliens, previously studied by Harlé, Gaudry and Boule, displaying a detailed knowledge of the questions concerning the classification of fossil hyenas.

4th letter. Pack 91 – Folder 5

Draft letter from Nery Delgado to Edouard Harlé copied by a secretary of Delgado.[1]

Lisbon, June 25, 1899

Mr. E. Harlé

I acknowledge receipt of your letter of the 15th of this month, to which I have the pleasure of replying.

First, I must tell you that there is no need to return the photographs; you may keep them and I will be happy that they can be useful to you. The remarks you make about my hyenas are very accurate and you have perfectly understood what I wanted to say to you. The tubercle of Es–Taliens is larger than any that I have in my collection, though it appears to be of a similar shape; moreover, it is implanted in the jaw in a manner quite different from that of the hyenas of Portugal, which are always placed perpendicularly to the carnassial. Now, I am inclined to believe, like you, that the hyena of Es-Taliens represents a particular race, if not a species different from those I know from my caves.

I must, however, declare that I have no upper jaw associated to the prisca *race; the identification of this species was based on mandible III. It is for this reason that I wrote in my previous letter that the hyena of Es-Taliens, which, considering the larger size of the tubercles than those of Montsaunés, and which in your text (p. 47) is indicated as also being a striped hyena, could possibly correspond to a* H. prisca.

Please accept, dear Sir, the expression of my respectful and very devoted feelings.

Following this correspondence, Nery Delgado sent to Harlé important mammal and bird remains from Portuguese Pleistocene deposits, including the Furninha material, thus enabling their study in France in optimal conditions. Harlé published a note on the fauna of the Fontainhas cave (Cercal cave) as mentioned in the missive from Delgado to Gaudry, where an intact skull of *Crocuta crocuta* was collected (note that in the old literature *Crocuta crocuta* was referred to as *Hyaena crocuta*). This was followed by the study of the fauna of the cave of Furninha and other caves (Harlé 1908; 1909). Such studies led to an important review that was published shortly thereafter (Harlé 1910/1911). Harlé attributed all the hyena remains from Furninha to *H. striata*, after discussing the *H. brunnea*'s alternative, except for a lower third premolar, in which he could not see, unlike other eight third premolars, 'a well-marked salience at the bottom of the anterior ridge' (translated from French) (Harlé 1910/1911, 33).

This tooth, was not identified in a studied set from Furninha (Cardoso 1993), so it was not possible to verify Harlé's observations. However, the absence of the morphological criteria may simply be explained by damage, as the size of the tooth is consistent with the known range of variation for *H. striata* (Cardoso 1993).

Discussion

Hyaena hyaena (aka *H. vulgaris* and *H. striata*) is a species currently known in the northern part of the African continent, as well as several Asian regions. The comparative study carried out by the author on the remains from Furninha allowed him to conclude that they all belong to a single form, *Hyaena hyaena prisca* or, alternatively, *H. prisca*, whose characteristics are close to those of the Lunel-Viel and L´Escale material (Cardoso 1993, 414), which is generally considered to represent the Pleistocene form of the current *H. hyaena*. More recently, some authors, influenced by an ethological perspective, have interpreted the Furninha's hyenas as belonging to a different species (*Hyaena prisca*) (Brugal, 2010), perhaps inspired by the monograph of M.-F. Bonifay (1971).

From a biometric point of view, the four skulls of Furninha (Fig. 6.5) are longer and especially narrower in the orbits than the 17 skulls of the current subspecies of *H. hyaena* studied at the Museum of Paris by the author (Cardoso 1993). These differences are even more pronounced for the two skulls of Lunel-Viel (Bonifay 1971). The nine mandibles of Furninha are also more robust than in the current species, this feature being even more obvious in the specimens from Lunel-Viel (2) and L'Escale (1) (Cardoso 1993).

The ratio between the height of the mandibular horizontal branch between the third and fourth premolars and the distance between the condyle border and the distal part of the first molar, indicates that the mean values of the Furninha and modern data sets (based on samples kept at the MNHN Paris) to be statistically different, as indicated by a Student's t-test (Cardoso 1993, 414). In addition, the comparative biometric study using Principal Component Analysis (PCA) to compare the three living species of hyenas – *Crocuta crocuta, Hyaena hyaena* and *Parahyaena brunnea* – and applied to skulls and jaws (Brugal *et al.* 2012, fig. 10) confirm these differences, as well as the uniqueness of the Furninha dataset.

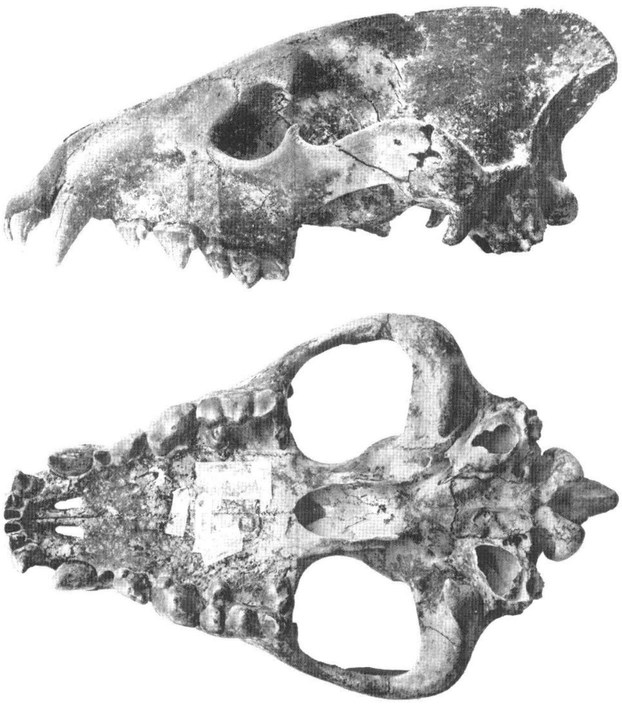

Figure 6.5. Skull of Hyeana prisca from Furninha cave. In Harlé (1910/1911), modified. Photo by the author.

Most European Pleistocene deposits with fossil hyaenids belong to the Middle Pleistocene: Montsaunes (Harlé 1894), Es-Taliens (Harlé 1895), Montmaurin (Boule 1902) and L´Escale (Bonifay 1971). Lunel-Viel has been dated to the beginning of the Mindel-Riss interglacial, but M.-F. Bonifay (1971) questioned the survival of *Hyaena prisca* at this time. The writer suggests the replacement of *Hyaena prisca* during the Mindel-Riss with an archaic form of cave hyena. For this author, the *Hyaena prisca* of Lunel-Viel, which is particularly old in origin, represents the last individual of a lineage that died out during the Middle Pleistocene (Bonifay 1971, 178). However, the remains of Verzé, probably from the beginning of the Riss (Argant 1991) and especially those of Hollabrunn (Austria), attributed to the Riss-Würrn period (Thenius 1965), suggest a later survival of the species in Europe. In Palestine, Kurtén (1965) observed the survival of *H. prisca* at the beginning of the Würm period, when it is replaced by *Crocuta crocuta spelaea*, which agrees with the age of the Furninha cave. In the context described, the Furninha assemblage assumes an ever increasing importance. Dated probably to the early Würm period (OIS 5), the Furninha hyenas probably represent a late surviving population of *H. prisca*, in the most western area of the Iberian Peninsula. Like in Palestine, the favourable ecological conditions may have led to this late survival, while the species had become extinct throughout the rest of Europe (Cardoso 1993; 1996).

Conclusions

Alongside Lunel-Viel (France), where the species was originally described, Furninha cave, whose age can be attributed to the beginning of the last glaciation (OIS 5), has produced the most important European collection of *Hyaena prisca* remains,. Nery Delgado, responsible for the 1879/1880 excavations, proposed in 1884 the coexistence of two types of hyenas, *Hyaena vulgaris* and *H. prisca*, based on the difference in size. In order to obtain confirmation of his evaluation, he consulted Albert Gaudry with two letters he sent in April and June 1885, which have been reported here. From the study carried out by the author of all the Furninha specimens preserved in the Geological Museum (LNEG) and its comparison with other collections held at the MNHN (Paris), differences in size and robustness within the assemblage were confirmed. Such variation, observed in both fossil and current species can be interpreted within the range of variation of each respective species. Thus, the suggestion proposed by Nery Delgado, when evaluating the variation in size of the Furninha remains, that two distinct species of hyenas occurred at the site, cannot be regarded as valid, as already stated by Albert Gaudry. Gaudry thought that the size variation was due to intraspecific variation. However, the author has identified significant size differences between the Furninha specimens and modern *H. hyaena*. Such differences could thus potentially support differentiation at a specific level and the recent application of PCA methodology to skulls and jaws confirmed the existence of clear differences between Furninha and the living hyaenids. In summary, the points of view of both Delgado, on the one hand, and Gaudry and Harlé, on the other, needed the integration of further analytical evidence.

The merit of Nery Delgado's research must be emphasised, as he had identified correctly, for the first time in the Pleistocene of the Iberian Peninsula, the presence of the striped hyena, using only the scanty information and analytical techniques he had available at the time. Also, the bold but at the same time careful approach he took to examine the research question, provides us with a good idea of his scientific rigour. The research question discussed in this paper, like many others, must be evaluated within the context of the historic times in which it was investigated. It is also likely that new methods and techniques will, in the future, allow us to refine our understanding of the intriguing problem of the identification of the Furninha remains, as well as the broader issue of which hyena species occurred in the Pleistocene of Europe.

Acknowledgements

I am grateful to A. Ávila de Melo for her support, during the writing of this paper, since the identification of the original documents in the Historical Archive of the former Instituto Geológico e Mineiro in Lisbon. I would also like to thank Dr.

Roz Gillis (ICArEHB, Algarve University) for the revision of the English translation, and Prof. Umberto Albarella for the scientific revision of the manuscript. Naturally all insufficiencies eventually identified in this contribution are exclusively the responsibility of the author.

Note

1 Original text in French.

References

Argant, A. (1991) Carnivores quaternaires de Bourgogne. *Documents du Laboratoire de Géologie de Lyon* 115, 301.

Bicho, N. and Cardoso, J.L. (2010). Paleolithic occupations and lithic assemblages from Furninha cave, Peniche (Portugal). *Zephyrvs* 66, 17–37.

Bonifay, M.-F. (1971) Carnivores quaternaires du Sud-Est de la France. *Mémoires du Museum national d'histoire naturelle* N.S.(C) 21(2), 43–377.

Boule, M. (1902) La caverne à ossements de Montmaurin (Haute-Garonne). *L'Anthropologie* 13, 305–319.

Breuil, H. and Zbyszewski, G. (1945) *Contribution à l' étude des industries paléolithiques du Portugal et de leurs rapports avec la géologie du Quaternaire. 2 – Les principaux gisements des plages quaternaires du littoral d'Estremadura et des terrasses fluviales de la basse vallée du Tage.* Lisboa, Serviços Geológicos de Portugal (Comunicações dos Serviços Geológicos de Portugal. Lisboa, 25).

Brugal, J.P. (2010) Carnivores pléistocènes (Hyenidés, Canidés, Félidés) dans les grottes du Portugal. *1.ª Reunión de científicos sobre cubiles de hiena (y otros grandes carnívoros) en los yacimientos arqueológicos de la Península Ibérica.* Zona Arqueológica 13, 93–106.

Brugal, J.-P., Argant, J., Crispim, J.A., Figueiredo, S., Martín Serra, A. and Palmqvist, P. (2012) The complex carnivore-rich assemblages from Furninha (Peniche, Portugal): a multidiciplary approach. *Journal of Taphonomy* 10 (3/4), 417–438.

Cardoso, J.L. (1993) *Contribuição para o conhecimento dos grandes mamíferos do Plistocénico superior de Portugal.* Oeiras, Câmara Municipal de Oeiras.

Cardoso, J.L. (1996) Les Grands Mammifères du Pléistocène Supérieur du Portugal. Essai de synthèse. *Geobios* 29(2), 235–250.

Cardoso, J.L. (2008) Joaquim Filipe Nery Delgado, arqueólogo. In *Nery Delgado (1835–1908), Geólogo do Reino*, 65–75. Lisboa, Museu Geológico/INETI/Centro de História e Filosofia das Ciências/FCT/UNL.

Cardoso, J.L. (2013) Carlos Ribeiro, a 'Breve notícia acerca do terreno quaternario de Portugal' e a questão do Homem terciário em Portugal. *Estudos Arqueológicos de Oeiras* 20, 27–88.

Cardoso, J.L. (2020) A primeira escavação arqueológica metodologicamente moderna foi realizada em Portugal em 1879/1880: a intervenção de Nery Delgado na gruta da Casa da Moura (Óbidos, Portugal). *Estudos Arqueológicos de Oeiras* 26, 123–242.

Cardoso, J.L. and Carvalho, A.F. (2010/2011) A gruta da Furninha (Peniche): estudo dos espólios das necrópoles neolíticas. *Estudos Arqueológicos de Oeiras* 18, 333–392.

Cardoso, J.L. and Melo, A.A. (2001) Correspondência anotada de Carlos Ribeiro e de Nery Delgado: contribuição para a história da arqueologia em Portugal. *Comunicações do Instituto Geológico e Mineiro* 88, 309–346.

Choffat, P. (1908) Notice nécrologique sur J.F. Nery Delgado (1835–1908). *Jornal de Sciencias Mathematicas, Physicas e Naturaes*, S. II, 7 (28), 14.

Christol, J. de and Bravard, A. (1828) Nouvelles espèces d'hyènes fossiles. *Mémoires de la Société d' Histoire Naturelle de Paris* 5, 368.

Delgado, J.F.N. (1884) La grotte de Furninha a Peniche. In *Congrès International d'Anthropologie et d'Archéologie Préhistoriques, Compte- Rendu de la 9ème Session (Lisboa, 1880)*, 207–278. Lisboa, Typografia da Academia.

Gaudry, A. and Boule, M. (1892) Les oubliettes de Gargas. *Matériaux pour l'Histoire des Temps Quaternaires* 4, 130.

Harlé, E. (1894) Découverte d'ossements d'Hyènes rayées dans la grotte de Monsaunès (Haute-Garonne). *Bulletin Société Géologique France* S. III 22, 44–49.

Harlé, E. (1895) Restes d'Hyènes rayées de la brèche d'Es-Taliens, à Bagnères-de-Bigorre (Hautes-Pyrenées). *Bulletin Société Géologique de France* S. IV 8, 460–466.

Harlé, E. (1908) Faune de la grotte das Fontainhas (Portugal). *Bulletin Société Géologique de France* S. IV 8, 460–466.

Harlé, E. (1909) Faune de la grotte à Hyènes rayées de Furninha et d'autres grottes du Portugal. *Bulletin Société Géologique de France* S. IV 9, 85–99.

Harlé, E. (1910/1911) Les mammifères et oiseaux quaternaires connus jusqu'ici en Portugal. *Comunicações da Comissão do Serviço Geológico de Portugal* 8, 22–85.

Kurtén, B. (1965) The carnívora of the Palestine caves. *Acta Zoologica Fennica* 107, 1–74.

Kurtén, B. (1968) *Pleistocene Mammals of Europe*. London, Weidenfeld & Nicolson.

Relatório (1881) *Relatorio do anno económico de 1880/1881.* Direcção Geral dos trabalhos Geologicos, Topographicos, hydrographicos e Geologicos do Reino. Lisboa, Imprensa Nacional 13, 14.

Relatório (1886) *Relatorio dos trabalhos geodesicos, topographicos, hydrographicos e geologicos do Reino pertencente ao anno economico de 1884–1885.* Lisboa, Imprensa Nacional.

Serres, M., de Dubreuil, J. and Jean-Jean, B. (1828) Mémoire sur les diverses espèces d'Hyènes fossiles découvertes dans les cavernes de Lunel-Viel (Hérault). *Mémoires Muséum Histoire Naurelle Paris* 17, 269–311.

Thenius, E. (1965) Uber das Vorkommen von Streifenhyänen (Carnivora, Mammalia) im Pleistozän Niederösterreichs. *Ann. Naturhist. Mus. Wien* 68, 263–268.

Zilhão, J. (1993) As origens da arqueologia paleolítica em Portugal e a obra metodologicamente precursora de J.F. Nery Delgado. *Arqueologia e História* S. X 3, 111–125.

7

Sheepish bones, sheepish dates, sheepish logic and the neolithization of Iberia

João Zilhão

Use of radiocarbon to date the human past depends on the accuracy of age measurements and the association of the sample with the events or processes one intends to date. With AMS (Accelerator Mass Spectrometry), samples that are themselves directly informative can be used, which adds correct specimen classification, typological or taxonomic, to the list of critical requirements for successful dating. These issues underpin to a large extent ongoing controversy relating to the spread of farming across west Mediterranean Europe. Good illustration is provided by different interpretations of a site in Spain, Cueva de Nerja. Appropriate consideration of potential sources of dating error – site formation process, bone chemistry, sheep/goat determination – shows that no support exists for the notion that the earliest Neolithic occupants of the site were bearers of a pre-Cardial, Impressed Ware tradition of Italian or North African origin. Based on current knowledge, the null hypothesis for Mediterranean Iberia is that the culture associated with its earliest Neolithic is the Cardial. Tested against the Nerja evidence using falsification logic instead of arguments from authority, the hypothesis stands.

Keywords: Iberia, Neolithic, sheep, early farming, radiocarbon chronology

Introduction

To use radiocarbon as a clock (for a recent review, see Wood 2015), we assume that, in the past, the radioactive isotope ^{14}C and the stable isotope ^{12}C were found in the Earth's atmosphere in the same proportion as today, and, by extension, that such is the case with the living organisms our samples of animal and plant material derive from. Additionally, the accuracy of radiocarbon dating results depends on (a) the instrument returning a $^{14}C/^{12}C$ ratio that is indeed the analysed sample's true ratio, and (b) the difference in ^{14}C content between samples and atmosphere being entirely due to the samples' post-mortem loss of ^{14}C by radioactive decay.

We now know that atmospheric $^{14}C/^{12}C$ ratios underwent secular and millennial variation, and that old carbon may be present in the oceans, rivers and underground aquifers; thus, if calculated against an atmospheric $^{14}C/^{12}C$ standard, the ages of samples derived from such environments can be erroneously old. But we also know that calibration and appropriate consideration of reservoir effects can by-pass these problems. Likewise, (a) systematic measurement errors can be detected by inter-laboratory blind testing, *i.e.* comparison of the results obtained for sub-samples of a single sample whose age is independently known (via, *e.g.* dendrochronology), and (b) rigorous pre-treatment of the material makes it possible to prevent the occurrence of errors caused by the presence in the sample of exogenous carbon acquired post-mortem (via, *e.g.* chemical exchange with the burial environment).

Gradually introduced over the years since the method was invented, these correction and safety mechanisms do not prevent the occasional production of results that deviate from expectations. The deviation may result from stochastic error or occasional malfunction (of the instrument of measurement, or of the sample-preparation chain). Alternatively, it can be that the expectation was based on interpretations of previously available evidence that turn out to be not as sound as one took them for. Re-dating of the same sample, or of comparable ones, will solve the inconsistency if the problem is strictly a 'dating' one (*e.g.* inappropriate sample pre-treatment). But, if it can be shown that there is nothing wrong with pre-treatment, measurement, or calibration, then it must be the case that (a) our interpretation of the historical event the sample was intended to date was flawed, and needs revision, or (b) our assessment of the sample as significant for the dating of the event or process in question was flawed, and needs revision. Ultimately, it this 'association' problem that lies at the heart of the interpretation of radiocarbon dates, as Waterbolk (1971; 1983) already pointed out nearly half a century ago.

In addition, all radiometric dating methods are accurate only within given uncertainty parameters. Put another way, their precision does not compare with that provided by historical sources, which often let us know about the day or month, or at least the year during which the event or process of interest took place. This represents a particularly difficult problem when our interest resides in epochs of rapid technological, cultural, economic or social change. Prehistory's so-called 'transitional' periods have therefore been especially prone to radiocarbon-related controversy. The ongoing debates on the nature of the Middle-to-Upper Palaeolithic transition in Europe, or of the Mesolithic-Neolithic transition in the Mediterranean, are good examples (Zilhão 2013; Higham *et al.* 2014; Martins *et al.* 2015; Isern *et al.* 2017; Manen *et al.* 2019).

The half-life of ^{14}C is 5730±40 years. Therefore, in the case of the Middle-to-Upper Palaeolithic transition, 35,000 to 45,000 years ago, the events of interest occurred very close to or at the limit of applicability of radiocarbon dating, and issues of accuracy, ones that deal with aspects of chemistry (the removal of contaminants) and physics (the measurement of the residual amount of ^{14}C atoms remaining in the samples), are of paramount importance in the assessment of results. This is not the case, however, with the Mesolithic-Neolithic transition in western Europe, which occurred at a time when half or almost half of the samples' original ^{14}C content is still there. Because sites, remains, and, hence, samples, are much more abundant and therefore easier to find or procure, it is also the case that the standard scientific approach to validation – experimental replication, in this case repetition of the dating – almost always suffices to detect laboratory error, whether systematic or stochastic. Thus, for the time of the Mesolithic-Neolithic transition, when repetition fails to solve inconsistencies between results and expectations, it is only reasonable to conclude that the problem is more likely to originate in the archaeological, not the physics/chemistry side of things.

A classic example is the 'Mesolithic sheep' hypothesis put forth in the 1970s by D. Geddes and J. Guilaine. At a few sites in southern France (*e.g.* Grotte Gazel, Abri Dourgne), levels defined as Mesolithic – by their artefact content and concordant radiocarbon dating on stratigraphically associated charcoal – yielded significant amounts of caprine remains that were zooarcheologically identified as sheep (*Ovis aries*). Based on this evidence, it was proposed that the regional Mesolithic-Neolithic transition had begun with the addition of sheep herding to the hunter-gatherer economy. The novel resource would have been acquired via long-distance exchange with coeval, fully farming societies established to the East, in the Italian peninsula and neighbouring regions of France. Eventually, it was acknowledged that the hypothesis relied on post-depositional mixing and the misidentification of the faunal remains of wild ungulates of similar size (ibex/wild goat, chamois, roe deer) – ones whose bones, especially when fragmentary and of juveniles, are very difficult, if not impossible, to differentiate from those of sheep (Geddes 1993; Guilaine and Manen 2007).

With the advent of AMS dating, it became possible to by-pass the impact of these classification issues via the individual dating of morphologically unambiguous samples. With regards to the Mesolithic-Neolithic transition, this meant that one could now directly date single cereal grains and single diagnostic bones of domesticated animals – *i.e.* samples that are straightforward indicators of the presence of components of the Neolithic package in the time/space unit represented by their age and site of provenience. Even if it represented huge progress, this approach, however, was not fool proof either – as was to be very clearly shown by the Tardiglacial result obtained for a bovid bone from the Basque site of Arenaza that had been zooarcheologically determined as cattle (Arias and Altuna 1999). Given the age returned by radiocarbon, there could be no doubt that the dated specimen was an aurochs bone, probably misidentified due to the overlap in size between male cattle and female aurochsen.

Bearing these cautionary tales in mind, a recent study of temporal patterning in the emergence of farming in Iberia used ZooMS (Zooarchaeology by Mass Spectrometry) to confirm that samples made available to the project as 'sheep' or 'domestic sheep/goat' belonged indeed to sheep and not to the native ibex species, *Capra pyrenaica* (Martins *et al.* 2015). When biomolecular confirmation could not be obtained, either because ZooMS identified the bone as *Capra* or because ZooMS was inconclusive, the samples were excluded from the chronological model. For consistency, the model also excluded previously published results for which a 'sheep' identification was not supported by biomolecular analysis or rigorous, differential morphological diagnosis.

García-Borja *et al.* (2018) have since questioned Martins *et al.*'s approach, particularly with regards to how the zooarchaeologically determined sheep/goat samples from Cueva de Nerja (Málaga, Spain) were dealt with. I find their points are well worth addressing because I believe that they provide good illustration of the errors of logic and fact that one frequently encounters in these debates.

Validation logic

The first problem with García-Borja *et al.*'s argument is that they treat as criticism directed specifically at their zooarchaeological determinations what is simply a general validation criterion applied to all the samples from all the sites used in Martins *et al.*'s study. That criterion was: if one cannot exclude the possibility that a given sample is ibex, then the domesticate status of the sampled specimen ought to be considered unproven and an open issue. In addition, García *et al.* claim that this criterion was used to reject the oldest dates, while, in fact, it was applied across the board. A few of the samples whose domesticate status Martins *et al.* considered unproven did return old dates (but they were not just from Nerja; Guixeres de Vilobí, in Catalonia, is another instance), while others returned rather young ones (*e.g.* Cova del Frare, where the Beta-325689 result for a ZooMS-undiagnostic caprine was 5010±40 BP).

To be consistent with their logic, Martins *et al.* could not but explain, and did explain, why the assignment to ibex of some of the samples in their study was a realistic alternative. Most such samples were fragmentary, represented skeletal parts that are morphologically undiagnostic, and came from sites located in mountainous terrain where ibex existed through the time of Neolithic occupation (and still exists today). In addition, ibexes are known to take refuge in caves and rock shelters, sometimes penetrating deep inside the karst, as unambiguously shown, to give just an example, by the footprints found in the deepest section of Grotte Chauvet, a cave art site in France (Garcia 2005). Thus, in caves situated in ibex country, as is the case with Nerja, the skeletal remains of caprine can represent natural deaths of the wild species; one cannot assume that they are anthropogenically accumulated, and of domestic animals, simply because the levels they were retrieved in also contain ceramics or other evidence of Neolithic human activity.

To say that 'ibex cannot be excluded' is, however, a very different thing from saying 'it is ibex, not sheep.' By rendering Martins *et al.*'s 'ibex cannot be excluded' statements as positive assignments to 'ibex' (Table 7.1), García-Borja *et al.* apparently fail to appreciate this distinction, which goes to the heart of the scientific method and concerns how validation operates. When testing contrasting interpretations of the same facts, the key question always is 'on which side does the burden of proof lie?' In this case, the issue being debated is whether (a) domesticates appear in Iberia as part of a 'Neolithic package' introduced at a definite point in time and in the context of the westward spread of farming populations, a process whose archaeological signature is, with current evidence, the spread of the Cardial culture, or (b) domesticates (and ditto for other elements of the package, *e.g.* ceramics) were acquired by local hunter-gatherers once they became available through long-distance exchange (as in the 1970s 'Mesolithic sheep' hypothesis), over a drawn-out period and with no demic diffusion of any significance being involved in the process. Both sides of the debate agree that the issue can be advanced via the direct dating of key samples. Obviously, this understanding is predicated on the domesticate status of such samples being certain; if reasonable doubt exists, then the sample's significance is compromised, and it cannot be considered evidence.

Table 7.1. What was actually said, and what is said to have been said.

Martins et al. *(2015)*	*García-Borja* et al. *(2018)*
'A somewhat different problem is raised by the sheep or sheep/goat samples for which (a) the biomolecular test failed to corroborate that they were of *Ovis* and, hence, domestic, or (b) the dating results fell outside the interval when domestication is securely attested in regional records. The prudent reading of the evidence in these cases is that the domesticate status of the sample remains unproven.' (p. 111) 'In order to eliminate any potential sources of error, only those samples from Mediterranean Spain for which biomolecular testing corroborated the zooarchaeological classification are considered in this study to be confirmed sheep (and, hence, unambiguous indicators of the presence of domesticates at the site of provenance and at the time indicated by their direct dating).' (p. 111) 'A corollary of our experiment is that the domestic status of all previous instances of directly dated remains classified as sheep or sheep/goat that come from Early Neolithic sites located in or adjacent to ibex terrain requires independent verification. Given the non-negligible probability of error in the taxonomic determination, the significance of such dates should be considered an open issue.' (p. 113)	'The authors did not consider valid the oldest dates on remains classified by one of us as domesticate, arguing that their identification might be erroneous […]. The remains would be of wild animals, not domesticated ones, and could have entered the cave through natural processes; therefore, they would not be representative of the initial stages of the Neolithic occupation.' (p. 14) 'The identifications of a fragment of tibia from NM-10 and another of vertebra from NM-12, both *Ovis/Capra* (OxA-26085 y OxA-26086), are questioned. It is said that they are skeletal parts of two ibex deposited in the cave by natural processes' (p. 14)

The above begs the question of what constitutes 'reasonable doubt'. In a blind test designed to assess the reliability of the morphological criteria used to identify post-cranial caprine bones as either sheep or goat, Zeder and Lapham (2010) found that the frequency of mistakes varied according to the analyst's experience and the skeletal part examined. However, all analysts made mistakes, even the most experienced, with individual error rates ranging between 4% and 16%. Given these figures, it stands to reason that the morphology-based identification of critical samples must be rigorous, *i.e.* supported by differential diagnosis, adequate illustration of the key features upon which the classification is based, and, if possible, biometric evidence. When such is not the case, independent corroboration is required – from context (*e.g.* ibex can be excluded because the species did not exist in the space/time unit under consideration), or from the biomolecular or genetic makeup of the dated bone.

To put it bluntly, few people would find it acceptable that someone be sentenced in a court of law based on a verdict acknowledging that the person had a 4–16% chance of being innocent. The same logic applies here.

Arguing from authority

One could object that the error rates in Zeder and Lapham (2010)'s study were rather high for some skeletal parts (as much as 32–44% for proximal radius, tibia, and calcaneus) but low for others – and in fact nil across all analysts for astragalus and first phalanx. Based on this experiment, one might therefore be led to conclude that 'sheep astragali' are a safe bet: for assessing the time and mode of the emergence of agro-pastoral economies in Iberia, any samples identified as such by a trained zooarchaeologist, even if a relatively inexperienced one, ought to be considered reliable 'domesticates' with no need for independent corroboration. Among the samples discussed by García-Borja *et al.*, that would be the case with the bone from the Vestíbulo hall of Nerja dated to 6590±40 BP (Beta-131577) (more will be said below about this sample); as they put it (p. 14), 'the determination of *Ovis aries* based on an astragalus in good condition is not prone to doubt, given the differential morphological traits, which have been precisely described.'

Such absolute trust that astragali can always be assigned to either sheep or goat would be, however, misplaced. The 0% error in Zeder and Lapham's (2010) experiment is flattering for the participants involved, but other experiments have failed to replicate such a success. Using a biometric approach, Davis (2017), for instance, tested his own previously published, morphology-based, sheep vs. goat astragali determinations. For the Early Neolithic site of Khirokitia, in Cyprus, Davis concluded that 12 of the 894 that he had studied (*i.e.* 1.3%) had been incorrectly classified. For a set of four archaeological sites in Portugal spanning the Neolithic and Early medieval periods, the number of erroneous identifications was significantly higher – between 6 and 9 out of 127, *i.e.* 4.7%–7.1% (which, as hypothesised by Davis, may have been due, at least in part, to change through time in the morphology of sheep astragali, due to breed improvement, *i.e.* artificial selection).

In another example, Martins *et al.* used sheep/goat astragali to date a mixed Holocene context from the Portuguese site of Galeria da Cisterna (Almonda karst system, Torres Novas). In this case, ibex could be excluded because, at the end of the Last Glacial, the species went extinct in the area. The idea was to better define the intervals of time during which use of the locus as a burial site ended up producing the palimpsest encountered by the archaeological excavation. To this end, all the ten sheep/goat right astragali giving the bone assemblage's highest MNI (Minimum Number of Individuals) were submitted for direct dating. Most samples (six out of 10) were failed due to insufficient collagen, but the experiment remains of interest for the issue at stake here because of the difficulty encountered when trying to apply Boessneck *et al.*'s (1964) and Zeder and Lapham's (2010) criteria to classify the specimens as either *Ovis aries* or *Capra hircus*. Including the left astragali, the total sample was 13, of which, combining those criteria with Davis' (2017) biometric ones, five could be classified as *Ovis* and three as *Capra*; the other five remained unidentified, even though it was possible to say that one was likely to be *Ovis* and another likely to be *Capra* (Figs 7.1–3).

García-Borja *et al.* (p. 16) use the following argument to pre-empt objections such as the above: 'in all instances the team of University of Valencia-trained zooarchaeologists classified the bones as corresponding to either domesticated or wild species, the rate of coincidence between the taxonomic identification based on morphological and biometric data and the molecular analysis was 100%.' Indeed, of the 14 such samples from three sites (Cova Fosca de la Vall d'Ebo, Cova de la Sarsa, and Cueva de Nerja/Mina) classified by that team and analysed by Martins *et al.*, ZooMS identified six as *Ovis*, one as probably *Ovis*, and none as *Capra* (the other seven were deemed undetermined). From this outcome, García-Borja *et al.* conclude that challenging their assignment of domesticate status to Nerja's ZooMS-undetermined samples is unreasonable.

However, none of the samples classified by the University of Valencia-trained zooarchaeologists was an astragalus or a first phalanx, the skeletal parts that participants in Zeder and Lapham's (2010) experiment always got right. One is therefore entirely within reason to expect the Cueva de Nerja/Mina identifications to be affected by Zeder and Lapham's (2010) overall error rate of 4–16%, *i.e.* that a significant probability exists that the University of Valencia-trained zooarchaeologists were wrong in one or two cases – which is not contradicted by them being shown to have

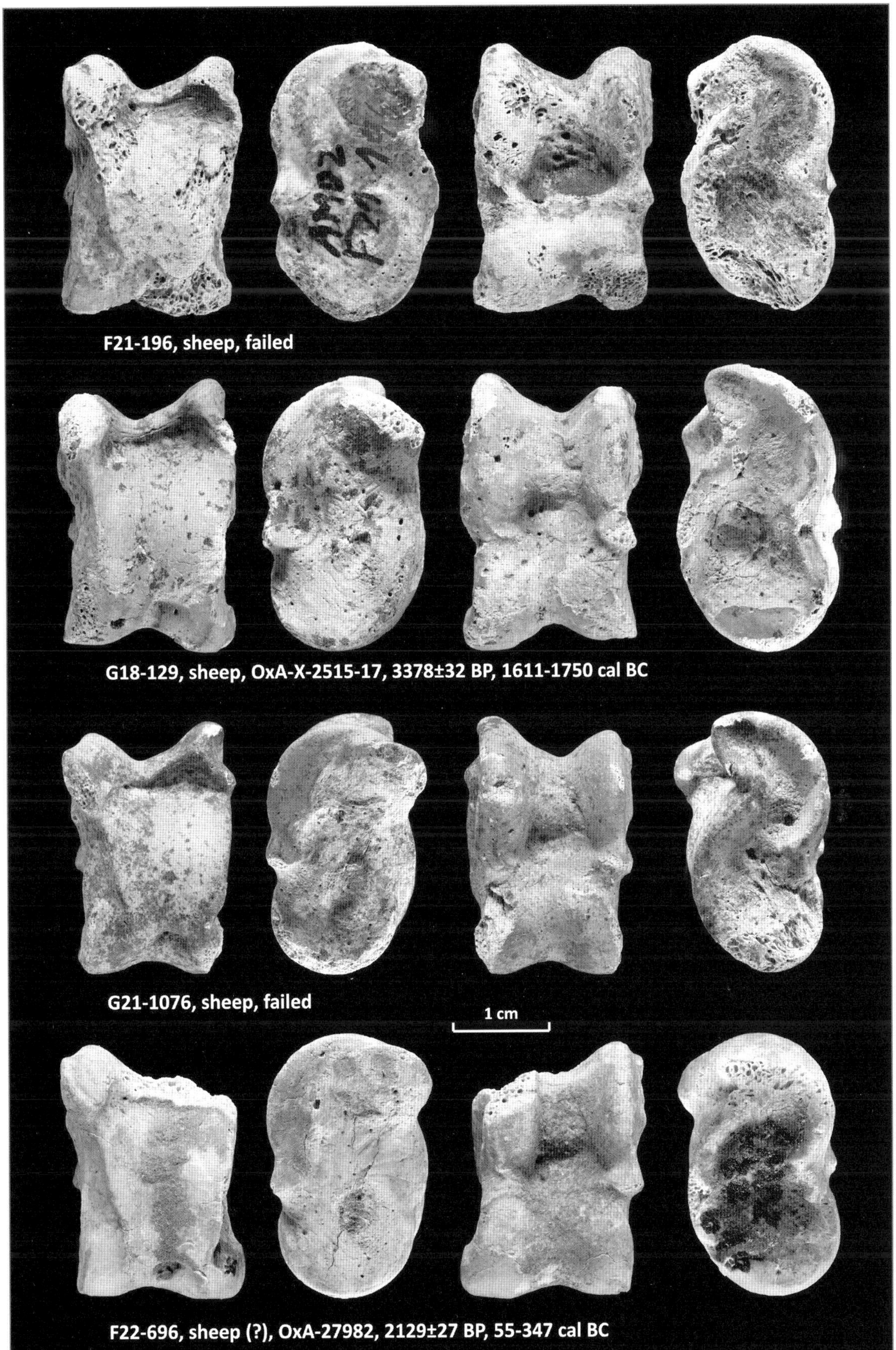

Figure 7.1. Sheep and sheep (?) astragali from the Galeria da Cisterna (Almonda karst system, Portugal).

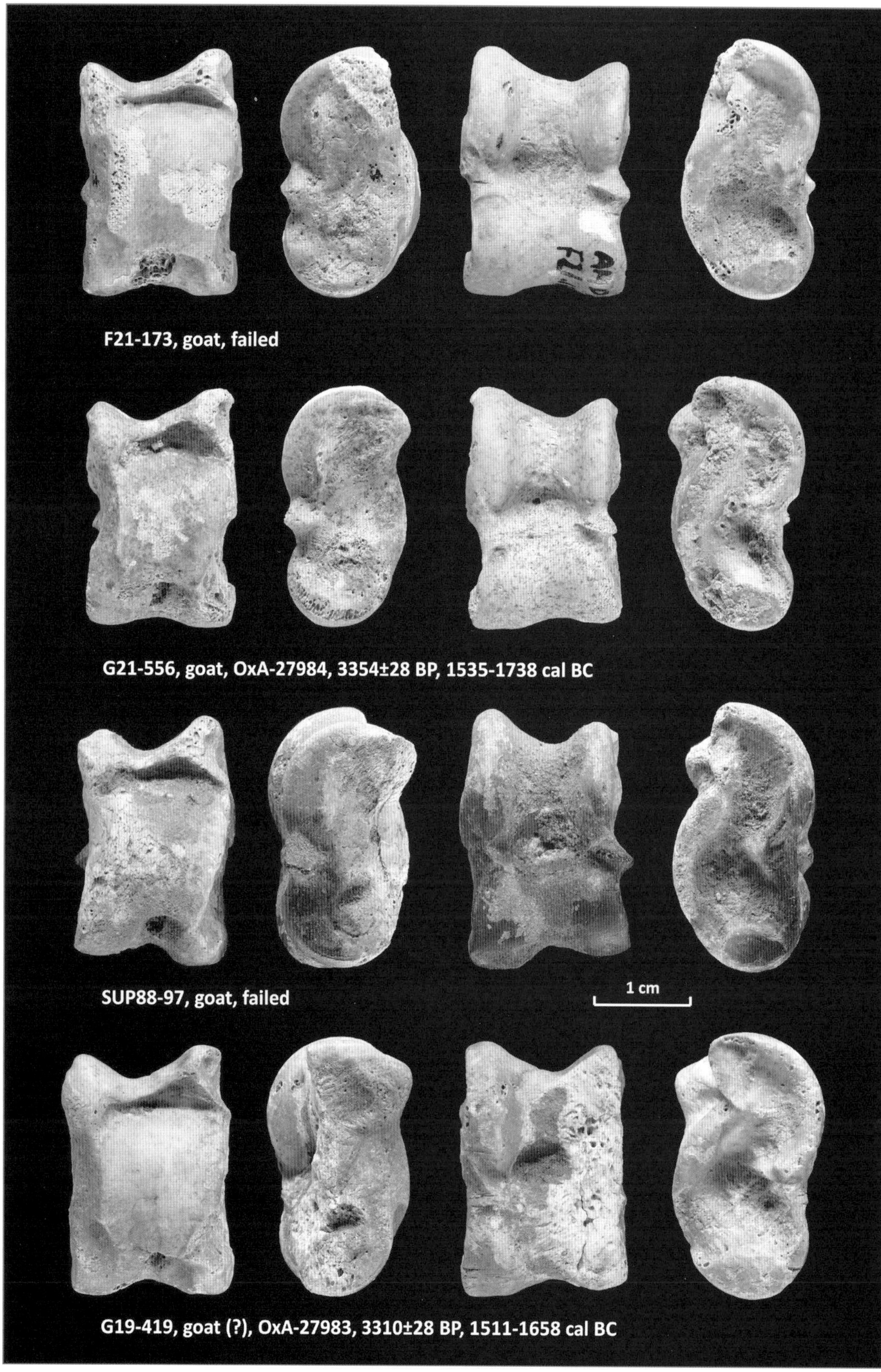

Figure 7.2. Goat and goat (?) astragali from the Galeria da Cisterna (Almonda karst system, Portugal).

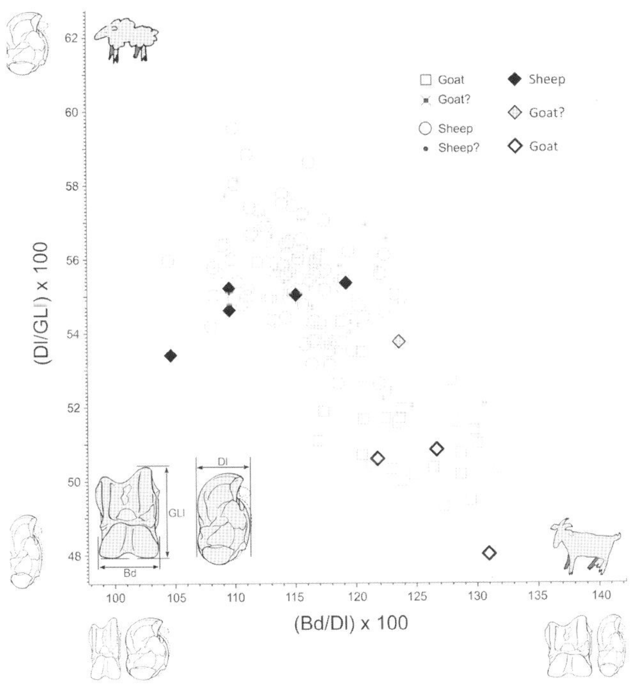

Figure 7.3. Biometric classification of the Galeria da Cisterna sheep/goat astragali (diamonds) against the background provided by four Neolithic to Early Medieval archaeological assemblages from Portugal (modified after Davis 2017, figs 9.3–9.4).

been right in six or seven out 14. This is the more so because ten of those 14 samples were only identified as 'sheep/goat' – meaning, as García-Borja *et al.* (p. 16) acknowledge, that they were deemed (a) to belong to either sheep or domestic goat and (b) definitely not to belong to ibex. The claim here is that not only can sheep be differentiated from domestic goat 100% of the time, but that such success is also possible when differentiating between the domestic (goat) and the wild (ibex) forms of *Capra* – even in the case of fragmentary remains such as those constituting the samples in Martins *et al.*'s study.

García-Borja *et al.*'s defence is an argument from authority. There is no question that, when it comes to assessing the results of an experiment, the level of expertise of the expert(s) involved carries weight. The fact that the zooarchaeologists among García-Borja *et al.* get their identifications right most of the time bespeaks of the fact that Zooarchaeology is a scientific discipline and that they are well trained and competent practitioners. What needs to be explained, however, is why, in face of the rest of the field's more prosaic success rates, one should place any faith in their infallibility.

Burnt bone dates

One of García-Borja *et al.*'s 'domestic sheep/goat' samples from Nerja that ZooMS had failed to identify dates to 7255±65 BP (OxA-X-2457-57). Given this age, Martins *et al.* argued that the sample had to belong to an ibex and, therefore, represented a case of mistaken zooarchaeological identification – evidence that the expectation was indeed not unreasonable, that one or two of the 14 samples identified as domesticate by García-Borja *et al.*'s team might well correspond to ibex. García-Borja *et al.* counter that they are right that the sample comes from a domesticate; the dating result would be erroneously old because the bone in question, a radius (Fig. 7.4), was charred and, hence, a chemically problematic sample upon whose age measurement no confidence is warranted.

Figure 7.4. The charred diaphyseal fragment of a caprine radius from level NM-8 of Cueva de Nerja, Mina hall. Scale bar: 1 cm. Photo by P. García-Borja.

García-Borja *et al.* use the sample's C:N ratio of 6.8 (outside the 2.9–3.4 acceptable range for pure collagen) to support their contention. Indeed, that ratio is why the dating lab used an OxA-X number format instead of the standard OxA format. In and of itself, however, that ratio does not necessarily imply that the age measurement is unreliable – as shown by DeNiro (1985), the C:N ratio of charred bone always falls outside the range for collagen. The issue with charred bone dates resides in the fact that the origin of the dated material (*i.e.* which part of the bone tissue the dated carbon comes from) is difficult to assess. This represents a problem because, in bone, carbon exists in an organic (collagen) and a mineral phase (bioapatite) and the latter is, in nature, susceptible to post-mortem acquisition of carbon present in mineral accretions (*e.g.* calcite, authigenic apatite) or derived from the burial environment (*e.g.* via ionic adsorption and exchange with enveloping sediment and percolating groundwater). If the C:N ratio of a charred bone sample indicates that the dated material was not pure collagen, or if the laboratory used the ABA (acid-base-acid) treatment typically applied to other types of charred substances, such as wood charcoal, then the measured $^{14}C/^{12}C$ ratio almost certainly includes a non-negligible amount of carbon derived from the mineral bone. Consequently, the possibility that the calculated age deviates from the true age of the sample must also be deemed non-negligible – and this regardless of whether the deviation is due to environmental contamination or to the use of glues or preservatives in the preparation of the specimen.

For the sake of the argument, let us therefore posit the worst-case scenario for the reliability of OxA-X-2457-57: namely, that the dated carbon all came from the mineral phase. Could it be that such carbon was at least in part exogenous and from an older source, thereby making the age measurement too old? To assess this possibility, we can model, using radiocarbon's decay equation, what percentage of a contaminant would be required to obtain an age of 7255±65 BP from a bone of the same age as the ZooMS-identified sheep radius with identical provenience (level NM-8 of Nerja/Mina) that Martins *et al.* (2015) dated to 6214±35 BP (OxA-26082). In the Mina hall of the site, the base of the excavated sequence consists of Holocene deposits containing a mix of Epipalaeoltihic and Mesolithic artefacts, and the dated samples underwent no preparation. Knowing this, any contaminant carbon must have come from the burial environment, and we can restrict the exercise to three cases: contamination by early Holocene, 10,000- or 7500-years-old carbon present in the sediments or the groundwater, and contamination by dead carbon derived from the cave's limestone walls.

As shown in Fig. 7.5, for the age measurement to be too old due to the post-mortem introduction of exogenous material, the contaminant, if 7500 years old, would have to represent 82% of the sample's carbon. For an even older

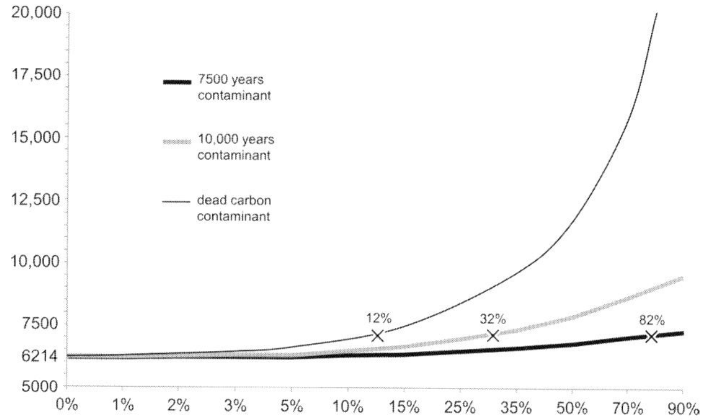

Figure 7.5. The amount of contamination required to produce a shift from 6214 BP, the age of a ZooMS-confirmed sheep radius from level NM-8 of Cueva de Nerja, to 7255 BP, the radiocarbon age obtained for the charred bone from the same level illustrated in Fig. 7.4.

contaminant, the percentage is smaller: 32% in the case of a 10,000 year-old contaminant, 12% if the latter is dead carbon. Note, however, that nitrogen was present in the sample and, therefore, part of the dated carbon must have come from endogenous collagen. Therefore, these numbers are minima: the outcome predicated by García-Borja *et al.* would have required even higher levels of contamination. Bearing this is mind, we can nonetheless compare our theoretical exercise to the results of real experiments conducted to assess the contamination of the mineral phase of archaeological bone by environmental sources.

Fernandes *et al.* (2013) report on six mid-Holocene human bone samples from the German site of Zauschwitz with various degrees of preservation and for which collagen dates ranging between 3025±30 BP (KIA-37855) and 6255±30 BP (KIA-37856) were obtained. Using untreated subsamples as well as subsamples treated with hydrochloric acid (HCl), the mineral phase of the same samples was also dated. The results showed that the treatment achieved significant removal of contaminants and, for well-preserved samples, returned results in agreement with the collagen dates; for the untreated subsamples and the treated subsamples from poorly preserved specimens, however, there was significant disagreement with the collagen date. The strongest disagreement was found between the collagen result for KIA-37854 – 46.06±0.19 pMC (percent Modern Carbon), 6225±30 BP – and its subsample of untreated bone mineral – 64.88±0.23 pMC, *i.e.* almost 3000 radiocarbon years younger.

Using radiocarbon's decay equation, we can gain an idea of how much contamination would have been necessary to cause the differences observed in Fernandes *et al.*'s (2013) experiment: some 35% if we model a 100 year-old contaminant, some 44% if the contaminant is modelled as 1000 years old. Thus, significant amounts of exogenous carbon, in the range of those required to bring about an age shift of the scale implied by García-Borja *et al.*'s interpretation of the OxA-X-2457-57 result, can indeed be found in the mineral phase of bone. However, in Fernandes *et al.*'s (2013) experiment, the shifts are never in the required direction – the contamination is always by younger carbon resulting in younger ages, not the other way around.

We can see the same pattern in Zazzo's (2014) compilation of instances in which bone apatite and bone collagen dates were obtained from the same sample. Good agreement exists in general for samples that are less than 8000 years old, even though the apatite date is statistically somewhat younger in a few cases; but disagreement becomes systematic for samples of Palaeolithic age, with the difference reaching up to 10,000 years in the case of Gravettian and Aurignacian samples from Abri Pataud (France). The key point, however, is that, in Zazzo's (2014) examples, the contamination caused by the presence of exogenous carbon always results in bone mineral ages that are younger, not older than the specimen's collagen age. Indeed, no examples exist in the literature of dates on the mineral phase of bone that are older than the collagen dates for the same sample, or older than expected – much less a millennium older, as implied by García-Borja *et al.*'s interpretation of OxA-X-2457-57.

To support their case, García-Borja *et al.* would need to reveal what geochemical processes known to science can explain the magnitude of the diagenetic processes involved, which, if the exogenous carbon is close in age to the sample, amounts to almost complete 'replacement', not just 'contamination'. In addition, they would need to explain how a bone found in primary position within a thick, largely coherent Neolithic sequence could have ended up in contact with the significantly older contaminating environment responsible for the age shift (in the Mina hall, the Mesolithic/Epipaleolithic level NM-12 is found 0.5–1.0 m below the NM-8 level where the OxA-X-2457-57 sample comes from). Instead, they claim that this Nerja anomaly is akin to that identified when dating human bone from Cova de la Sarsa, in Valencia. Here, an Early Neolithic human mandible was dated to 6800±30 BP (Beta-384725), while a subsequently dated cranial sample from the same individual yielded the much younger, and within expectations, result of 6309±36 BP (OxA-31629). As García-Borja *et al.* acknowledge, the reason for the difference is that the remains had been treated with preservatives – eliminated by a solvent, prior to pretreatment, in the case of the OxA result but not in the case of Beta's.

García-Borja *et al.* present the Sarsa example as a cautionary tale that contamination may indeed result in ages much older than expected and, therefore, that such is the parsimonious explanation in the Nerja case. This argument is a textbook example of the non-sequitur fallacy. In Sarsa, we are dealing with the artificial addition of chemicals undertaken in a restoration laboratory by a human agent. In Nerja, the outcome would have to result from the action of a natural agent in the burial environment. The two cases are not comparable. In addition, their explanation requires that we accept two empirically untenable premises. The first is that analysts can determine 100% of the time whether a given bone belongs to the domestic (goat) or the wild (ibex) species of *Capra*, even when the bone in question is a diaphyseal fragment lacking the diagnostic proximal epiphysis (Fig. 7.4) – recall that, in Zeder and Lapham's (2010) experiment, the error rate in distinguishing proximal radii at the genus, *Ovis* vs. *Capra*, not even the species level, was the highest, as much as 32–44%. The second is that a mid-Holocene bone's endogenous carbon can be replaced in a significant amount, if not almost completely, by exogenous carbon derived from an immediate burial environment that is significantly younger than the age of the putative contaminant.

The correct scientific interpretation of the OxA-X-2457-57 age measurement is that, if inaccurate, it is so on the younger, not the older side of the error. Put another way, the worst case 'accuracy scenario' for this result is that it represents a minimum age only. But this scenario makes no difference to the conclusion that the sample dates to a time range during which the only caprine extant in the region was the ibex.

As acknowledged by García-Borja *et al.*, the interface between the Neolithic and the underlying Early Holocene and Tardiglacial deposit is, at Nerja, significantly affected by post-depositional disturbance, including 'pits' containing Neolithic material that penetrate deeply into the Mesolithic/Epipaleolithic levels. The corollary of such disturbances is that, conversely, upwardly displaced Mesolithic/Epipaleolithic bones must be present in the overlying Neolithic levels. Clearly, the charred caprine radius dated by OxA-X-2457-57 is one such bone, and therefore an instance of ibex misidentified as domestic sheep/goat.

Association issues

The radiocarbon age of the dated astragalus from the Vestíbulo hall of Nerja is statistically identical to the earliest so far obtained for confirmed domesticates in the adjacent regions of Valencia and Catalonia: sheep from Guixeres de Vilobí and La Sarsa, wheat from El Cavet and La Falguera, and barley from Mas d'Is (Martins *et al.* 2015, tables 4–5). Accepting that the bone is sheep, domesticates would be present in eastern Andalucía by 5481–5616 cal BC (the 95.4% probability interval of the date), and the Neolithic would have begun there at about the same time as in Valencia and Catalonia. Given the 135-year width of the interval, such penecontemporaneity does not contradict the general west Mediterranean pattern of rapid, directional, time-transgressive coastal dispersal, from East to West and North to South (Isern *et al.* 2017). Add that the probability of mistaken identifications is, for sheep astragali, low, and such consistency with context makes it legitimate to use the Nerja/Vestíbulo result, with due caution, to assess the mode and rhythms of the process. In such a theoretical context, one may be justified in taking the Nerja bone, with appropriate caveats, as a space/time proxy for the presence of farming.

What García-Borja *et al.* do, however, is something entirely different: they use the Nerja/Vestíbulo astragalus result to assess the cultural affinities of the site's Early Neolithic and to date the corresponding ceramic assemblage. They argue that 'the comparison with assemblages of similar chronology from Valencia and Catalonia shows little affinity, as in these regions the percentage of impressed Cardial ceramics hovers around 40% between 5500 and 5200 cal BC, and that is not the case at Nerja' (p. 28). The astragalus result would instead support the notion that the region's earliest Neolithic was an Impressed Ware, pre-Cardial tradition that diffused to Iberia from either Liguria, along a northern Mediterranean coastal route, or southern Italy, via the Maghreb and the Straits of Gibraltar.

Because the advent of the Neolithic did not mean that people abandoned hunting, whether a given bone stands for the slaughtering of sheep from a herded flock or of ibex from a successful hunt is, in the context of assessing cultural tradition, largely irrelevant. What matters is the integrity of the deposit the bone comes from: the bone will date the artefacts found alongside it if, and only if, their stratigraphic association is reliable. If it is, the sample, whether wild or domesticate, dates the presence of the Neolithic tradition defined by the associated material culture. If it is not, the sample dates itself and nothing else; its significance for Neolithic archaeology will then depend entirely on the issues of taxonomic identification and age measurement discussed in the preceding sections.

The validity of García-Borja *et al.*'s argument therefore depends not on the status – domestic or wild, sheep or ibex – of the caprine individual the dated Nerja/Vestíbulo astragalus belonged to, but on whether it came from an archeologically homogeneous, stratigraphically intact context. Did it? No. It came from a major post-depositional disturbance feature: a pit containing some Neolithic finds that penetrated deep into Epipaleolithic level NV-4 (Aura *et al.* 2005). In short, García-Borja *et al.* use a sample from a disturbed context excavated in the Vestíbulo hall of the site to date the ceramic assemblage retrieved in the Mina hall. Nothing could be farther from date-by-association logic.

The cultural significance that García-Borja *et al.* attach to the dated Nerja/Vestíbulo astragalus thus comes not from its empirical context but from the apposition of meaning derived from an aprioristic view: that the bone represents the same Neolithic activity that resulted in the on-site discard of the Nerja ceramic assemblage as a whole, which would therefore be of the same early age (or belong to a tradition stretching back in time to that early age). Under such a *modus operandi*, it would be equally legitimate, however, for others to conclude that the sample stands for a Late Mesolithic acquisition of domesticates prior to the subsequent Neolithic occupation represented by the ceramics. Dismissing the latter claim as speculation that does not qualify as a scientific assessment of the evidence implies doing the same for the former. Both represent no more that different forms of what Binford (1981) called a 'post-hoc accommodative argument' and Smith (2015, 19) describes as 'the scientific equivalent of the farmer who paints bulls-eyes around the bullet holes in his barn in order to show his superior shooting skill.'

Instead, we can look at Nerja under the falsification logic used in science. To do this, we need to recall the stratigraphic framework of Nerja's Early Holocene sequence (Aura *et al.* 2005; 2013; García Borja *et al.* 2014; 2018; Table 7.2). In the Vestíbulo hall, the *in situ* Early Neolithic context is level NV-2; underlying level NV-3 contains a mix of Neolithic, Mesolithic and Epipaleolithic finds, and basal level NV-4 is an Epipaleolithic deposit cut by disturbance features that contain Neolithic material. In the Mina hall, *in situ* Early Neolithic contexts are found in levels NM-6 to NM-9. Level NM-10 is subsumed with level NM-11, described as mixed deposits containing both Neolithic and Mesolithic material, while underlying level NM-12 contains a scatter of lithics, ornaments and bone tools of both Epipaleolithic and Neolithic affinities, but no ceramics.

Given this stratigraphic framework, two dating scenarios and four site formation hypotheses can be advanced (Table 7.3). Scenario 1 accepts that García-Borja *et al.*'s Neolithic samples, whether domestic or wild, are all indeed Neolithic and archeologically associated with the material culture found alongside. Scenario 2 only considers the samples whose domesticate status is secure because of the possibility that the others represent background noise (*e.g.* natural accumulations of ibex dead at the site during a period of human abandonment, or upwardly displaced remains of ibex hunted and consumed in the context of preceding Mesolithic occupations). In Hypotheses B, C and D, the site's first Neolithic settlement would not have been intensive enough to result in the constitution of separate, recognisable stratigraphic units; thus, the remains left behind would have ended-up mixed in the Mesolithic/Epipaleolithic deposit that, at the time, formed the two halls' ground floor. Under these three hypotheses, the Vestíbulo sequence – phases 'in NV-4' and 'mixed within NV-3' at the bottom, followed by phase '*in situ* within NV-2' at the top – would have stratigraphic value and represent a true succession of different Neolithic occupation phases; and ditto for the different units of the Mina hall. Under Hypothesis A, no such phasing exists; despite the vertical scatter, the assemblage would have been deposited in the framework of a single, relatively short period of site use.

In Scenario 1 (all dated samples are Neolithic and their artefact associations are genuine), Hypotheses A and C are rejected because the samples are of statistically different

Table 7.2. AMS radiocarbon dates for the Cueva de Nerja deposits that yielded Early Neolithic remains, calibrated under Calib 7.0.1 with INTCAL13 (Reimer et al. *2013). For additional details on sample chemistry, specimen identification, and sample provenience, see Martins* et al. *(2015) and García-Borja* et al. *(2010; 2014; 2018).*

Level	*Specimen*	*Lab #*	*Date BP*	*Cal BC 2σ*	*ZooMS*
Vestíbulo hall					
NV-2	*Hordeum vulgare* grain	Beta-284149	5050±40	3958–3715	n/a
NV-2	*Ovis/Capra* bone	MAMS-20437	6185±21	5215–5057	Not done
NV-3	*Ovis aries* radius	Beta-369357	6300±40	5362–5213	Not done
Pit	*Ovis aries* astragalus	Beta-131577	6590±40	5616–5481	Not done
NV-3	*Pinus pinea* wood	GifA-102010	7610±90	6639–6257	n/a
NV-4	*Capra pyrenaica* pelvis	Beta-156020	10040±40	10011–9305	n/a
Mina hall					
NM-6	*Ovis/Capra* tibia	OxA-26079	6207±32	5291–5055	*Ovis*
NM-7	*Ovis/Capra* metacarpal	OxA-26080	6196±31	5287–5045	Undetermined
NM-7	*Ovis/Capra* radius	OxA-26081	6219±33	5299–5061	Undetermined
NM-8	*Ovis/Capra* radius	OxA-26082	6214±35	5297–5058	*Ovis*
NM-8	*Ovis/Capra* radius (a)	OxA-X-2457-57	7255±65	6235–6009	Undetermined
NM-9	*Ovis/Capra* femur	OxA-26083	6252±33	5312–5078	*Ovis*
		OxA-26084 (b)	6254±33	5313–5078	
NM-10	*Ovis/Capra* tibia	OxA-26085	6342±37	5464–5221	Undetermined
NM-12	*Ovis/Capra* vertebra	OxA-26086	6466±33	5484–5366	Undetermined
NM-11	*Lathyrus* sp. seed	Beta-284146	7150±40	6080–5923	n/a
NM-11	*Pinus pinea* bract	Beta-284148	7490±40	6435–6251	n/a

(a) charred
(b) repeat date

ages and the dating results respect the stratigraphic order of their provenience. No sherd-refitting data being available, choosing between B and D must be done on the strength of the typological evidence alone. According to García-Borja *et al.* (2014, tables 2 and 8–9), of the 13 fragments in Vestíbulo's 'in NV-4' phase that one might associate with the astragalus dated to 6590±40 BP (Beta-131577), only one is decorated, and its decoration is made with the edge of a shell. And, of the 283 fragments in Mina's NM-10/NM-11 phase that one might associate with the sheep/goat dates of 6466±33 BP for NM-12 (OxA-26086) and 6342±37 BP (OxA-26085) for NM-10, only 25 are decorated, 18 of which featuring the cordons, knobs and impressions that are typical of the Cardial (Guilaine and Manen 2007); in addition, the instrument used to decorate three out of the eight impressed fragments was a shell. More to the point, no such shell-impressed sherds are found higher up in the Mina sequence. In this scenario, therefore, only Hypothesis D is not rejected, and the Nerja evidence remains consistent with a model in which the site's earliest Neolithic occupation is a fleeting, Cardial-related episode.

In Scenario 2 (only the securely domestic samples are considered), Hypotheses B, C and D are rejected because the three dates that can be used range between 6207±32 BP (OxA-26079) for NM-6 and 6254±33 BP (OxA-26084) for NM-9 (Table 7.2) and are statistically the same. With regards to Hypothesis A, support might be sought in the fact that, together, levels NM-6 to NM-9 yielded 2469 ceramic fragments (450 decorated), while the underlying mixed/reworked levels NM-10 and NM-11 yielded 283 (25 decorated). Put another way: 89.7% of the ceramics found in the levels of the Mina hall that García *et al.* define as Early Neolithic, and 94.7% of the decorated fragments in that assemblage, are associated with dates comprised between 5055 and 5313 cal BC. Adding level NV-3 of the Vestíbulo hall on the basis that its unconfirmed sheep date (6300±40, Beta-369357; Table 7.2) is statistically the same as those for levels NM-6 to NM-9 of the Mina hall, we get even higher numbers: of the 699 decorated fragments in both halls, no less than 673 (450 in NM-6 to NM-9, 165 in NV-2, and 58 in NV-3), *i.e.* 96%, come from levels in which the Neolithic component post-dates 5362 cal BC, the lower limit of the calibrated range of the Beta-369357 result (6300±40 BP).

These Scenario 2 percentages, however, only allow treatment of the ceramics retrieved in the basal Neolithic levels of Nerja as a single, homogeneous assemblage (Hypothesis A) if the concentration of Cardial diagnostics found therein is devoid of statistical significance. To assess

Table 7.3. Different site formation hypotheses capable of explaining the composition of Nerja's Early Neolithic ceramics, and their test implications.

		Artefact implications	*Dating implications*
A	The assemblage is homogeneous; it represents a single period of occupation, the remains of which are found *in situ* in levels NM-6 to NM-9 and NV-2, and as intrusions in underlying Mesolithic/Epipaleolithic levels	The same types of decorated ceramics are found from top to bottom of the sequence; refits between fragments found *in situ* and as intrusions are common	Samples are all statistically the same age
B	The assemblage is homogeneous; it represents different, stratigraphically ordered occupations belonging to a single tradition, the earliest represented by the ceramics intruded in the Mesolithic/Epipaleolithic levels, the latest by those found *in situ* in levels NM-6 to NM-9 and NV-2	The same types of decorated ceramics are found from top to bottom of the sequence; refits between fragments found *in situ* and as intrusions are spatially related to post-depositional disturbance features and limited in number	Samples are of statistically different age and the dating results agree with the samples' stratigraphic order
C	The assemblage is heterogeneous; it represents different occupations belonging to different traditions, the stratigraphic precedence of the earliest having been erased by post-depositional mixing within the disturbance features found in the Mesolithic/Epipaleolithic levels	The same types of decorated ceramics are found from top to bottom of the sequence; refits between fragments found *in situ* and as intrusions are common	Samples show little agreement between age and stratigraphic order
D	The assemblage is heterogeneous; it represents different, stratigraphically ordered occupations belonging to different traditions, the earliest represented by the ceramics intruded in the Mesolithic/Epipaleolithic levels, the latest by those found *in situ* in levels NM-6 to NM-9 and NV-2	Certain ceramic types are only found in the basal levels; refits between fragments found *in situ* and as intrusions are spatially related to post-depositional disturbance features and limited in number	Samples are of statistically different age and the dating results agree with the samples' stratigraphic order

Table 7.4. Vertical distribution of impression types across the Nerja sequence of decorated ceramics.

Stratigraphic provenance of the ceramics	*Shell impressions*	*Other impressions*	*Total*
In situ in the Early Neolithic (NV-2, NM6 to NM-9)	2	190	192
Intrusive in the underlying Mesolithic/Epipaleolithic (NV-3, NV-pit, NM-10, NM-11)	6	20	26
Total	8	210	218

whether such is the case, we can apply the chi-square test to the vertical distribution of impression types, using the matrix in Table 7.4. The p-value obtained, 2.04E-08, rejects the null hypothesis that the stratigraphic distribution of shell impressions does not change across the sequence – and, therefore, rejects Hypothesis A too. Thus, under Scenario 2, this hypothesis – that the Nerja ceramics are a homogeneous assemblage belonging to a single period or a single tradition – can be salvaged only if one rejects the validity of the definition as Cardial diagnostics of the modes of ceramic decoration that archaeologists have so far used, something for which the evidence is non-existent and in any case not provided by García-Borja *et al.*

With current evidence, the only model that is consistent with the data is a two-phase sequence in which, initially, there is a fleeting Cardial occupation of the site, followed by a much more substantial, post-Cardial or Epicardial one – as, indeed, Martins *et al.* already had suggested. García-Borja *et al.*'s rejection of this model resorts to an argument from authority that is also a non-sequitur: 'our experience with the study of some of the richest Neolithic sites with Cardial impressed ceramics in western Europe leads us to question that hypothesis, as one of these groups' characteristics is precisely the high visibility of their occupations in the archaeological record.' Surely, one should not need to point out that the hazards of García-Borja *et al.*'s experience need not imply that all Cardial occupations must be highly visible, or that low-visibility Neolithic occupations cannot be Cardial by definition.

Be it as it may, García-Borja *et al.* base their interpretation of the site's Early Neolithic ceramics in a non-existent inconsistency. The fact that Cardial ceramics are rare at

Nerja can be fully explained in a way that is fully consistent with current knowledge of the Early Neolithic of west Mediterranean Europe: Scenario 1/Hypothesis D, where the ceramic assemblage is heterogeneous and bearers of Cardial wares archaeologically imprinted the basal levels of the Vestíbulo's Holocene sequence in a manner that, by comparison with the imprint left by later, post-Cardial occupations of the site, was very light.

Conclusions

Compounded by limited precision, radiometric dating errors have a potentially huge impact in our interpretation of the human past. On the archaeological side of the dating chain, the two most important and commonest sources of error are the mistaken identification of dated specimens and the disassociation between the dated sample and the event or process the sample was intended to date. In west Mediterranean Europe, two such dating-related, decades-old problems with profound consequences for our understanding of the emergence of farming persist. One is the apparent-but-false association between Neolithic diagnostics and samples of an older age resulting from the fuzzy nature that stratified cave and rock shelter sites ubiquitously show at the interface between their first Neolithic and their last hunter-gatherer (Palaeolithic, Epipaleolithic, or Mesolithic) occupations. The other is the misidentification as sheep or domestic goat of the bone remains of wild ungulates of similar size and skeletal morphology. In Iberia specifically, Nerja provides good illustration of these problems, but it is not an isolated case, far from it.

The Nerja/Mina radius from level NM-8 whose radiocarbon age (7255±65 BP; OxA-X-2457-57) implies Mesolithic ibex rather than Neolithic sheep/goat is simply another instance of a rather extended phenomenon; in no way is it a rare, exceptional occurrence. Another unambiguous instance of the upward displacement into thick and coherent Neolithic deposits of specimens derived from levels of an earlier age can be found in Martins *et al.*'s (2015) study: the human bone from the upper part of the Neolithic levels of Cova Fosca de la Vall d'Ebo (Valencia) dated to 10060±45 BP (OxA-26073). Given the age measurement, this bone clearly derived from the site's much deeper Epipaleolithic levels. Contra García-Borja *et al.*, invoking this type of processes to explain dating anomalies is not special pleading.

Likewise, it is quite often the case that the Neolithic occupation of a site is represented by ceramics, stone tools, beads, human bones or the remains of domesticates found as scattered intrusions in a deposit whose content is mostly, if not overwhelmingly, of an earlier age (Mesolithic, Epipaleolithic or Upper Palaeolithic). A case in point is Gruta do Caldeirão (Portugal), whose earliest Neolithic context consists of an ensemble of finds retrieved from the Magdalenian deposit (layer Eb) that caps the site's Pleistocene succession (Zilhão 1992; 1993). Based on artefact typology and the bones' surface condition, tested by relative (fluorine) and absolute (radiocarbon) dating, layer Eb's Neolithic-related finds could be sorted out and defined as a separate site occupation event (Horizon NA2) – albeit one that did not exist at the site as a differentiated stratigraphic unit (much as, within Hypotheses C-D above, would be the case with Nerja's earliest Neolithic too).

Perusal of the literature also shows that mistaking ibex for sheep or domestic goat and making the Neolithic at a given site begin 'earlier than expected' on the back of the dating of such samples are not as infrequent occurrences as one might have hoped for. This is a distinct possibility in the case of the Nerja astragalus, but examples exist at other sites. Cases in point are the putative domestic sheep/goat specimens from the Middle Neolithic of Cova de la Carihuela (Granada) and the Early Neolithic of the Peña Larga rock-shelter (Euskadi), dated to 6950±40 BP (COL-1563.1.1; Medved, 2013) and 6720±40 BP (Beta-242783; Fernández-Eraso 2011), respectively.

The occurrence of these problems does not detract from the value of zooarchaeology as a scientific discipline, or from the competence of the individual analysts responsible for the determinations. It simply goes to show that no one is infallible, that one should be prepared to admit that errors have been made, and that seeking additional diagnostic rigour and independent corroboration (namely, via such newly developed scientific tools as ZooMS and ancient DNA), is mandatory when the samples play a critical role in the testing of hypotheses of major significance.

Arguments from authority should have no place here. In this, archaeologists and zooarchaeologists alike ought to be inspired by the example of a leading figure of our twin disciplines: the author of the seminal textbook *The Archaeology of Animals* (Davis 1987), whose career the book this chapter appears in intends to honour. Faced with the same domestic vs. wild issue at the heart of the Nerja debate, Davis (2017) did not invoke his decades of experience with hundreds of thousands of sheep/goat bones from the Eastern Mediterranean, the British Isles and Iberia to assert the accuracy of his own past determinations. Instead, he developed biometric tools designed to capably assist the traditional morphology-based criteria, and then used them to find where and to what extent those determinations were wrong. It turns out that he had been almost always right. But that's how science works, and that's the true measure of a scientist.

Acknowledgements

I thank the organisers of the 2017 symposium 'From the Mediterranean to the Atlantic: Themes in Old World Zooarchaeology' for the invitation to participate and contribute to this book. The identification of the sheep/goat astragali from Galeria da Cisterna was carried out with the expert help of Montserrat Sanz, Jordi Nadal and Simon Davis. Mostly, I want to acknowledge Simon Davis for his friendship and for the trust he placed in my country and in myself when, 20 years ago, he decided to leave the UK and take up his current position in Lisbon. Simon's commitment and hard work made it possible that zooarchaeology caught up in Portugal, from the incipient basis it then had to the flourishing discipline it is today. *Obrigado!*

References

Arias, P. and Altuna, J. (1999) Nuevas dataciones absolutas para el Neolítico de Arenaza (Bizkaia). *Munibe* 51, 161–171.

Aura-Tortosa, J.E., Badal-García, E., García-Borja, P., García-Puchol, O., Jordá, J.F., Pascual-Benito, J.LL., Pérez-Jordá, G., Pérez-Ripoll, M. and Jordá-Prado, J. (2005) La Cueva de Nerja (Málaga). Los niveles neolíticos de la sala del Vestíbulo. In P. Arias, R. Ontañón, C. García-Moncó (eds) *Actas del III Congreso del Neolítico en la Península Ibérica*, 975–987. Santander, Instituto Internacional de Investigaciones Prehistóricas de Cantabria.

Aura-Tortosa, J.E., Jordá-Prado, J., García-Borja, P., García-Puchol, O., Badal-García, E., Pérez-Ripoll, M., Pérez-Jordá, G., Pascual-Benito, J. LL., Carrión-Marco, Y. and Morales-Pérez, J.V. (2013) Una perspectiva mediterránea sobre el proceso de neolitización. Los datos de la cueva de Nerja en el contexto de Andalucía (España). *Menga. Revista de Prehistoria de Andalucía* 4(3), 53–77.

Binford, L.R. (1981) *Bones: Ancient Men and Modern Myths*. New York, Academic Press.

Boessneck, J., Müller, H. and Teichert, M. (1964) Osteologische unterscheidungmerkmale zwischen Schaf (*Ovis aries* Linné) und Ziege (*Capra hircus* Linné). *Kühn-Archiv* 78(1–2), 1–129.

Davis, S.J.M. (1987) *The Archaeology of Animals*. London, Routledge.

Davis, S.J.M. (2017) Towards a metrical distinction between sheep and goat astragali. In P. Rowley-Conwy, D. Serjeantson and P. Halstead (eds) *Economic Zooarchaeology*, 50–82. Oxford, Oxbow Books.

DeNiro, M.J., Schoeninger, M.J., and Hastorf, Ch.A. (1985) Effect of heating on the stable carbon and nitrogen isotope ratios of bone collagen. *Journal of Archaeological Science* 12(1), 1–7.

Fernandes, R., Hüls, M., Nadeau, M.-J., Grootes, P.M., Garbe-Schönberg, C.-D., Hollund, H.I., Lotnyk, A. and Kienle, L. (2013) Assessing screening criteria for the radiocarbon dating of bone mineral. *Nuclear Instruments and Methods in Physics Research Section B: Beam Interactions with Materials and Atoms* 294, 226–232.

Fernández-Eraso, J. (2011) Las cerámicas neolíticas de la Rioja alavesa en su contexto: los casos de Peña Larga y los Husos I y II. In J. Bernabeu, M.A. Rojo and L. Molina (eds) *Las primeras producciones cerámicas: el VI milenio cal AC en la Península Ibérica,* 117–130. València, Universitat de València.

Garcia, M.-A. (2005) Ichnologie générale de la grotte Chauvet. *Bulletin de la Société Préhistorique Française* 102(1), 103–108.

García-Borja, P., Aura-Tortosa, J.E., Bernabeu-Aubán, J. and Jordá-Pardo, J.F. (2010) Nuevas perspectivas sobre la neolitización en la Cueva de Nerja (Málaga-España): la cerámica de la Sala del Vestíbulo. *Zephyrus* 66, 109–132.

García-Borja, P., Aura-Tortosa, J.E., Jordá-Pardo, J.F. and Salazar-García, D.C. (2014) La cerámica neolítica de la Cueva de Nerja (Málaga, España): salas del Vestíbulo y la Mina. *Archivo de Prehistoria Levantina* 30, 81–131.

García-Borja, P., Salazar-García, D.C., Jordá-Pardo, J.F., Pérez-Ripoll, M. and Aura-Tortosa, J.E. (2018) El inicio del Neolítico en la cueva de Nerja y la Cova de la Sarsa. Contexto arqueológico y dataciones radiocarbónicas. *Pyrenae* 49(2), 7–36.

Geddes, D. (1993) La faune de l'abri de Dourgne: paléontologie et paléoéconomie. In J. Guilaine, M. Barbaza, J. Gasco, D. Geddes, J. Coularou, J. Vaquer, J.-E. Brochier, F. Briois, J. André, G. Jalut and J.-L. Vernet (eds) *Dourgne. Derniers chasseurs-collecteurs et premiers éleveurs de la Haute-Vallée de l'Aude*, 365–397. Toulouse/Carcassonne, Centre d'Anthropologie des Sociétés Rurales/Archéologie en Terre d'Aude.

Guilaine, J. and Manen, C. (2007) From Mesolithic to Neolithic in the western Mediterranean. *Proceedings of the British Academy* 144, 21–51.

Higham, T., Douka, K., Wood, R., Ramsey, C.B., Brock, F., Basell, L., Camps, M., Arrizabalaga, A., Baena, J. and Barroso-Ruíz, C. (2014) The timing and spatiotemporal patterning of Neanderthal disappearance. *Nature* 512, 306–309.

Isern, N., Zilhão, J., Fort, J. and Ammerman, A.J. (2017) Modeling the role of voyaging in the coastal spread of the Early Neolithic in the West Mediterranean. *Proceedings of the National Academy of Sciences USA* 114(5), 897–902.

Manen, C., Perrin, T., Guilaine, J., Bouby, L., Bréhard, S., Briois, F., Durand, F., Marinval, P., Vigne, J.-D. (2019) The Neolithic transition in the Western Mediterranean: a complex and non-linear diffusion process – the radiocarbon record revisited. *Radiocarbon* 61, 531–571.

Martins, H., Oms, F.X., Pereira, L., Pike, A.W.G., Rowsell, K. and Zilhão, J. (2015) Radiocarbon dating the beginning of the Neolithic in Iberia: new results, new problems. *Journal of Mediterranean Archaeology* 28(1), 105–131.

Medved, I. (2013) *Continuity vs. Discontinuity: Epipaleolithic and Early Neolithic in the Mediterranean Southeast of the Iberian Peninsula*. Unpublished thesis, University of Cologne.

Reimer, P.J., Bard, E., Bayliss, A., Beck, J.W., Blackwell, P.G., Bronk Ramsey, C., Buck, C.E., Cheng, H., Edwards, R.L., Friedrich, M., Grootes, P.M., Guilderson, T.P., Haflidason, H., Hajdas, I., Hatté, C., Heaton, T.J., Hoffmann, D.L., Hogg, A.G., Hughen, K.A., Kaiser, K.F., Kromer, B., Manning, S.W., Niu, M., Reimer, R.W., Richards, D.A., Scott, E.M., Southon, J.R., Staff, R.A., Turney, C.S.M. and van der Plicht. J. (2013) IntCal13 and MARINE13 radiocarbon age calibration curves 0–50000 years calBP. *Radiocarbon* 55(4), 1869–1887.

Smith, M. (2015) How can archaeologists make better arguments? *The SAA Archaeological Record* 15(4), 18–23.

Waterbolk, H.T. (1971) Working with radiocarbon dates. *Proceedings of the Prehistoric Society* 37, 15–33.

Waterbolk, H.T. (1983) The integration of radiocarbon dating in archaeology. *Radiocarbon* 25, 639–644.

Wood, R. (2015). From revolution to convention: the past, present and future of radiocarbon dating. *Journal of Archaeological Science* 56, 61–72.

Zazzo, A. (2014) Bone and enamel carbonate diagenesis: A radiocarbon prospective, *Palaeogeography, Palaeoclimatology, Palaeoecology* 416, 168–178.

Zeder, M.A. and Lapham, H.A. (2010) Assessing the reliability of criteria used to identify postcranial bones in sheep, *Ovis*, and goats, *Capra*. *Journal of Archaeological Science* 37, 2887–2905.

Zilhão, J. (1992) *Gruta do Caldeirão. O Neolítico Antigo*. Lisboa, Instituto Português do Património Arquitectónico e Arqueológico.

Zilhão, J. (1993) The spread of agro-pastoral economies across Mediterranean Europe: a view from the Farwest. *Journal of Mediterranean Archaeology* 6(1), 5–63.

Zilhão, J. (2013) Neandertal-modern human contact in Western Eurasia: issues of dating, taxonomy, and cultural associations. In. T. Akazawa, Y. Nishiaki and K. Aoki (eds) *Dynamics of Learning in Neanderthals and Modern Humans* 1, 21–57. Tokyo, Springer.

8

Astragali and their archaeological contexts in the Iberian Peninsula. Significance, meanings and historical implications

Ana Margarida Arruda

To Simon, who welcomed me into his Noah's Ark.

The functions of astragali in Antiquity have been studied and evaluated, mainly taking stock of their contexts but also Greek and Roman iconography.

Studies have chiefly approached the Levant and Eastern and Central Mediterranean islands. We now describe data referring to this type of bone in Iberia.

Recurrently present in necropolises and sanctuaries of the Iron Age, astragali also appear in habitat sites of the same chronology. An analysis of contexts, gathering information on the Far West of Iberia, is important. Also, the adoption of exogenous models in ritual practices and aspects of daily life, which talus bones illustrate, must be discussed in terms of the phenomena of the mobility allowed by the Phoenician colonisation, giving historical perspective to animal bones.

Keywords: Astragali, Iberian Peninsula, Iron Age religious systems, rituals and religion activities

Astragali in Antiquity – uses, functions, contexts

Astragalus reproduction, in different materials and sizes, was common practice in Antiquity. Having begun, at least, in the Bronze Age, it has lasted until today (Fig. 8.1). Their reproduction was the most common among all kinds of animal bones – at least since the second millennium BCE – and these copies performed several functions. We have found astragalus reproductions in bronze, glass, ceramic, stone (onyx, marble, hyaline quartz) and ivory (Fig. 8.2). But astragali, the bones themselves, were often carved and/or polished and used in different daily activities. They are almost always small-sized, collected from *Ovis*/*Capra* or *Sus*. Larger bones from bovines are rarer.

The use of these small-sized bones, or their reproductions, in recreation, religious and devotional activities has long been recognised (Elia and Carè 2004; Lebo 2016). The researcher to whom we pay tribute in this paper also had the opportunity to recall this (Davis 2004, 3).

The history of astragali use in Antiquity has already been traced (Guilmour 1997; Lebo 2006; De Grossi Mazzorin and Minniti 2013). Even so, let me briefly recall that many researchers take for granted their association with the children's world, due to some iconography, particularly of the Hellenistic and Roman periods (Fig. 8.3), but also because their presence is frequent in many children's tombs

Figure 8.1. Bronze astragalus monument. Atyrau, Kazakstan. Tolegen et al. *2018, fig. 4.*

Figure 8.2. Attic pottery. Sotades painter. 5th century BCE. British Museum (London), 1860, 1201. Courtesy of British Museum.

Figure 8.3. Part of a marble group of two boys quarrelling over a game of knucklebones. 2nd or 1st century BCE. British Museum (London), 1805,0703.7. Courtesy of British Museum.

in Italy and Greece (Elia and Carè 2004; Carè 2012), in which they could have performed an apotropaic function. Such function was clearly proved by the way in which the nearly 9,000 astragali found at the necropolis of Lucifero, at Locri Epize (Southern Italy), were distributed around the bodies in graves (Fig. 8.4), seemingly forming a protection barrier (Elia and Carè 2004). Also, in grave 101 of the

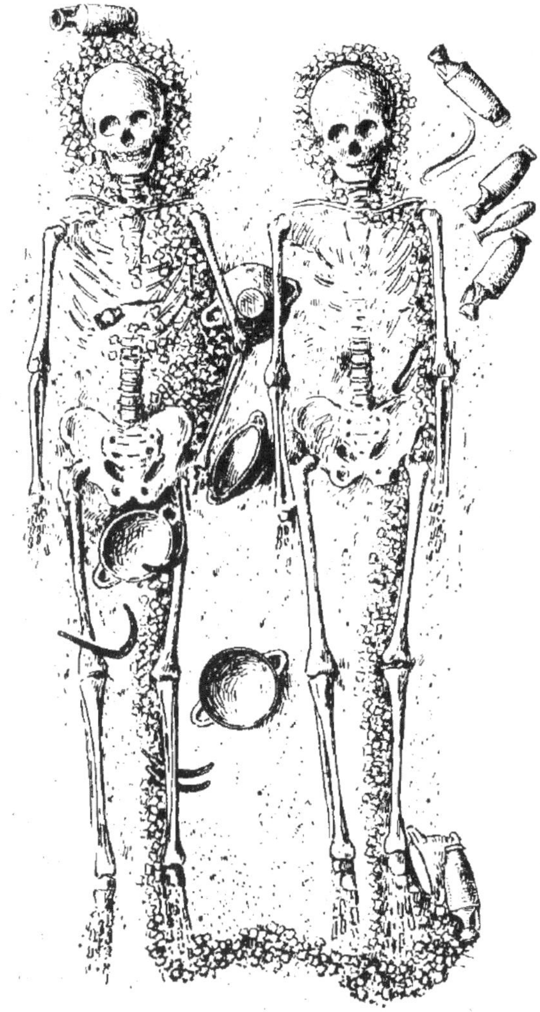

Figure 8.4. Tomb 348 of Lucifero cemetery. De Grossi and Minniti 2013, fig. 4.

Varranone necropolis (Fig. 8.5), at Poggio Picenze, astragali were placed between the body and the entrance (De Grossi Mazzorin and Minniti 2013).

In this context, we must also remember that the association of these objects with recreational activities is also unquestionable – particularly (but not exclusively) in the children's sphere, as evidenced by statues, mural paintings and coroplast work (Fig. 8.3), and referred to in the Greek and Roman texts: 'Do not put my bones apart from yours, Achilles, but together, just as we were raised in your house, when Menoitios led me, still little, from Opus to your house under [the ban of] grievous manslaughter, on that day when I killed the son of Amphidamas, foolishly, unwillingly, angered over a game of astragali; there receiving me in his house, the horseman Peleus raised me carefully and named me your henchman'. (Homer, Iliad, XXIII, 83–92)

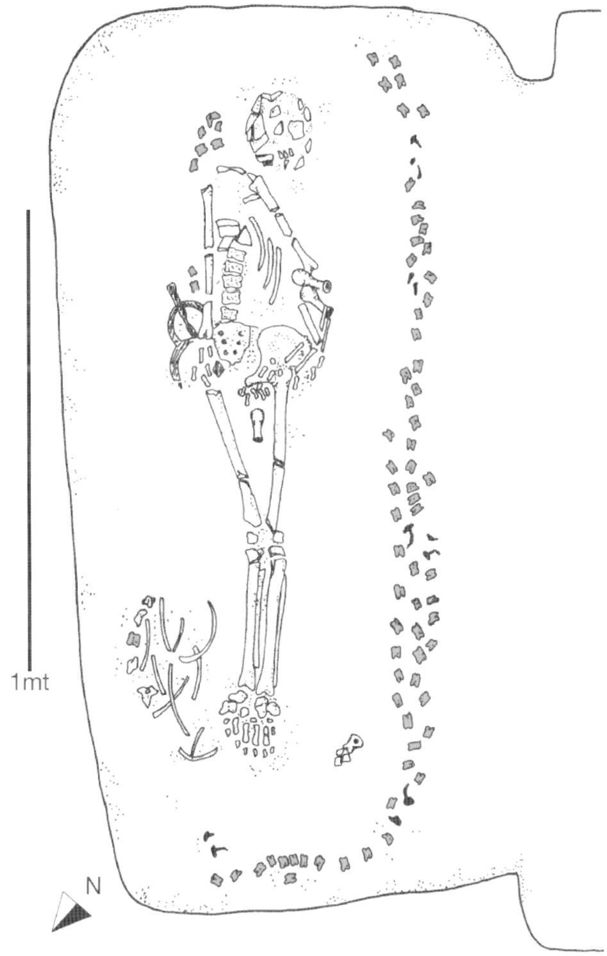

Figure 8.5. Tomb 101 of Varranone cemetry. De Grossi and Minnit 2013, fig. 3.

Figure 8.6. Set of three astragal-shaped bronze weights from Pera Museum (Turkey). Courtesy of Pera Museum.

Figure 8.7. Attic pottery. Painter of London D 12. 5th BCE. Metropolitan Museum (New York), 40.11.22. Courtesy of Metropolitan Museum.

Astragali were also frequently placed in graves in other cultural (namely Semitic) settings, such as Anatolia, the Syrian-Palestinian corridor (Guilmor 1997; Minniti and Peyronel 2005), Cyprus (Minniti and Peyronel 2005), Carthage and Utica, in North Africa (Lancel 1982, figs 579–580; Cardoso *et al.* 2016). Their use as gaming pieces, which per se could explain their reproduction in different raw materials, should also take into account the fact that some bone-made astragali were filled with lead, as found at Lapis Niger, the *forum* sanctuary in Rome (Blanc and Blanc 1958; *apud* in De Grossi Mazzorin and Minniti 2013), but also in Spain in Huelva (González de Canales *et al.* 2004, 169) and in grave 86G/26-11 of the Medellín necropolis (Almagro Gorbea 2008, 482–483). Lead-filled astragali have been considered weighting artefacts by certain authors; one should remember that some metal pieces, in bronze, were certainly used for this purpose, and reproduced the astragalus shape (De Grossi Mazzorin and Minniti, 2013, 372), like the those deposited at the Turkey Pera Museum (Fig. 8.6), an excellent testimony for Roman times. Others, ceramics, with identical morphological characteristics, are earlier, from Iron Age (5th century BCE (Fig. 8.7).

In my opinion, attributing the same connotation to images of women using this type of artefact (Figs 8.8–9) is more controversial. Women are represented in pairs in these cases, usually in scenes painted on Greek vases. Perhaps they depict practices potentially associated with divination – although domestic and almost popular, without the oracle features found in sanctuaries.

Figure 8.8. Attic pyxis with women playing with knucklebones, and astragalus as handle of the cover. 425–400 BCE. Metropolitan Museum (New York), 06.1021.119a, b. Courtesy of Metropolitan Museum.

Figure 8.9. Women playing Knucklebones. Hellenistic terracotta. Capua or Apulia, 330–300 BCE. British Museum (London), 1867,0510.1. Courtesy of British Museum.

In fact, in places of worship, astragali and their reproductions were artefacts directly associated with the rituals performed in those sanctuaries of public worship. An astounding number of pieces have been found in certain places, the best example being the cave of Corycian, near Delphi (Guilmour 1997), in which 22,780 items were found (Larson 2011).

In Antiquity, of both Greco-Latin and Semitic matrix, astragali, either originals or reproductions, were systematically related to the magical-religious universe. Also, we can (and should) associate them with merely recreational functions.

Similarly, this type of artefact was found outside the direct influence of the Mediterranean, in the context of the European continent. In fact, similar pieces were found in the necropolises of urn fields of Central Europe (Wiesner 2013), but also in Egypt, namely in Tutankhamun tomb (in amber), and the Balkans, in Romania (Kogalniceanu *et al.* 2014) and Bulgaria, which are dated from the Chalcolithic (Sidéra and Vernicu 2016). This extended geographical and chronological scope shows that the use of astragali is common in different cultural contexts.

Astragali and their contexts in the Iberian Peninsula

The link between these bones and the magical-religious universe is well known for Eastern and Central Mediterranean regions, with plenty of data published and studied. As regards the Western Mediterranean, only recently were these pieces deemed important.

Their presence is none the less significant. It deserves to be underscored and interpreted in the framework of the Iron Age, during which this region was greatly influenced by Mediterranean contributions, in terms of genetics and demography, but also culturally (Fig. 8.10).

First, we should state that astragali used as ritual objects are well documented in necropolises and sanctuaries of Iberia, mainly in coastal territories – more specifically in the South-west, the Levant and the South-east. In the first two areas, their presence is remarkable in necropolises of the so-called Iberian World – *e.g.* Coimbra del Barranco Ancho, Cigarralejo or Albufareta (Cuadrado 1987; García Cano 1997, 251–253; Martín-Poveda 2014). The presence of Phoenician and Greek colonies profoundly influenced these areas and is evidenced by funerary furniture and rituals. As regards the last area, we must remember that astragali are very frequent in the necropolises of Ampurias (Almagro Basch 1953) – a Greek colony which obviously replicated the rituals of the metropolis.

Astragali are rarer in the Iberian Far West, a fact possibly explainable by the small number of necropolises studied so far. Still, they were found at Cruz del Negro, in Carmona, Spain (Bernáldez-Sánchez *et al.* 2013), dated from the 8th century BCE, a site whose links to Western Phoenician colonisation have been demonstrated. Three-hundred and eight astragali were found here (307 from goats and 1 from a red deer), unevenly distributed in several graves – varying from 1 to 210 pieces per grave.

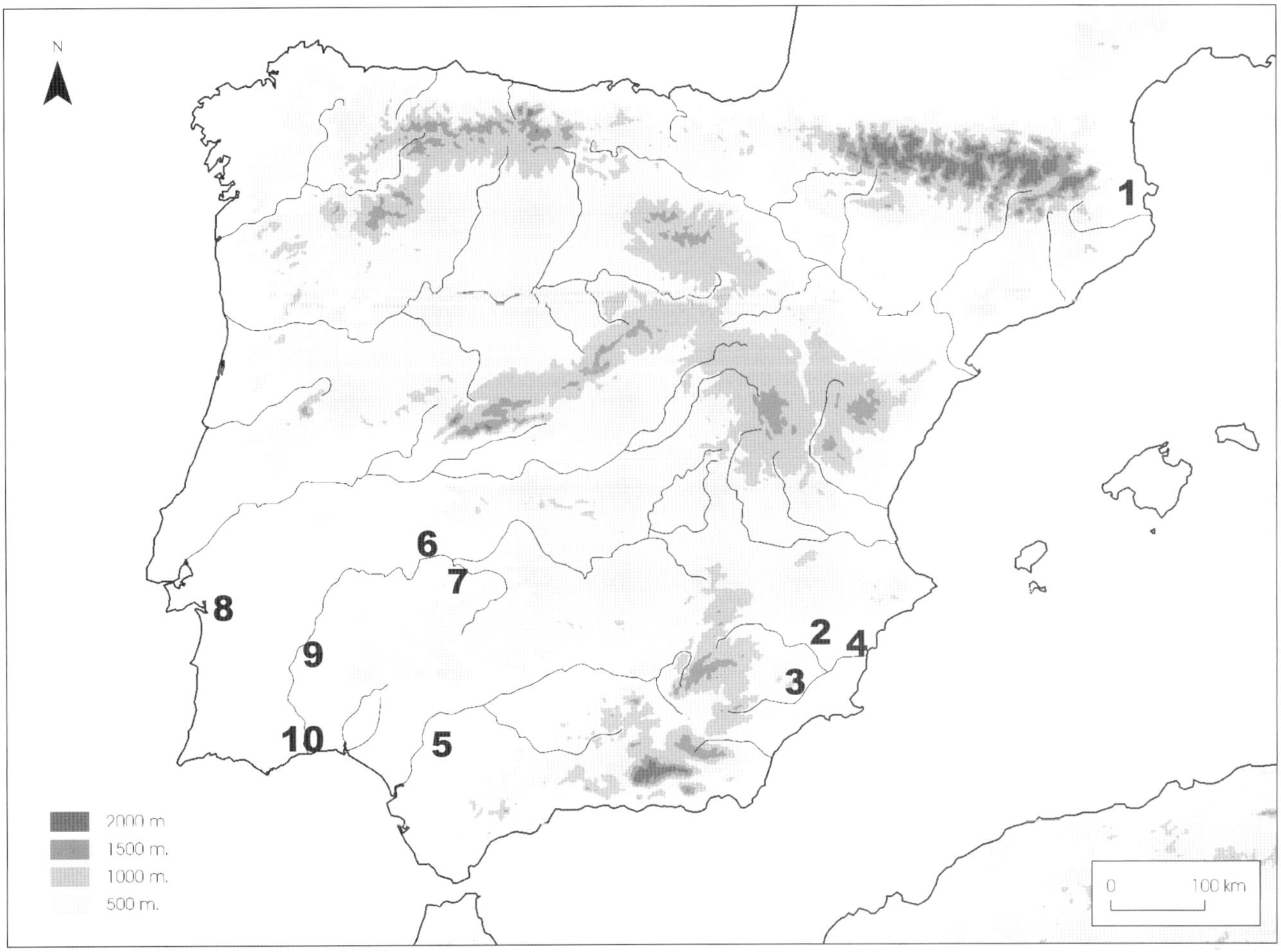

Figure 8.10. Astragali in Iberian Peninsula. 1. Ampúrias; 2. Coimbra del Barranco Ancho; 3. Cigarralejo; 4. Albufareta; 5. Cruz del Negro; 6. Medellín; 7. Cancho Roano; 8. Alcácer do Sal; 9. Azougada; 10. Castro Marim.

In Medellín (Spanish Extremadura), they were found in smaller numbers, *i.e.* 35 pieces in eight graves of all kinds, in terms of age, sex and time frame – as they appeared in all chronologies of the necropolis, from the 7th to the 5th century BCE (Almagro Gorbea 2008, 481–483). The number of astragali per grave is variable, some having 23 pieces (grave 83G/26) and others only one (*e.g.* grave 86G/42B).

Also related to the funerary world, the 14 astragali from Alcácer do Sal (Portugal) were found inside graves (Cardoso and Arruda 2016). Unfortunately, we have no data concerning their distribution within the necropolises, although they undoubtedly belong to contexts of the Iron Age (7th to 4th century BCE). Most of them belonged to sheep and goats, but some *Bos* astragali were identified – sometimes longitudinally cut, or polished.

Their absence from other funerary spaces – namely those excavated in recent decades in the region of Beja/Pedrogão (Portugal), is bewildering, on one hand because Oriental-type worship practices and funerary rituals were identified in these necropolises (Arruda 2016a; 2016b). On the other hand, the remains unearthed in such spaces find their most similar counterparts precisely in Medellín and Alcácer do Sal. Such absence cannot be attributed to soil conditions, as the human remains found were particularly well preserved. On the other hand, child burials are represented, as well as women burials.

In the Iberian Far West, other types of contexts also provided significant collections of astragali. In Spanish Extremadura, we may mention those of Cancho Roano (Maluquer 1983, 365) and La Mata (Rodríguez Diaz 2004). Although the function performed by these two sites remains under discussion (sanctuary *versus* palace), highly expressive ritual activities were developed – at least in the first.

At Alcácer do Sal, apart from the abovementioned necropolis of Senhor dos Mártires, nine astragali with polished sides – *e.g. Bos taurus* (four), *Cervus elaphus* (one), *Ovis capra* (three), and *Ovis aries* (one) – were found at Rua do Rato (Arruda *et al.*, in press) (Fig. 8.11).

During the archaeological intervention undertaken at this site, near the river, we found a remarkable collection

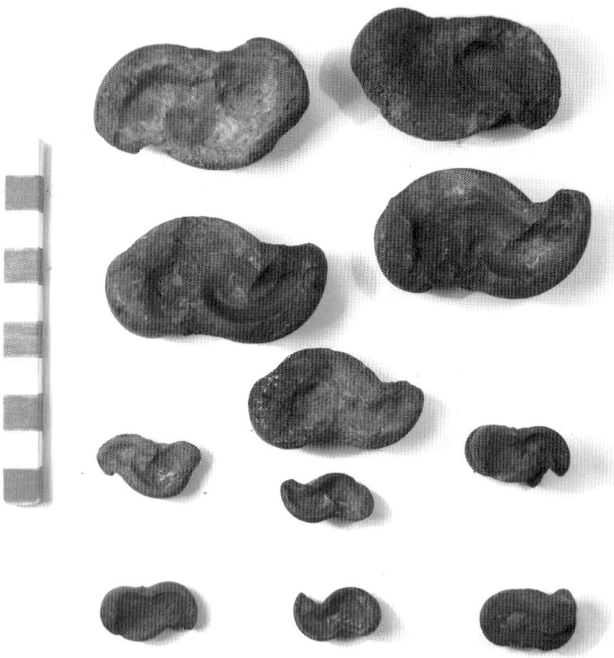

Figure 8.11. Astragalus from Rua do Rato (Alcácer do Sal). Photo by author.

of archaeological materials from the Iron Age in primary association – including ceramic vases, some in miniature, a large group of metal artefacts and bone artefacts, plus some mammal fauna. The characteristics of these materials – including pieces of furniture (*e.g.* a lion protome and several fragments of *diphroi*), silversmith work (including anvils, a possible spindle and fragments of a weighing scale with its weights), a *thymiaterion*, a Cypriot shell, and bone plates decorated with Oriental symbols (a tree of life) suggest that this context was formed, *grosso modo*, in the 6th to 5th century BCE – although more likely in the second half of the latter. Most of the unearthed artefacts were found *in situ*. Their quality is exceptional and their state of preservation excellent. These characteristics allow us to hypothesise that this might have been a place of worship, of the harbour type.

Further south and inland, an astragalus of sheep/goat were found at Azougada, in the municipality of Moura (Nabais and Soares 2017). This site was also a place of worship, as demonstrated on more than one occasion (Antunes 2009a; 2009b; 2017). The piece was flattened, and all the trochlea removed (Nabais and Soares 2017). The presence of a single astragalus may be due to the selective way in which materials were collected at this important site, excavated in the 1940s. In those days, faunal remains were not collected, unless they showed obvious signs of processing into identifiable artefacts. The team of Manuel Heleno picked this astragalus because it had been processed into a small rectangular plate. Perhaps they did not pick other astragali that were not so carefully processed.

The same connection with votive and cult activity can be advanced to the 89 astragalus collected at Iron Age levels of the Castro Marim Castle (Davis 2007, app. C), since the excavated area at this site functioned as a space of worship, from the 7th century to, at least, the 3rd BCE (Arruda 2005; Arruda *et al.* 2009; Arruda *et al.* 2017).

Final remarks

Astragali found in the Iberian Peninsula belong to a vast magical-religious universe, deeply rooted in the Mediterranean – as evidenced by the type of contexts, of burial and worship, from which they were recovered. This should be associated with the strongly Orientalising trend that influenced the Iberian South-west since, at least, the 9th century BCE. Also, it is related to the foundation of the Greek colony of Ampurias, insofar as the findings in the South-east and Levante are concerned.

Colonisation processes triggered the transfer of technological knowledge (*e.g.* potter's wheel, rotating mill, iron reduction, orthogonal plans, 'bunker' walls, etc.) and deep social and political changes. They also gave rise to changes in the religious universe, as previously demonstrated by the architectural characteristics of sanctuaries in the South-west, as well as the entities worshipped in such places. The ones at El Carambolo (Fernández Flores and Rodriguez Azouge 2007) and Castro dos Ratinhos (Berrocal Rangel and Silva 2010) are fine examples of such reality.

The use of astragali in necropolises and sanctuaries located in vast areas of Iberia evidences the full adoption of an exogenous religious system, as the cult and its specific rituals cannot be separated from theology and belief. We must, however, acknowledge the curious total absence of astragali from the necropolises of the Beja and Ourique regions, among others, although their funerary objects are clearly Orientalising. The fact that the funerary architecture itself has obvious indigenous roots, contrary to what was found in Alcácer do Sal and Medellín, combined with the absence of astragali, should be interpreted in the context of a local framework of memory and resistance.

Acknowledgements

This work was financed by Portuguese funds through FCT – Fundação para a Ciência e a Tecnologia in the framework of the project UID/ARQ/00698/2019.

Bibliography

Almagro Basch, M. (1953) *Las necrópolis de Ampurias. 1. Introdución y necrópolis griegas*. Monografias Ampuritanas. Barcelona, Seix y Barral.

Almagro Gorbea, M. (2008) *La necrópolis de Medellín. II. Estudio de los hallazgos*. Madrid, Real Academia de la História.

Antunes, A.S. (2009a) O 'Castro' da Azougada (Moura, Portugal): sacralidade e dinamismo comercial no Baixo Guadiana durante o Pós-Orientalizante, In P. Mateos, S. Celestino, A. Pizzo and T. Tortosa (eds) *Santuarios, oppida y ciudades: arquitectura sacra em el origen del desarrollo urbano del Mediterráno ocidental*, 131–152. Mérida, IAM/CSIC.

Antunes, A.S. (2009b) *Um conjunto cerâmico da Azougada. Em torno da Idade do Ferro Pós-Orientalizante da margem esquerda do Baixo Guadiana*, Suplemento 5 de O Arqueólogo Português. Lisboa, Museu Nacional de Arqueologia.

Antunes, A.S. (2017) A Azougada (Moura) e o sistema metrológico da Idade do Ferro Pós-Orientalizante do Baixo e Médio Guadiana, In J.M. Arnaud and A. Martins (eds) *Arqueologia em Portugal-Estado da Questão*, 907–928. Lisboa, AAP.

Arruda, A.M. (2005) O 1º milénio a.n.e. no Centro e no Sul de Portugal: leituras possíveis no início de um novo século. *O Arqueólogo Português* 4, 23, 9–156.

Arruda, A.M. (2016a) O touro da necrópole de Cinco Reis 8 (Beringel, Beja, Portugal). *Rivista di Studi Fenici* 44, 347–356.

Arruda, A.M. (2016b) À vol d'oiseau. Pássaros, passarinhos e passarocos na Idade do Ferro do Sul de Portugal. In A.C. Sousa, A. Carvalho and C. Viegas (eds) *Terra e água. Escolher sementes, invocar a deusa*, 403–423. Lisboa, Uniarq.

Arruda, A.M., Ferreira, M., Sousa, E., Lourenço, P., Lima, J., Carvalho, A. (in press) *Contributos para o conhecimento da Idade do Ferro de Alcácer do sal: os dados da Rua do Rato*.

Arruda, A.M., Freitas, V., Oliveira, C., Sousa, E., Lourenço, P. and Carretero, P. (2009) Castro Marim: um santuário pré-romano na foz do Guadiana. In P. Mateos, S. Celestino, A. Pizzo and T. Tortosa (eds) *Santuarios, Oppida y Ciudades: Arquitectura Sacra en el origen y desarrollo urbano del Mediterráneo Occidental (Anejos de AEspA, XLV)*, 79–88. Mérida, Instituto de Arqueología de Mérida/CSIC).

Arruda, A.M., Oliveira, C. and Freitas, V.T. (2017) Castro Marim entre indígenas, fenícios e tartésicos. In J.J. Ávila (ed.) *Sidereum Ana III. El río Guadina y Tartessos*, 445–466. Mérida, Consórcio de la Ciudad Monumental.

Bernáldez-Sánchez, E., García-Viñas, E., Gamero-Esteban, M., Amores-Carredano, F. and Ocaña-García De Veas, A. (2013) Knucklebones and other animal deposits in the 'Cruz del Negro' necropolis: Possible Phoenician funerary rituals in SW Spain? *Anthropozoologica* 48(2), 323–340.

Berrocal Rangel, L. and Silva, A.C. (2010) *O Castro dos Ratinhos (Barragem do Alqueva, Moura). Escavações num povoado proto-histórico no Guadiana*. Lisboa, Museu Nacional de Arqueologia.

Cardoso, J.L. and Arruda, A.M. (2016) Faunas domésticas e rituais funerários em Alcácer do Sal (Idade do Ferro). In R. Vilaça and M. Serra (eds) *Matar a fome, alimentar a alma, criar sociabilidades*, 193–217. Coimbra, FLUC, CEPBA, Palimpsesto.

Cardoso, J.L., López Castro, J.L., Ferjaoui, A., Mederos Martín, A., Martínez Hahnmuller, V. and Jerbania, I. (2016) What the people of Utica (Tunisia) ate in the 9th century BC Zooarchaeology of a North African early Phoenician settlement. *Journal of Archaeological Science* 8, 314–322.

Carè, B. (2012) L'astragalo in tomba nel mondo greco: un indicatore infantile? Vecchi problemi e nuove osservazioni a proposito di um aspetto del costume funerário. In A. Hermary and C. Dubois (eds) *L'enfant et la mort dans l'Antiquité III; Le matériel associé aux tombes d'enfants*, 403–416. Paris/Aix-en-Provence, Errance/Centre Camile Julien (BIAMA), 12.

Cuadrado, E. (1987) *La necrópolis ibérica de 'El Cigarralejo' (Mula, Murcia)*. Madrid, Consejo Superior de Investigaciones Científicas.

Davis, S. (2004) The *astragalus*: why is this bone important to zoo-archaeologists? *Rubber Band Society Gazette* 4(1), 3.

Davis, S. (2007) *Mammal and Bird Remains from the Iron Age and Roman Periods at Castro Marim, Algarve*. Trabalhos do CIPA, 107. Lisboa, Instituto Português de Arqueologia.

De Grossi Mazzorin, J. and Minniti, C. (2013) Ancient use of the knucklebone for rituals and gaming piece. *Anthropozoologica* 48(2), 371–380.

Elia, D. and Carè, B. (2004) Ancora sull' astrolalomania a Locri Epizeferi. La documentazione dalla necropolis incontrada Lucifero. *Orizzonti: Rassegna di Arqueologia* V, 77–90.

Fernández Flores, A. and Rodriguez Azouge, A. (2007) *Tartesos desvelado. La colonización fenícia del suroeste peninsular y el origen y ocaso de Tartesos*. Cordóva, Almuzara.

Garcia Cano, J.M. (1997) *Las necrópolis ibéricas de Coimbra del Barranco Ancho (Jumilla. Murcia). I. Las excavaciones y estudio analítico de los materiales*. Murcia, Universidad.

Gilmour, G.H. (1997) The nature and function of astragalus bones from archaeological contexts in the Levant and Eastern Mediterranean. *Oxford Journal of Archaeology* 16(2), 167–175.

González de Canales, F., Serrano, L. and Llompart, J. (2004) *El emporio fenicio precolonial de Huelva, ca. 900–770 a.C.* Madrid, Biblioteca Nueva.

Lancel, S. (1982) *Byrsa II. Mission Archéologique française à Carthage. Rapports préliminaires des fouilles de 1977–1978 (niveaux et vestiges puniques)*. Rome, École Française de Rome.

Larson, J. (2011) The Corycian Nymphs and the Bee Maidens of the Homeric. *Hymn to Hermes, Greek, Roman and Byzantine Studies* 36, 341–357.

Lebo, P. (2016) The astragalomancy in Etruria: an analysis of the astragali at Cetamura del Chianti. In *Proceedings of the National Conference of Undergradute Research*, 420–430. Asheville, University of North Carolina.

Kogălniceanu, R., Ile, A., Mărgărit, M., Simalcsik, A. and Dumitraşcu, V. (2014) A hoard of astragals discovered in the Copper Age settlement at Iepureşti, Giurgiu County, Romania. *Documenta Praehistorica* 41, 283–304.

Maluquer de Motes, J. (1983) *El Santuario Protohistórico de Zalamea de la Serena, Badajoz. II 1981–1982* (Programa de Investigaciones Protohistoricas V). Barcelona, Consejo Superior de Investigaciones Científicas, Universidad de Barcelona.

Martín-Poveda, J.G. (2014) Uso y significado de los astrágalos en los ajuares funerários ibéricos: la necrópolis del poblado de Coímbra del Barranco Ancho (Jumilla, Múrcia). *Alberca* 12, 43–57.

Minniti, C. and Peyronel, L. (2005) Symbolic or Functional Astragali from Tell Mardikh-Ebla (Syria). *Archeofauna* 14, 7–26.

Nabais, M., and Soares, R. (2017) Os ossos trabalhados do Castro da Azougada (Moura, Portugal). In J.M. Arnaud and A. Martins (eds), *Arqueologia em Portugal-Estado da Questão*, 263–301. Lisboa, Associação dos Arqueólogos Portugueses.

Rodriguez Diaz, A. (ed.) (2004) *El Edificio Protohistorico de 'La Mata' (Campanario, Badajoz) y Su Estudio Territorial.* Cáceres, Universidad.

Sidéra, I. and Vornicu, A. (2016) The archaeology of games. Playing with knucklebones in the early chalcolithic of the Balkans. In K. Bacvarov and R. Gleses (eds) *Southeast Europe and Anatolia in Prehistory, Essays in Honor of Vassil Nokolov on his 65th Anniversary*, 378–388. Bonn, Velag Dr. Rudolf Habelt GmbH.

Tolegen, Z., Moldabekov, M., Koshenov, K. and Mugzhanova, G. (2018) Roles of public ethnocultural spaces in Kazakhistan. *Asta Solvensis* 6(11), 761–774.

Wiesner, N. (2013) Astragali in Gräbern der mitteleuropäischen Urnenfelderkultur. *Germania* 91, 89–113.

9

Origins of metallurgy in the southern Levant: microscopic examination of butchering marks on animal bones at Tel Yarmuth, Israel

Haskel J. Greenfield, Annie Brown and Pierre de Miroschedji

It is commonly assumed that with the introduction of bronze metallurgy, there is a concomitant shift in the type of raw material technology used in daily activities – i.e. *from stone to metal. In this paper, we test that assumption with microscopic examination of butchering marks from the Early Bronze Age (EB) site of Tel Yarmuth, Israel, where Simon Davis has long been analysing the faunal assemblage. The deposits mostly date to the EB II–III (*c. *3000–2500 BCE). Two types of analyses are conducted. First, microscopic butchering marks on faunal remains from the various deposits at Tel Yarmouth are first evaluated to determine if the primary butchering tools were made of chipped stone or metal. Although bronze technology is first utilised for prestige objects (*e.g. *ornaments, mace heads), it is unclear when this material is adopted for daily activities, such as animal carcass processing (*e.g. *butchering). Second, this study tests whether the adoption of metal technology for quotidian activities, such as butchering, occurred differentially across sites – did the entire settlement adopt metallurgy all at once or did the status of the occupants in different parts of the site have any effect upon the adoption of the new and more expensive butchering technology. This is the only major urban settlement in the region from the EB where many different parts of the settlement have been excavated and the fauna collected. Data from each of the excavation areas are compared through the EB to test this proposition. This study follows up on Simon Davis' pioneering work on Early Bronze Age urban settlements in the southern Levant (Israel) and the relationship between changing animal exploitation practices and the evolution of early complex societies.*

Keywords: Early Bronze Age, butchering; zooarchaeology, southern Levant, stone-tool technology, scanning electron microscopy, Tel Yarmuth, Near East

Introduction

This paper is written in honour of Simon Davis, who has had a long and illustrious career that has influenced many in the discipline, but more specifically the authors of this manuscript in so many inspiring ways. Simon has been an ever-present figure in each of our academic lives, with his huge quantity and variety of publications. He is one of the pioneers of the discipline of faunal analysis (archaeozoology/zooarchaeology).

Haskel accidentally met Simon for the first time one day in 1981/2 at the Graduate Center of the City University of New York City when he was struggling with the taxonomic identification of sheep and goat phalanges from a site in Serbia. Simon walks in unannounced through the door of the Archaeology Lab, introduces himself, and sets about teaching Haskel how to distinguish between the two taxa. In short order, he put Haskel on the straight, and sometimes not so narrow, path of sheep/goat taxonomic identification. Annie first met Simon during the ASWA 2013 conference in Haifa where he had presented a paper on zooarchaeology. Pierre has known Simon for over 30 years since he became the senior archaeozoologist of the Tel Yarmuth project.

When Haskel began his study on the evolution of metallurgy, Simon graciously allowed him to include the material from Tel Yarmuth. This paper is the result of that study. Simon has been the archaeozoologist for Pierre's excavations at Tel Yarmuth since its inception in the 1980s. It is one of the largest Early Bronze faunal assemblages from the region and its analysis is only now wrapping up. The assemblage is available for further study since it is being archived at the Israel Antiquity Authority.

The origins of metallurgy lie in the distant past. Most studies of early metallurgy have focused on the display and ceremonial items from the Neolithic to the Chalcolithic into the later periods (Hayden 1998; Birch *et al.* 2013). However, very few studies have been concerned with the origins and spread of the quotidian metallurgy, otherwise known as tools used in daily life activities. The most common tools used in daily life are those that involve subsistence activities and food production, such as the axes, hammers, blades and scrapers used for butchering and processing animals.

Technological change is often assumed to occur because of perceived increased effectiveness. Older technologies should be quickly abandoned and their functions replaced with the new technology because they are supposedly more effective (*e.g.* Childe 1951; Bamforth 1986; Bleed 1986; van der Leeuw and Torrence 1989; Torrence 1989; Hayden *et al.* 1996). In this paper, we test this assumption with respect to the shift from a chipped stone- to a metal-based technology for butchering animals. The assumption is that metal tools, particularly knives, replace their stone counterparts quickly because they are sharper and would last longer since they can be sharpened. Also, metal objects can be recycled by melting them down and molding them anew. In some ways, this is true for expedient chipped stone tools, such as simple blades and flakes that may have been used in butchering activities. Such tools would require less time, effort and skill to resharpen on the spot since they would simply need a new flake to be struck off the core or broken tool. In general, chipped stone tools made from local stone (*e.g.* flint) would be generally cheaper to produce than metal objects imported from distant sources. Furthermore, chipped stone blade tools can be equally or sharper.

In this paper, we examine the assumption that people switched from a stone to metal (*i.e.* bronze) technologies as soon as metal knives initially appear in the archaeological record. Archaeologically, bronze metal objects (such as knives) appear early in the 3rd millennium BCE of the Near East (Moorey 1988) or Early Bronze (EB) III of the southern Levant (Shalev 1994), but they don't appear to be made of tin bronze. Hard tin bronze, when tin begins to be mixed with copper to make bronze implements, only appears during the second half of the 3rd millennium BCE (Moorey 1994; Miroschedji 2014). But, in the southern Levant, it only becomes significant during the Middle Bronze (MB) II at the advent of the 2nd millennium BCE (Shalev 2009; Bourke 2014; Cohen 2014).

In reality, this is not what has been observed in faunal assemblages which enables us to monitor the effect that the introduction of bronze technology has on butchering, technology and other related activities involving animal remains. We utilise zooarchaeological data to investigate this issue.

Few sites from the region exhibit the range of deposits and quantity and quality of zooarchaeological remains from the Early Bronze to be able to test the above assumptions. Tel Yarmuth stands alone among Early Bronze Age (henceforth EB) sites from the region in the scale of excavations, and quantity and quality of remains, and variety of contexts. Hence, it is an apt assemblage to investigate what we can learn from zooarchaeology about the origins of metallurgy in the southern Levant.

The site

Tel Yarmuth is a large multi-period site located in central Israel. Geographically, it sits within the central Shephelah (a rolling low hill country forming the foothill of the Judean highland) at the southern edge of the modern city of Beth Shemesh (Fig. 9.1). The site covers 18 hectares and occupies

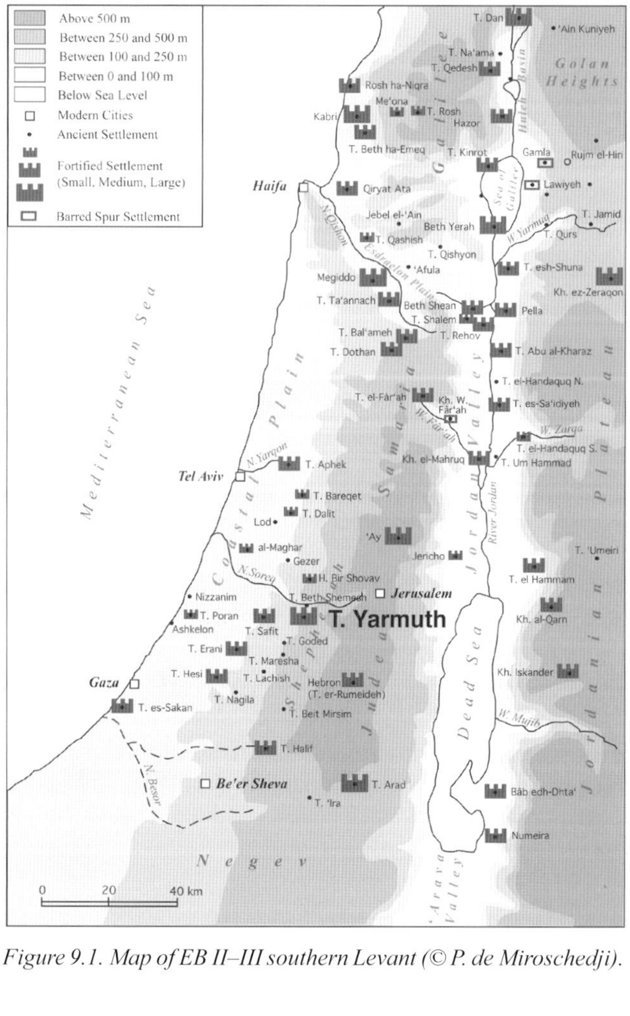

Figure 9.1. Map of EB II–III southern Levant (© P. de Miroschedji).

a natural hill overlooking the Yarmuth River valley. It was excavated for 18 seasons since 1980 under the direction of Pierre de Miroschedji under the auspices of the *Centre de recherche français à Jérusalem* (CNRS and French Ministry of Foreign Affairs) (Miroschedji 1988a; 1988b; 2000; 2003; 2006; 2013; 2014). The site consists of a vast lower city (16.5 ha) and a much smaller acropolis (1.5 ha). Tel Yarmuth was settled first during the second half of the fourth millennium BCE, during the EB IB period (*c.* 3400 BCE) and was continuously inhabited until the end of the EB IIIB period (*c.* 2450/2400 BCE), when it was abandoned. The site was later reoccupied, but on the acropolis only, which was inhabited more or less continuously from the Late Bronze Age (henceforth LB) II until the early Byzantine period, when it was finally abandoned (see Jasmin and Miroschedji 2020). Hence, there is a long gap in occupation from the EB III to LB II, which is unfortunate since this is the time period when it is likely that hard tin-bronze metallurgy was introduced into and adopted for use with cutting implements (Greenfield 2013).

During the EB, which is the only period of occupation in the Lower City, Yarmuth was a major regional centre, with massive stone fortifications and monumental buildings, including two successive palaces, the last of which covered 6000 m^2, surrounded by extensive domestic quarters (Fig. 9.2).

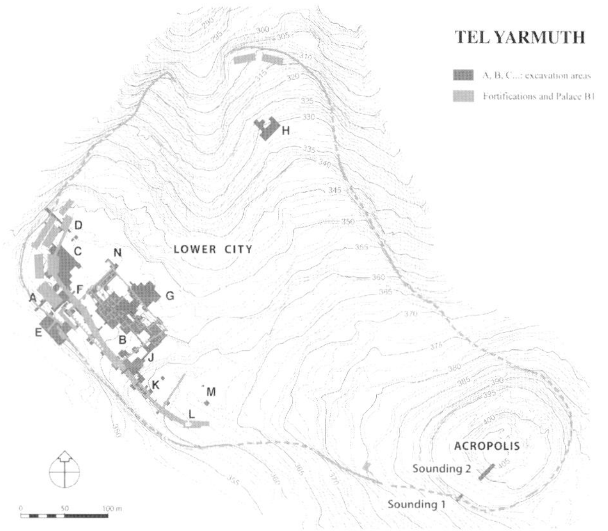

Figure 9.2. Map of Tel Yarmuth showing location of the excavation areas (© P. de Miroschedji).

The Yarmuth assemblage is particularly apt for this analysis since it is has a long chronological sequence that extends from early in the EB until the Byzantine period (Fig. 9.3); there is a diverse range of depositional contexts,

PERIODS	GENERAL STRATA	LOWER CITY												ACROPOLIS	
		City, areas							Fortifications, areas					Sounding 1	Sounding 2
		B	C	G	H	Ja-b	K	M	A	Bf	D	E	Jc		
BYZANTINE	N	B-0	C-0	G-0	H-1	J-0	K-0	M-0		+	D-9*	E-5*		Acr.-6*	Acr.-IA
ROMAN	M														Acr.-IB
HELLENISTIC	L														Acr.-IC
PERSIAN	K														Acr.-IIA
IRON II	J						ABANDONMENT							Acr.-5*	? Acr.-IIB
IRON I	I														Acr.-III Acr.-IV Acr.-V
LATE BRONZE	H														Acr.-VIA ?
MIDDLE BRONZE	G														Acr.-VIB
INTERMEDIATE BRONZE (EB IV)														ABANDONMENT	
EARLY BRONZE IIIC	F	B-1	C-1	G-1 ? G-2	H-2 ? H-3	J-1 J-2	K-1	M-1	A-7*	Bf-6*	D-8*	E-4* & 5*	Jc-4*		Acr.-VII
EARLY BRONZE IIIB	E	B-2	C-2	G-3	H-4	J-3	K-2	M-2	A-6*	Bf-5*	D-7*	E-3*		Acr. 4*	Acr.-VIII
EARLY BRONZE IIIA	D	B-3	C-3 C-4 C-5	G-4	H-5	J-4	Unexcavated	Unexcavated	A-5*	Bf-4*	D-6*	E-2*			
EARLY BRONZE II	C	B-4 B-5	C-6 C-7	G-5 Bedrock	Unexcavated	J-5 Unexcavated			A-4* A-3* A-2* A-1*	Bf-3* Bf-2* Bf-1*	D-5* D-4* D-2* & 3* D-1*	E-1b* E-1a* E-0*	Jc-3* Jc-2* Jc-1*	Acr.-3* Acr.-2* Acr.-1*	Acr.-IX
EARLY BRONZE IB	B A	B-6 Bedrock	C-8 C-9 Bedrock						A-0* Bedrock	Unexcavated	D-0* Bedrock	Bedrock	Unexcavated	Bedrock	Bedrock

Figure 9.3. Stratigraphic table of Tel Yarmuth (© P. de Miroschedji).

including both high and lower status deposits; and a broad spatial exposure for especially the EB that is missing on most other EB sites from the region.

Simon Davis is the primary archaeozoologist for the Yarmuth excavations and his final report on the fauna of 18 seasons of excavations is forthcoming. Unfortunately, we could not compare the results of our analyses because of analytical differences – *e.g.* he did not analyse most vertebrae and ribs since they are difficult to identify to the more specific taxonomic level (Davis 1988). In contrast, we included such remains in our analysis since many relevant specimens have butchering marks on them.

Analytical method

Experiments have identified the relationship between the raw material of the butchering implement (metal or stone knife or other tools) and the shape of slicing butchering mark on bone (*e.g.* Walker 1978; Shipman 1981; Olsen 1988; Greenfield 1999; 2006; 2013; Bello and Soligo 2008). Metal knife blades tend to have a flat edge that produces a symmetrically balanced V-shaped groove when examined in profile because the blade edge is symmetrical. Both sides of the groove rise steeply away from the apex at a similar angle (Fig. 9.4A). As blades become dull over time, the bottom of the groove may become relatively flat (Fig. 9.4B). The depth of the groove tends to be deeper and cleaner with metal tools.

The grooves made by chipped stone tool blades and flakes cannot be distinguished from one another. Chipped stone tool slicing marks tend to be shallower and coarser than those of metal tools. The slices appear as asymmetrical V-shaped grooves. One side descends rapidly, while the other side descends much more gradually (Fig. 9.4D). Production style can also be differentiated. Unifacially produced blades/flakes create an uneven dorsal face that contains one or more parallel ridges that run the length of the groove. The ventral side of the tool's surface leaves a relatively smooth and steeply descending surface. Unretouched blades/flakes leave a couple of (1–2) lateral striations on the dorsal face of the groove (Fig. 9.4E). Unifacial retouching leaves more lateral ridges on the dorsal slope of the groove. Bifacial retouching leaves many lateral ridges on both sides of the groove (Fig. 9.4F).

In the same manner, chops marks can be distinguished between stone and metal materials. Metal axe marks leave clean and straight marks, while chipped stone or ground stone axes leave faces with more irregular shapes or as bashes.

In each table, the NISP (Number of Individual Specimens) is provided. In addition, an additional measure is provided = NI (Number of Incidences) (Greenfield *et al.*, 2016; Greenfield and Brown, 2016). The NISP only counts a single specimen once. The number of incidences may reflect the number of times that a butchering activity may have occurred on the same bone fragment. If there are separate butchering grooves on two different side of the bone or on two different clusters on the same face, the NISP would be 1, but the NI would be 2. This avoids double counting of elements and taxa, but allows a more subtle analysis of butchering patterns.

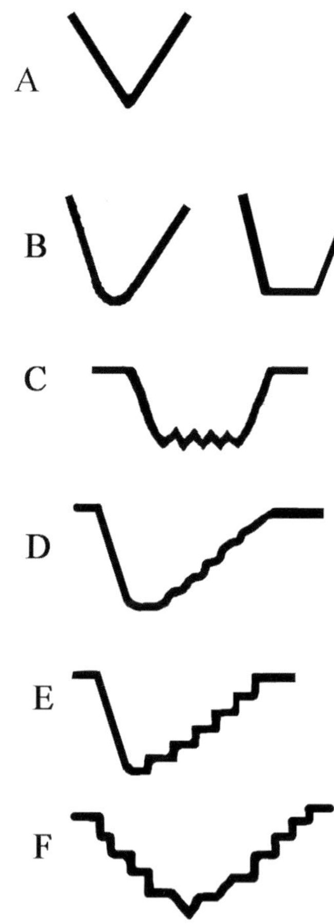

Figure 9.4. Profile of typical metal and chipped stone slicing marks. A. profile of metal blade – sharp flat-edged; B. profile metal blade – dulled flat-edged; C. profile of chipped stone scraper; D. profile of chipped stone blade – unretouched; E. profile of unifacially produced chipped stone blade with unifacial retouch; F. profile of chipped stone blade – bifacial retouch.

All of the faunal remains are currently being relocated to the major curatorial facilities of the Israel Antiquity Authority in the city of Beth Shemesh.

The assemblage

The faunal assemblage from Tel Yarmuth derives from both the soundings dug on the acropolis and from several areas of excavations in the Lower City, which were much more extensive but chronologically limited to the EB (Fig. 9.3). Based on our experience with contemporary assemblages from the region, we estimate that only about 3–5% of bones had preserved butchering marks on them. Almost a thousand

Table 9.1. Frequency distribution of all specimens by period from all parts of site (NISP).

Time period	*Sum of NISP (butchered)*	*Sum of NISP (all)*	*%*
EB	15	20	2.15%
EB II	266	289	31.04%
EB III	420	556	59.72%
EB I-II	3	3	0.32%
EB III+Byzantine	4	5	0.54%
EB II-III	11	12	1.29%
EB II-IIIA	1	1	0.11%
EB+Persian+Byzantine	1	1	0.11%
Iron I	10	12	1.29%
Iron IB-II+Persian	1	1	0.11%
Iron II	1	1	0.11%
Iron II+Persian	2	2	0.21%
Iron I-II+Persian	5	6	0.64%
LB III to Iron IA	1	1	0.11%
LB III+Iron I	2	2	0.21%
Roman+Byzantine	3	4	0.43%
Byzantine	4	9	0.97%
Byzantine+later	1	1	0.11%
Surface	1	1	0.11%
Cat. # does not exist	0	1	0.11%
Cat. # lost	0	1	0.11%
Unknown	1	1	0.11%
Not relevant	0	1	0.11%
Grand Total	753	931	100.00%

(N=931) specimens were identified to have butchering marks on them (Table 9.1). Once the problematic specimens were removed from the analysis, 738 specimens were available for the final analysis. The larger size of the lower city and its vast area of excavation dwarfs that of the acropolis (N=21; Table 9.2). This affects the sample size when comparing between periods.

The vast majority of butchered bones are from EB deposits – 97% (Table 9.3). Most come from the EB III (61%) and EB II (31.7%). Very few come from the other periods (LB, Iron Age and later deposits – 3%).

Analysis

Taxonomy

The butchered assemblage is overwhelmingly dominated by domestic animals (97.7%; Table 9.4). Caprines, when both domestic sheep and goat are combined, also dominate (64/79% if indeterminate mammals are included/excluded; Table 9.5a–b): sheep/goat (*Ovis/Capra* – 39.3/43.9%), goat (*Capra hircus* – 22.4/25.1%; Fig. 9.5), and sheep (*Ovis aries* – 12.7/14.2%). Cattle (*Bos taurus* – 12.3/13.7%) are the next most common. All other taxa, both domestic and wild are represented by few butchered remains (*e.g.* donkey/*Equus asinus* – 0.1%). All of the other common domestic taxa are absent from the butchered assemblage: dog (*Canis familiaris*), pig (*Sus domesticus*) and horse (*Equus caballus*). Many of the very fragmentary mammal rib, vertebrae, cranial and long bone shaft fragments could not be taxonomically identified and are listed as large mammal (3.4%), medium mammal (7.1%) and small mammal (probably hare – 0.6%). The large and medium sized mammals are probably dominated by domestic taxa. A variety of wild taxa were butchered – gazelle (*Gazella* sp. – 1.3%), fallow deer (*Dama dama* – 0.1%), roe deer (*Capreolus capreolus* – 0.1%) and ibex (*Capra ibex* – 0.4%).

Types of butchering marks over time

Several different types of butchering marks were identified in the assemblage: bash, chop, scrape and slice (Table 9.6). The vast majority of marks are slices (94%; Fig. 9.5), but that is probably because they are the easiest to identify. Chops in the EB tend to be very ambiguous since butchers do not appear to be using metal axes, which would leave clear chop marks, yet. Many of the butchering marks (*e.g.* chops and slices) are combined with other forms of processing, such as breakage. There is one sawn piece only. There is not much difference between the frequencies of each type in the Acropolis or the Lower City.

When the types are broken down by period, there are differences. Yet, most of the differences can be attributable to assemblage size variation. The larger the frequencies of remains in a period, the more likely more types of processing marks will appear (Table 9.7). For example, only in the large EB2 and 3 assemblages is there a variety of butchering types present. For example, only slices are present in the small EB I–II assemblage, while many more are present in the larger EB II and III assemblages (Fig. 9.6). In all the other periods, which have small assemblages, one or more types are missing.

In the EB II and III assemblages, all types of processing marks are present, and slices represent either 98–100% of the types of marks. All of the others are present in very small numbers. The small acropolis assemblage is heavily biased toward slicing marks. All of the marks are slices except for one in the Iron Age 2 assemblage.

Raw material over time

There is a dramatic change in the frequency of stone and metal marks over time at the site (Table 9.8). There is no evidence of metal tool marks in the EB. All are made by chipped stone tools during the EB. This is clear from the very asymmetrical

Table 9.2. Frequency distribution of all specimens by period and part of site (NISP).

	Sum of NISP (butchered)		*% column*		*% row*		*Total Sum of NISP (butchered)*	*Total % column*	*Total % row*
Time period	*Acropolis*	*Lower city*	*Acropolis*	*Lower city*	*Acropolis*	*Lower city*			
EB		15	0.00%	2.09%	0.00%	100.00%	15	2.03%	100.00%
EB I-II		3	0.00%	0.42%	0.00%	100.00%	3	0.41%	100.00%
EB II		263	0.00%	36.68%	0.00%	100.00%	263	35.64%	100.00%
EB II?		1	0.00%	0.14%	0.00%	100.00%	1	0.14%	100.00%
EB II-III		6	0.00%	0.84%	0.00%	100.00%	6	0.81%	100.00%
EB II-IIIA		6	0.00%	0.84%	0.00%	100.00%	6	0.81%	100.00%
EB III	2	5	9.52%	0.70%	28.57%	71.43%	7	0.95%	100.00%
EB IIIA	2	59	9.52%	8.23%	3.28%	96.72%	61	8.27%	100.00%
EB IIIA?		2	0.00%	0.28%	0.00%	100.00%	2	0.27%	100.00%
EB IIIA-B		3	0.00%	0.42%	0.00%	100.00%	3	0.41%	100.00%
EB IIIB		189	0.00%	26.36%	0.00%	100.00%	189	25.61%	100.00%
EB IIIB?		2	0.00%	0.28%	0.00%	100.00%	2	0.27%	100.00%
EB IIIC		147	0.00%	20.50%	0.00%	100.00%	147	19.92%	100.00%
EB IIIB-C		11	0.00%	1.53%	0.00%	100.00%	11	1.49%	100.00%
LB III+Iron IA	3		14.29%	0.00%	100.00%	0.00%	3	0.41%	100.00%
Iron IB	10		47.62%	0.00%	100.00%	0.00%	10	1.36%	100.00%
Iron II (1st half of 1st mill. BCE)	1		4.76%	0.00%	100.00%	0.00%	1	0.14%	100.00%
Roman+Byzantine	3		14.29%	0.00%	100.00%	0.00%	3	0.41%	100.00%
Byzantine		4	0.00%	0.56%	0.00%	100.00%	4	0.54%	100.00%
Byzantine+later		1	0.00%	0.14%	0.00%	100.00%	1	0.14%	100.00%
Grand Total	21	717	100.00%	100.00%	2.85%	97.15%	738	100.00%	100.00%

profile and lateral striations that is characteristic of chipped stone tools which is visible in the SEM photograph (Fig. 9.7).

In the mixed EB I–II deposits, all of the marks are made by stone tools. In the clear EB II and EB III deposits, a very small percentage (*c.* ≤1%) of marks might be made by metal tools. All the rest are clear and unambiguous chipped stone tool marks. We suspect that the so-called identifications of metal marks in the EB II and III assemblages are misidentifications since these data were collected in the early years of this project, when the method was less well-developed. Re-examination of some of those specimens allowed us to recognise that they were mistakenly identified as metal marks and should not be given much credence. In general, it can be said that the EB assemblage contains no clear and unambiguous evidence of metal butchering marks.

Metal marks only appear very late in time – during the LB III and Iron Age – where there is a clear symmetrical profile without any lateral striations (Figs 9.8 and 9.9). Surprisingly, stone butchering marks continue to be present during the Iron Age, when they range from 50–100% of marks in the various deposits. But, the small size of the assemblage from these later periods suggests that the percentages are unreliable. Other assemblages from the region and elsewhere in the Near East also show that stone tools continue to be an important part of animal processing in post-EB periods (Greenfield 2013; Greenfield and Brown 2016; Greenfield and Chaput in press).

Stone tool production technique

There is little or no evidence for bifacial production of tools. All of the marks are made by unifacially produced tools. Only two specimens in the entire assemblage could be identified positively as being made by bifacially retouched tools – and both derive from the EB II Stratum in the Lower City. All the rest were made by either unretouched or only unifacially retouched blades or flakes.

Butchering efficiency

Butchering efficiency (BE) is a useful measure for determining if there are changes in butchering efficiency when there is a

Table 9.3. Frequency distribution of all specimens by more finite periods from all parts of the site (NISP).

Period and sub-period	*Sum of NISP (problems removed)*	*%*	*Sum of NISP (butchered)*
EB	**20**	**2.20%**	**15**
EB	20	2.20%	15
Iron I	**12**	**1.32%**	**10**
Iron IB	12	1.32%	10
Roman+Byzantine	**4**	**0.44%**	**3**
Roman+Byzantine	4	0.44%	3
Byzantine	**9**	**0.99%**	**4**
Byzantine+later	**1**	**0.11%**	**1**
EB II	**289**	**31.76%**	**266**
EB II	286	31.43%	263
EB II?	1	0.11%	1
EB IIIB?	2	0.22%	2
EB II-III	**12**	**1.32%**	**11**
EB II-IIIA	6	0.66%	5
EB II-III	6	0.66%	6
EB III	**555**	**60.99%**	**420**
EB III	8	0.88%	7
EB IIIA	67	7.36%	61
EB IIIA-B	3	0.33%	3
EB IIIA?	3	0.33%	2
EB IIIB	251	27.58%	189
EB IIIB?	0	0.00%	0
EB IIIC	210	23.08%	147
EB IIIB-C	13	1.43%	11
Iron II	**1**	**0.11%**	**1**
Iron II (1st half of 1st mill. BCE)	1	0.11%	1
EB II-IIIA	**1**	**0.11%**	**1**
EB II-IIIA	1	0.11%	1
EB I-II	**3**	**0.33%**	**3**
EB I-II	3	0.33%	3
LB III+Iron I	**2**	**0.22%**	**2**
LB III+Iron IA	2	0.22%	2
LB III to Iron IA	**1**	**0.11%**	**1**
LB III+Iron IA	1	0.11%	1
Grand Total	910	100.00	738

change in technology or the organisation of the butchering process (Table 9.9). It can be calculated through the Butchering Incidence index (BI) which evaluates the average number of slices per butchering incidence on a bone. A butchering incidence is when there is either a single isolated mark or a cluster of marks with the same orientation on a bone. They likely reflect multiple strokes occurring during the same action. Each cluster is considered a single action set (*e.g.* removal of a specific tendon, or filleting flesh off a part of the bone) and is recorded as a separate incident. A single bone may have several butchering incidences. The BE index is calculated by summarising the number of strokes and dividing them by the number of incidences.

There is a clear increase in BE from the EB I–II and II (13 to 11.2 marks/incidence) to the EB III (7.7 marks/incidence). Possibly this reflects a reorganisation of the butchering process to service the palace more efficiently. Surprisingly, the BE does not change significantly between the EB III and the LB (8 marks/incidence) even though metal begins to play a major role in the butchering technology beginning in the Late Bronze. Given that stone tools continue to be used during the LB and Iron Age, this may be why butchering efficiency does not change all.

Butchering efficiency only marginally increases in the Iron 2 and Roman-Byzantine deposits (5.6–7 BE), but not into the final Byzantine + later assemblage, where it soars to its highest level (19 BE).

Conclusions

As Simon Davis has shown with his innovative and abundant publications, zooarchaeology can have an impact on our understanding of the past that extends beyond the traditional identification of zoological taxonomic status (Davis 1987). It can be used to identify all kinds of behavioural processes that help us to elucidate the nature of human behaviour. We build on his pioneering research by examining the adoption and spread of metal technology and its effect upon the butchering process at Tel Yarmuth.

It is clear that stone continued to be the major raw material for animal butchering and tool production technologies during the EB. The EB is a continuation of the earlier Stone Age patterns in terms of animal processing. Despite the presence of the only major palace structure in

Table 9.4. Frequency distribution of all butchered specimens by state of domestication (NISP).

	Sum of NISP (butchered)		*% column*		*% row*		*Total Sum of NISP (butchered)*	*Total % column*	*Total % row*
State	*Acropolis*	*Lower city*	*Acropolis*	*Lower city*	*Acropolis*	*Lower city*			
Domestic	21	565	95.45%	97.75%	3.58%	96.42%	586	97.67%	100.00%
Wild	1	13	4.55%	2.25%	7.14%	92.86%	14	2.33%	100.00%
Grand Total	22	578	100.00%	100.00%	3.67%	96.33%	600	100.00%	100.00%

Table 9.5a. Frequency distribution of all butchered specimens by taxon (NISP).

	Sum of NISP (butchered)		*% column*		*% row*		*Total Sum of NISP (butchered)*	*Total % column*
Taxon	*Acropolis*	*Lower city*	*Acropolis*	*Lower city*	*Acropolis*	*Lower city*		
?								
Mammal - large		23	0.00%	3.52%	0.00%	100.00%	23	3.41%
Mammal - medium		48	0.00%	7.35%	0.00%	100.00%	48	7.11%
Mammal - small		4	0.00%	0.61%	0.00%	100.00%	4	0.59%
Domestic								
Bos taurus	2	81	9.09%	12.40%	2.41%	97.59%	83	12.30%
Capra hircus	9	142	40.91%	21.75%	5.96%	94.04%	151	22.37%
Equus sp.	1		4.55%	0.00%	100.00%	0.00%	1	0.15%
Ovis aries	2	84	9.09%	12.86%	2.33%	97.67%	86	12.74%
Ovis/Capra	7	258	31.82%	39.51%	2.64%	97.36%	265	39.26%
Wild								
Capra ibex		3	0.00%	0.46%	0.00%	100.00%	3	0.44%
Capreolus capreolus		1	0.00%	0.15%	0.00%	100.00%	1	0.15%
Dama dama		1	0.00%	0.15%	0.00%	100.00%	1	0.15%
Gazelle sp.	1	8	4.55%	1.23%	11.11%	88.89%	9	1.33%
Grand Total	22	653	100.00%	100.00%	3.26%	96.74%	675	100.00%

Table 9.5b. Frequency distribution of all butchered specimens by taxon (NISP), after indeterminates removed.

	Sum of NISP (butchered)		*% column*		*% row*		*Total Sum of NISP (butchered)*	*Total % column*	*Total % row*
Domestication and taxon	*Acropolis*	*Lower city*	*Acropolis*	*Lower city*	*Acropolis*	*Lower city*			
Domestic									
Bos taurus	2	81	9.09%	14.01%	2.41%	97.59%	83	13.83%	100.00%
Capra hircus	9	142	40.91%	24.57%	5.96%	94.04%	151	25.17%	100.00%
Equus sp.	1		4.55%	0.00%	100.00%	0.00%	1	0.17%	100.00%
Ovis aries	2	84	9.09%	14.53%	2.33%	97.67%	86	14.33%	100.00%
Ovis/Capra	7	258	31.82%	44.64%	2.64%	97.36%	265	44.17%	100.00%
Wild									
Capra ibex		3	0.00%	0.52%	0.00%	100.00%	3	0.50%	100.00%
Capreolus capreolus		1	0.00%	0.17%	0.00%	100.00%	1	0.17%	100.00%
Dama dama		1	0.00%	0.17%	0.00%	100.00%	1	0.17%	100.00%
Gazelle sp.	1	8	4.55%	1.38%	11.11%	88.89%	9	1.50%	100.00%
Grand Total	22	578	100.00%	100.00%	3.67%	96.33%	600	100.00%	100.00%

Table 9.6. Frequency distribution of all butchered specimens by butchering mark frequency and location within site (NISP).

	Sum of NISP (butchered)		*% row*		*Total Sum of NISP (butchered)*	*Total % row*
Butcher type	*Acropolis*	*Lower city*	*Acropolis*	*Lower city*		
Bash		7	0.00%	0.97%	7	0.93%
Chop	1	20	3.57%	2.76%	21	2.79%
Chop + break		2	0.00%	0.28%	2	0.27%
Possible chop		1	0.00%	0.14%	1	0.13%
Saw		1	0.00%	0.14%	1	0.13%
Scrape		11	0.00%	1.52%	11	1.46%
Slice	27	681	96.43%	93.93%	708	94.02%
Slice + break		1	0.00%	0.14%	1	0.13%
Slice + chop		0	0.00%	0.00%	0	0.00%
Slice + scrape		0	0.00%	0.00%	0	0.00%
Slice?		1	0.00%	0.14%	1	0.13%
Grand Total	28	725	100.00%	100.00%	753	100.00%

*Figure 9.5. Butchering slice marks on juncture between base of horn core and frontal bone of goat (*Capra hircus*) from an EB II Stratum. Photo 8040; butchering sample 43; catalogue #7171 (Area C, Stratum C-6, L. 603-2).*

the region for the EB, all animal remains were processed with stone tools. Metal was used for making weapons and various kind of tools (axes, adzes, awls), but not for such mundane quotidian activities such as food processing. Soft metals, such as copper and arsenic-copper alloy, are present already in the EB, but a hard tin bronze is not yet developed or adopted in this region during the EB. During the LB and later periods, clear and unambiguous metal tool marks begin to appear. However, stone tool marks continue even in these later periods, as at many sites in the region and elsewhere. This is expected given that flint chipped stone tools, such as blades, continue to be produced until the Iron Age (Rosen 1984; 1996; 1997; Manclossi *et al.* 2019).

The data from Tell Yarmuth corresponds and substantiates the larger overall regional model that we have identified in other sites (Greenfield 2013; 2004), but it also adds in the new dimension to our analysis – that even the high status occupants of the site have limited access to bronze during the EB.

Butchering efficiency can also be monitored from our analysis. The BE from the later EB deposits at Yarmuth is very similar to what has been identified at the nearby sites of Nahal Tillah and Halif (*e.g.* 8–9 BE). The higher frequencies seen for the earlier phases of the EB are not characteristic of the other examined EB assemblages (Greenfield and Brown 2016; Greenfield *et al.* 2018), but they are much higher than at EB sites in Anatolia where obsidian was used instead of flint as the major raw material for butchering – *e.g.* Göltepe 2.5/incidence (Greenfield and Chaput in press).

This study confirms our observations from other assemblages across the region – that the EB is a period when stone continues to be the dominant technology for the processing of foods, such as animals. In the

Table 9.7. Frequency distribution of all butchered specimens by butchering mark frequency, location and period (NISP).

	Sum of NISP (butchered)		% row		Total Sum of NISP (butchered)	Total % row
Period	Acropolis	Lower city	Acropolis	Lower city		
EB		**15**	**0.00%**	**100.00%**	**15**	**100.00%**
Slice		15	0.00%	100.00%	15	100.00%
Iron I	**10**		**100.00%**	**0.00%**	**10**	**100.00%**
Slice	10		100.00%	0.00%	10	100.00%
Roman+Byzantine	**3**		**100.00%**	**0.00%**	**3**	**100.00%**
Slice	3		100.00%	0.00%	3	100.00%
Byzantine		**4**	**0.00%**	**100.00%**	**4**	**100.00%**
Scrape		1	0.00%	100.00%	1	100.00%
Slice		3	0.00%	100.00%	3	100.00%
Byzantine+later		**1**	**0.00%**	**100.00%**	**1**	**100.00%**
Slice		1	0.00%	100.00%	1	100.00%
Grand Total	13	20	39.39%	60.61%	33	100.00%

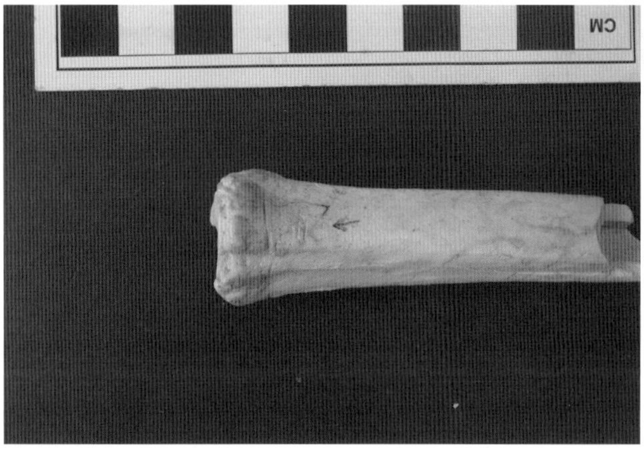

*Figure 9.6. Example of butchering slice marks on proximal end of sheep (*Ovis aries*) metatarsus from an EB IIIA Stratum. Note the use wear polish on the proximal epiphysis. Photo 7982; butchering sample 326; catalogue #9018 (Area Cne, Stratum C-3, Structure 1014).*

Chalcolithic, most (not all) metal objects were used for display and did not have other more practical or utilitarian uses. But in the EB, many metal objects types had, or could have had a practical use – such as tools and weapons (Philip 1989; 2001; Levy *et al.* 2001; Levy *et al.* 2002). Yet, it is clear that these metal objects were not yet used for quotidian activities, such as the processing of animal carcasses for their primary products (*i.e.* meat, skins, bones, etc.). It is clear from the analysis of the butchering marks on animal bones from Tel Yarmuth and from elsewhere in the region that the shift to a utilitarian metal butchering and carcass processing technology does not take place in the EB when bronze metal implements, such as knives, first appear (Greenfield 2013). It likely occurs a thousand or more years later, long after the end of the EB. The first hints of the change occur in the MB, but it is not until the LB that metal begins to become the dominant technology.

Table 9.8. Frequency distribution of all butchered specimens by raw material frequency, location in site and period (NISP).

	Sum of NISP (butchered)			% row			Total Sum of NISP (butchered)	Total % row
	Acropolis		Lower city	Acropolis		Lower city		
Period	Metal	Stone	Stone	Metal	Stone	Stone		
EB			15	0.00%	0.00%	100.00%	15	100.00%
Iron I	4	5		44.44%	55.56%	0.00%	9	100.00%
Roman+Byzantine		3		0.00%	100.00%	0.00%	3	100.00%
Byzantine			4	0.00%	0.00%	100.00%	4	100.00%
Byzantine+later			1	0.00%	0.00%	100.00%	1	100.00%
Grand Total	4	8	20	12.50%	25.00%	62.50%	32	100.00%

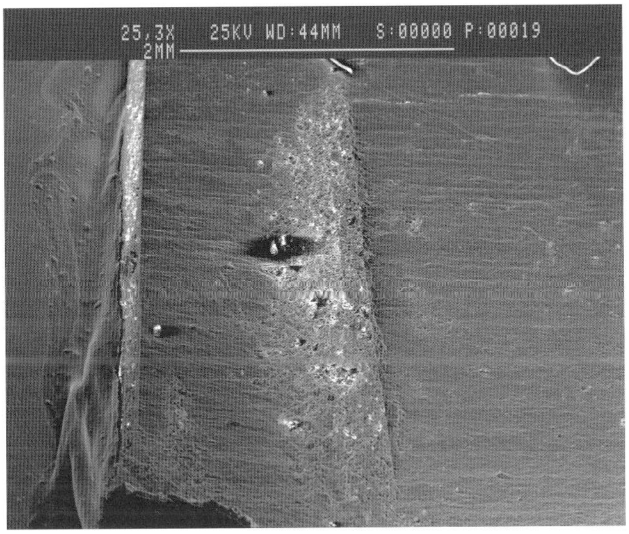

*Figure 9.7. Example of SEM photograph of silicone mold of the inverse of a groove made by a unifacially produced chipped stone tool blade/flake slicing mark metal butchering slice marks on caudal face of distal shaft of cattle (*Bos taurus*) anterior phalange 1 from the final EB III Stratum in the Lower City. Butchering sample 980A; catalogue # 17212 (Area Bb, Stratum B-1 [abandonment], Cistern 2683).*

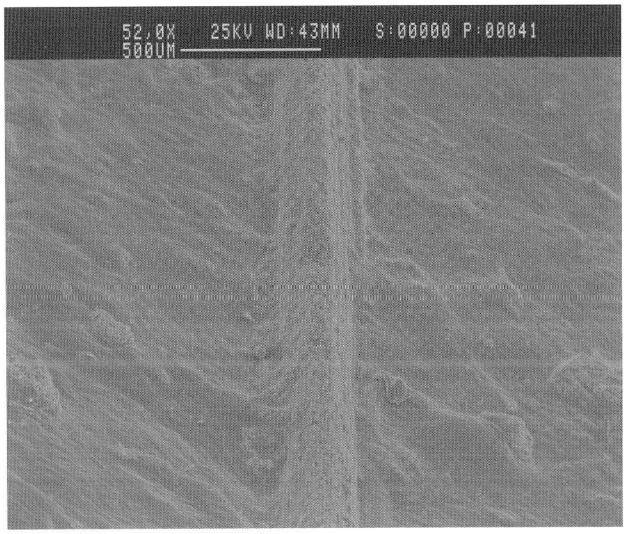

Figure 9.9. SEM photograph of silicone mold of the inverse of a groove made by a metal butchering knife blade at x52 magnification of Fig. 9.8.

*Figure 9.8. Example of metal butchering slice marks on caudal face of distal shaft of cattle (*Bos taurus*) anterior phalange 1 from an Iron IB+Iron II-Persian Stratum in the Acropolis. Photo DSCN9196; butchering sample 93; catalogue #9581 (Acropolis, Sounding 1, Stratum IIA, L. 1122).*

Table 9.9. Frequency distribution of all butchered specimens by butchering efficiency (BI) over time.

Period	*Sum of NBI (Number of butchering incidences)*	*Sum of # of slices under microscope*	*Efficiency*
EB	15	217	14.46666667
EB I-II	3	39	13
EB II	266	2992	11.2481203
EB II-III	11	52	4.727272727
EB II-IIIA	1	35	35
EB III	420	3264	7.771428571
EB mean			14.368915
LB III+Iron I	2	16	8
LB III to Iron IA	1	8	8
Iron I	10	102	10.2
Iron II	1	4	4
Roman +Byzantine	3	17	5.666666667
Byzantine	4	29	7.25
Byzantine+later	1	19	19
LB+later mean			8.8738095
Grand Total	738	6794	9.20596206
Total mean			9.20596206

Acknowledgements

We gratefully thank Simon Davis for allowing us to study and report on the Tel Yarmuth faunal assemblage. We also thank Tina L. Greenfield, Elizabeth R. Arnold, Matthew Singer, and Trent Cheney for contributing to the collection of the zooarchaeological specimens; Liora Horwitz for helping to select the assemblage for analysis; and Rivka Rabinovich (National Natural History Collections, Institute of Earth Sciences, Institute of Archaeology, The Hebrew University of Jerusalem) for providing essential access to the collection and laboratory space for the analysis in Jerusalem. Funding for this research was provided by the Social Science and Humanities Research Council of Canada, St Paul's College of the University of Manitoba and the University of Manitoba. We would like to thank them all for their support. Any errors are the fault of the authors, alone.

References

Bamforth, D.B. (1986) Technological efficiency and tool curation. *American Antiquity* 51, 38–50.

Bello, S.M. and Soligo, C. (2008) A new method for the quantitative analysis of cutmark micromorphology. *Journal of Archaeological Science* 35, 1542–1552.

Birch, T., Rehnen, T. and Pernicka, E. (2013) The metallic find from Çatalhöyük: a review and preliminary work. In I. Hodder (ed.) *Substantive Technologies at Çatalhöyük*, 307–318. Los Angeles, UCLA Press.

Bleed, P. (1986) The optimal design of hunting weapons: maintainability or reliability. *American Antiquity* 5, 737–747.

Bourke, S.J. (2014) The southern Levant (Transjordan) during the Middle Bronze Age. In M.L. Steiner and A.E. Killebrew (eds) *The Oxford Handbook of the Archaeology of the Levant c. 8000–332 BCE*, 465–481. New York, Oxford University Press.

Childe, V.G. (1951) *Man Makes Himself.* New York, Mentor.

Cohen, S.L. (2014) The southern Levant (Cisjordan) during the Middle Bronze Age. In M.L. Steiner and A.E. Killebrew (eds) *The Oxford Handbook of the Archaeology of the Levant c. 8000–332 BCE*, 445–458. New York, Oxford University Press.

Davis, S.J.M. (1987) *The Archaeology of Animals.* New Haven, Yale University Press.

Davis, S.J.M. (1988) The mammal bones from Tel Yarmuth 1980–1983. In P. de Miroschedji (ed.) *Yarmouth 1, rapport sur les trois premières campagnes de fouilles à Tel Yarmouth (Israel), 1980–1982*, 143–149. Paris, Association Pour la Diffusion de la Pensée Française, Editions Recherche sur les Civilisations.

Greenfield, H.J. (1999) The origins of metallurgy: distinguishing stone from metal cut marks on bones from archaeological sites. *Journal of Archaeological Science* 26, 797–808.

Greenfield, H.J. (2004) The butchered animal bone remains from Ashqelon, Afridar – Area G. *'Atiqot* 45, 243–261.

Greenfield, H.J. (2006) Slicing cut marks on animal bones: diagnostics for identifying stone tool type and raw material. *Journal of Field Archaeology* 31, 147–163.

Greenfield, H.J. (2013) 'The Fall of the House of Flint': a zooarchaeological perspective on the decline of chipped stone tools for butchering animals in the Bronze and Iron Ages of the southern Levant. *Lithic Technology* 38, 161–178.

Greenfield, H.J. and Brown, A. (2016) 'Making the cut': changes in butchering technology and efficiency patterns from the Chalcolithic to modern Arab occupations at Tell Halif, Israel. In N. Marom, R. Yeshurun, L. Weissbrod and G. Bar-Oz (eds) *Bones and Identity: Zooarchaeological Approaches to Reconstructing Social and Cultural Landscapes in Southwest Asia (Proceedings of the ICAZ-SW Asia Conference, Haifa, June 23–28, 2013)*, 273–291. Oxford, Oxbow Books.

Greenfield, H.J. and Chaput, T. (2021) Stone or metal butchering technology butchering technology at an early tin metal producing site at Göltepe, SE Turkey. In K.A. Yener (ed.) *Tin Production at Göltepe: Excavations at an Early Bronze Age Mining Town in the Central Taurus Mountains, Turkey. Volume 1, Parts 1 and 2*, 159–176. Philadelphia, Institute for Aegean Academic Press.

Greenfield, H.J., Beller, J.A. and Levy, T.E. (2018) Butchering technology during the Early Bronze Age I: an examination of microscopic cut marks on animal bones from Nahal Tillah, Israel. In I.Shai, J.R. Chadwick, L.Hitchcock, A. Dagam, C. Mckinny, and J. Uziel (eds) *Tell it in Gath: Studies in the History and Archaeology of Israel. Essays in Honor of A.M. Maeir on the Occasion of his Sixtieth Birthday, Ägypten und Altes Testament – Studien zu Geschichte, Kultur und Religion* Ägyptens *und des Alten Testaments, Band 90*, 20–40. Münster, Zaphon.

Greenfield, H.J., Cheney, T. and Galili, E. (2016) A taphonomic and technological analysis of the butchered animal bone remains from Atlit Yam, a submerged PPNC site off the coast of Israel. In N. Marom, R. Yeshurun, L. Weissbrod and G. Bar-Oz (eds) *Bones and Identity: Zooarchaeological Approaches to Reconstructing Social and Cultural Landscapes in Southwest Asia (Proceedings of the ICAZ-SW Asia Conference, Haifa, June 23–28, 2013)*, 89–112. Oxford, Oxbow Books.

Hayden, B. (1998) Practical and prestige technologies: the evolution of material systems. *Journal of Archaeological Method and Theory* 5, 1–55.

Hayden, B., Franco, N. and Spafford, J. (1996) Evaluating lithic strategies and design criteria. In G.H. Odell (ed.) *Stone Tools. Theoretical Insights into Human Prehistory*, 9–45. New York, Plenum Press.

Jasmin, M. and de Miroschedji, P. (eds) (2020) *Fouilles de Tel Yarmouth (1980–2009), Rapport final, Volume 1: Les fouilles sur l'acropole.* Ancient Near Eastern Studies Supplement 56. Leuven, Peeters.

Levy, T.E., Adams, R.B., Hauptmann, A., Prange, M., Schmitt-Strecker, S. and Najjar, M. (2002) Early Bronze Age metallurgy: a newly discovered copper manufactory in southern Jordan. *Antiquity* 76, 425–437.

Levy, T.E., Adams, R.B., Witten, A.J., Anderson, J., Arbel, Y., Kuah, S., Moreno, J., Lo, A. and Wagonner, M. (2001) Early metallurgy, interaction, and social change: the Jabal Ḥamrat Fīdān (Jordan) research design and 1998 archaeological survey, preliminary report. *Annual of the Department of Antiquities of Jordan* 45, 159–187.

Manclossi, F., Rosen, S.A. and Böeda, E. (2019) From stone to metal: the dynamics of technological change in the decline of chipped stone tool production: a case study from the southern Levant (5th–1st Millennia BCE). *Journal of Archaeological Method and Theory* 26(4), 1276–1326.

Miroschedji, P. de (1988a) Données nouvelles sur le Bronze ancien de Palestine: les fouilles de Tel Yarmouth. *Comptes rendus de l'Académie des Inscriptions et Belles-Lettres* 1988, 186–211.

Miroschedji, P. de (ed.) (1988b) *Yarmouth 1, Rapport sur les trois premières campagnes de fouilles à Tel Yarmouth (Israel).* Paris, Edition Recherche sur les Civilisations.

Miroschedji, P. de (2000) Fouilles de Tel Yarmouth: résultats des 11e, 12e et 13e campagnes de fouilles (1996–1999). *Comptes rendus de l'Académie des Inscriptions et Belles-Lettres*, 679–710.

Miroschedji, P. de (2003) The late EB III palace B1 at Tel Yarmuth: a descriptive summary. *Eretz-Israel* 27 (Tadmor Volume), 153–170.

Miroschedji, P. de (2006) At the dawn of history: sociopolitical developments in southwestern Canaan in Early Bronze Age III. In A.M. Maeir and P. de Miroschedji (eds) *'I Will Speak the Riddles of Ancient Times': Archaeological and Historical Studies in Honor of Amihai Mazar on the Occasion of His Sixtieth Birthday* 55–78. Winona Lake, IN, Eisenbraun.

Miroschedji, P. de (2013) Fouilles de Tel Yarmouth: résultats des travaux de 2003 à 2009 (14e–18e campagnes). *Comptes rendus de l'Académie des Inscriptions et Belles-Lettres* 2013, 759–796.

Miroschedji, P. de (2014) The southern Levant (Cisjordan) during the early Bronze Age. In M.L. Steiner and A.E. Killebrew (eds) *The Oxford Handbook of the Archaeology of the Levant c. 8000–332 BCE*, 308–328. New York, Oxford University Press.

Moorey, P.R.S. (1988) Early metallurgy in Mesopotamia. In R. Maddin (ed.) *The Beginning of the Use of Metals and Alloys*, 28–33. London, The MIT Press.

Moorey, P.R.S. (1994) *Ancient Mesopotamian Materials and Industries*. Oxford, Oxford University Press.

Olsen, S.L. (1988) The identification of stone and metal tool marks on bone artifacts. In S.L. Olsen (ed.) *Scanning Electron Microscopy in Archaeology*, 337–360. British Archaeological Reports ISS 452. Oxford, BAR.

Philip, G. (1989) *Metal Weapons of the Early and Middle Bronze Ages in Syria-Palestine*. Oxford, BAR.

Philip, G. (2001) The Early Bronze I–III Ages. In B. Macdonald, R. Adams and P. Bienkowski (eds) *The Archaeology of Jordan. Levantine Archaeology 1*, 163–232. Sheffield, Sheffield Academic Press.

Rosen, S.A. (1984) The adoption of metallurgy in the Levant: a lithic perspective. *Current Anthropology* 25, 504–505.

Rosen, S.A. (1996) The decline and fall of flint. In G.H. Odell (ed.) *Stone Tools. Theoretical Insights into Human Prehistory*, 129–158. New York, Plenum Press.

Rosen, S.A. (1997) *Lithics after the Stone Age: A Handbook of Stone Tools from the Levant*, Walnut Creek, CA, Altamira Press.

Shalev, S. (1994) The change in metal production from the Chalcolithic period to the Early Bronze Age in Israel and Jordan. *Antiquity* 68, 630–637.

Shalev, S. (2009) Metals and society: production and distribution of metal weapons in the Levant during the Middle Bronze Age II. In S.A. Rosen and V. Roux (eds) *Techniques and People: Anthropological Perspectives on Technology in the Archaeology of the Proto-historic and Early Historic Periods in the Southern Levant*, 69–80. Paris, Edition de Boccard.

Shipman, P. (1981) Application of scanning electron microscopy to taphonomic problems. *Annals of the New York Academy of Sciences* 375, 357–385.

Torrence, R. (ed.) (1989) *Time, Energy and Stone Tools*, Cambridge: Cambridge University Press.

Van Der Leeuw, S. and Torrence, R. (1989) *What's New? A Closer Look at the Process of Innovation*. London, Unwin Hyman.

Walker, P.L. (1978) Butchering and stone tool function. *American Antiquity* 43, 710–715.

Figure 10.1. Map of Egypt showing the location of Malqata. Map by Aidan Dodson.

mission carried out further excavations at the site between 1971 and 1977. This team was focussed on settlement archaeology, and archaeozoological material was collected (Kemp and O'Connor 1974; O'Connor 1979). Most of this material was left in Egypt, however, awaiting study and is now inaccessible. Fortunately, though, a small amount was exported as a sample collection to Cambridge University. In 2007 the Michael C. Carlos Museum (subsequently replaced by The Ancient Egyptian Heritage and Archaeology Fund) collaborated with the MMA and returned to the site. Although their main focus was the physical preservation of Malqata, which is increasingly under threat, new excavations were also carried out, including some in domestic spaces that yielded faunal material. This paper presents some of the results from both the new and the old excavations.

Materials and methods

The specimens that form the basis of this study are derived from the University of Pennsylvania excavations of the 1970s

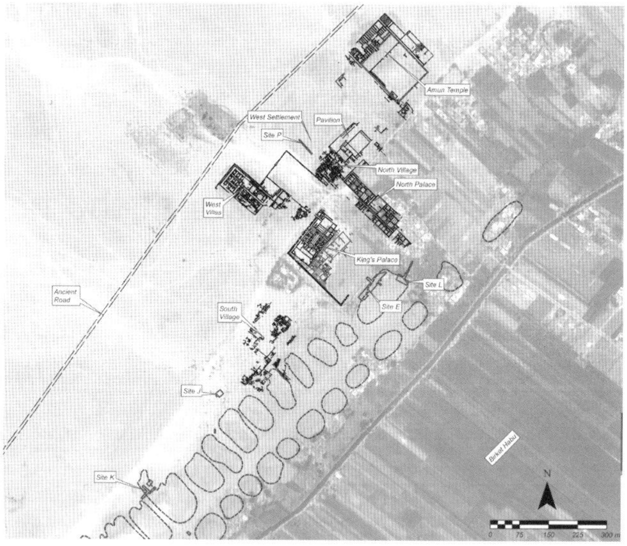

Figure 10.2. Map of Malqata showing areas discussed in the text. Courtesy of the Joint Expedition of the Metropolitan Museum of Art and the Ancient Egyptian Heritage and Archaeology Fund. Map by Joel Paulson.

kept in Cambridge University, together with those of the MMA/Carlos Museum, which started in 2007. The Cambridge material was studied in the University's bone laboratory using comparatives, while the later material was studied in the field, using comparatives loaned by the American University in Cairo's Ibrahim Helmy Archaeozoology Laboratory and published atlases for bones where there were no comparatives (Cohen and Serjeantson 1986).

Faunal material came from a variety of locations whose character was defined by the size and complexity of the surviving architecture, and the materials and texts found within them (Fig. 10.2). These spaces included palaces (Areas E and K), zones of elite residences (Area P and the North Village), mid- to low-socio-economic settlement areas (Area J, also known as the South Village), and poorer structures, near animal pens (Area L). The West Settlement is a northern part of Area P, and is probably similar in nature to it, though the architecture is slightly different; the faunal remains can potentially contribute to a better definition of its character. Material from the North Village, West Settlement and South Village was all sieved, although this procedure was not carried out in the other areas, which is why Area P is not merged with the West Settlement in this paper, though probably they are similar in character. It is apparent that a great deal of material, particularly fish bones, was lost as a result, which has skewed the overall data.

For the vertebrate remains, the following information was recorded for each specimen: taxon, anatomical element, portion (similar to diagnostic zone), side, age (based on epiphyseal fusion, level of porosity, tooth eruption and wear patterns), processing (butchery marks, signs of being worked, gnawed or burnt) and erosion or breakage patterns.

Whenever possible measurements were made following von den Driesch (1976). When it was not totally clear if a fragment belonged to a specific species, particularly in the case of limb bones, ribs, and vertebrae, it was identified only by mammal size: Large (cow size), Large-Medium (calf size), Medium (sheep size), Medium to Small (dog size), and Small (rabbit size). The ageing systems used follow Barone (1965), Silver (1969), Grant (1982), Payne (1973) and Hambleton (2001).

Results

A total of 9056 specimens were identified (Table 10.1, which excludes fragmentary bones identified by mammal size), with the majority coming from the South Village (Area J), followed by the palace edge (Area E), closely followed by the West Settlement. The smallest number of bones came from the edge of the Small Palace (Area K).

As a general rule, the majority of bones derived from animals that were consumed in ancient Egypt come from river fish, followed by birds, then cattle (*Bos taurus*), sheep/goat (*Ovis/Capra*), and finally, pigs (*Sus scrofa domesticus*) (Table 10.2; Fig. 10.3). Although the bones of fish and birds are plentiful, it should be remembered, particularly for the fish, that vertebrae, spines and ribs are part of the count, giving an exaggerated impression of their number. This is not the case for the identified mammals; with the inclusion of counts for mammals identified by size, the distribution of bones is more equitable. Also noteworthy is the fact that most birds and riverine fish, with some exceptions, provide far less meat than cattle, sheep/goat and pigs, so that their contribution to the overall volume of food supplied is less.

Cattle are the most common large-scale meat-providing animals found at Malqata, dominating the domesticated species almost to the exclusion, in some cases, of all others (Figs 10.3 and 10.4). The North Village is the only area where sheep/goat outnumber them, but just barely. Most cattle were killed before they reached the age of two years, thus providing succulent meat. They are followed by sheep/goat. The greatest number of sheep/goats occurs in Area J, then the North Village, West Settlement, and finally, the palace. The other locations have negligible or no

*Table 10.1. The Number of Identified Specimens (NISP) from each area at Malqata, including to mammal size (large, medium, small). * denotes sieving.*

Area	*NISP*
E = Palace Edge	1982
J = South Village*	4805
K= Small Palace Edge	134
L=Mangers/poor huts	271
P=Town, elite	466
West settlement*	1205
North village*, elite	193
Total	9056

*Table 10.2. The distribution of Number of Identified Specimens from each area at Malqata, excluding fragments only identified to mammal size (large, medium, small). * denotes sieving.*

Taxa	*E*	*J**	*K*	*L*	*P*	*W Set.**	*N Vill.**	*Total*
Cattle	146	92	38	27	31	63	36	433
Sheep/ Goat	16	80	0	1	1	22	40	160
Pig	26	40	4	3	2	0	0	75
Equid	2	8	0	0	0	0	0	10
Dog	6	1	0	0	0	0	0	7
Crocodile	0	0	0	1	0	0	0	1
Gazelle	2	3	0	0	0	0	0	5
Bird	50	75	11	11	14	268	28	457
Fish (riverine)	16	80	0	4	6	483	45	634
Fish (marine)	0	0	0	1	0	1	1	2
Turtle	0	1	0	0	0	0	0	1
Total	264	380	53	48	54	837	150	1785

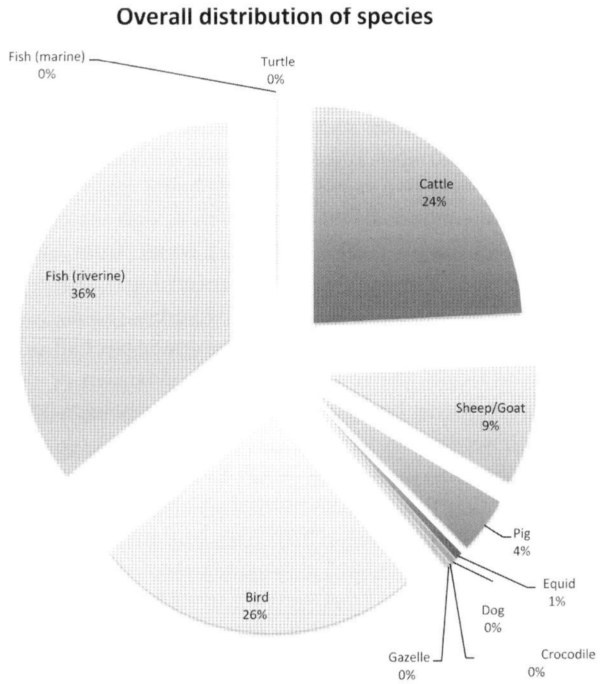

Figure 10.3. Chart showing the overall distribution of animals in the whole assemblage.

on the Malqata assemblage possible: for the Cambridge material I am grateful to Barry Kemp, Marsha Levine, Jessica Rippengal and Anthony J. Legge for access to, and advice with this assemblage. For access to the on-site material and permission to publish the maps and plans I am indebted to Diana Craig Patch and Peter Lacovara.

References

Barone, R. (1965) *Anatomie comparée des mammifères domestiques*. Paris, Vigot Frères.

Bertini, L. and Ikram, S. (2020) Food for the Forces: An investigation of military subsistence strategies in New Kingdom border regions. In K.L. Younger, Jr. and R.E. Averbeck (eds) *Festschrift for James Hoffmeier*. University Park, PA, Eisenbrauns.

Brewer, D. and Friedman, R. (1989) *Fish and Fishing in Ancient Egypt*. Warminster, Aris and Phillips.

Cohen, A. and Serjeantson, D. (1986) *A Manual for the Identification of Bird Bones from Archaeological Sites*. London, Archetype.

Daressy, G. (1903) Le palais d'Aménophis III et le Birket Habou. *Annales du Service des Antiquités de l'Égypte* 4, 164–70.

von den Driesch, A. (1976) *A Guide to the Measurement of Animal Bones from Archaeological Sites*. Cambridge, MA, Peabody Museum/Harvard University.

Ghoneim, W. (1977) *Die Ökonomische Bedeutung des Rindes im Alten Agypten*. Bonn, Habelt.

Grant, A. (1982) The use of tooth wear as a guide to the age of domestic ungulates. In B. Wilson, C. Grigson, and S. Payne (eds) *Ageing and Sexing Animal Bones from Archaeological Sites*, 91–108. BAR British Series 109. Oxford, British Archaeological Reports.

Griffith, F. Ll. (1891) The Account Papyrus no. 18 of Boulaq. *Zeitschrift für Ägyptische Sprache und Altertumskunde* 29, 102–116.

Hambleton, E. (2001) A method for converting Grant mandibular wear stages to Payne style wear stages in sheep, cow, and pig. In A. Millard (ed.) *Archaeological Sciences, Proceedings of the Conference Held at the University of Durham, 2nd–4th September 1997*, 103–108. BAR Reports International Series 939. Oxford, Archaeopress.

Hayes, W.C. (1951a) Inscriptions from the Palace of Amenhotep III 1. *Journal of Near Eastern Studies* 10(1), 35–56.

Hayes, W.C. (1951b) Inscriptions from the Palace of Amenhotep III 2. *Journal of Near Eastern Studies* 10(2), 82–112.

Hayes, W.C. (1951c) Inscriptions from the Palace of Amenhotep III 3. *Journal of Near Eastern Studies* 10(3), 156–183.

Hecker, H.M. (1982) A zooarchaeological inquiry into pork consumption in Egypt from prehistoric to New Kingdom times. *Journal of the American Research Center in Egypt* 19, 59–71.

Hecker, H.M. (1984) Preliminary report on the faunal remains from the Workmen's Village. In B.J. Kemp (ed.,) *Amarna Reports* I, 154–164. Occasional Publications 1. London, Egypt Exploration Society.

Hekkala, E., Shirley, M.H., Amato, G., Austin, J.D., Charter, S., Thorbjarnarson, J., Vliet, K.A., Houck, M.L., Desalle, R. and Blum, M.J. (2011) An ancient icon reveals new mysteries: mummy DNA resurrects a cryptic species within the Nile crocodile. *Molecular Ecology* 20(20), 4199–4215.

Ikram, S. (1995) *Choice Cuts: Meat Production in Ancient Egypt*. Leuven, Peeters.

Kemp, B.J. and O'Connor, D. (1974) An ancient Nile harbour: University Museum excavations at the 'Birket Habu'. *International Journal of Nautical Archaeology and Underwater Exploration* 3, 101–136.

Kozloff, A., Bryan, B. and Berman, L.M. (1992) *Egypt's Dazzling Sun Amenhotep III and his World*. Cleveland, Cleveland Museum of Art.

Lacovara, P. (1994) Realm of the Sun King: the palace city of Amenhotep III at Malkata. *Amarna Letters* 3, 6–21.

Lacovara, P. (2012) Malkata/Malqata. In R.S. Bagnall, K. Brodersen, C.B. Champion, A. Erskine and S.R. Huebner (eds) *The Encyclopedia of Ancient History*, 1st edition, 4241–4243. Malden, MA, Wiley/Blackwell.

Legge, A. (2010) The mammal bones Form Grid 12. In B. Kemp and A. Stevens (eds) *Busy Lives at Amarna: Excavations in the Main City (Grid 12 and the House of Ranefer, N49.18). Vol. 1: The Excavations, Architecture, and Environmental Remains*, 445–452. London, Egypt Exploration Society.

O'Connor, D. (1979) The University Museum Excavations at the Palace-City of Malkata. *Expedition* 21.2, 52–53.

Payne, S. (1973) Kill off patterns in sheep and goats: the mandibles from Asvan Kale. *Anatolian Studies* 23, 281–305.

Payne, P. (2004) Tell el-Amarna, 2004. Appendix III: the faunal remains. *Journal of Egyptian Archaeology* 90, 24–26.

Payne, P. (2006) Recovering animal bone at the house of the high priest Panehsy. In B. Kemp, The 2005/6 season at Tell el-Amarna, *Journal of Egyptian Archaeology* 92, 45–52.

Pendlebury, J.D.S. (1951) *The City of Akhenaten*, 3.i–ii. London, Egypt Exploration Society.

Peyster Tytus, R. de (1903) *On the Re-excavation of the Palace of Amenhotep III*. New York, Winthrop Press.

Posener-Kriéger, P. (1976) *Les Archives du Temple Funéraire de Neferirkare-Kakai*, 1–2. Cairo, Institut Français d'archéologie orientale.

Redding, R. (2010) Status and diet at the Workers' Town, Giza, Egypt. In D. Campana, P. Crabtree, S.D. de France, J. Lev-Tov and A. Choyke (eds). *Anthropological Approaches to Zooarchaeology*, 65–75. Oxford, Oxbow Books.

Silver, I.A. (1969) The ageing of domestic animals. In D. Brothwell and E.S. Higgs, (eds). *Science in Archaeology*, 282–302. London, Thames and Hudson.

Te Velde, H. (1967) *Seth God of Confusion*. Leiden, E.J. Brill.

Uphill, E. (1999) Tell el-Amarna (city). In K. Bard (ed.). *Encyclopedia of the Archaeology of Ancient Egypt*, 932–937. New York, Routledge.

Volokhine, Y. (2014) *Le porc en* Égypte *ancienne. Mythes et histoire à l'origine des interdits alimentaires*. Liège, Presses Universitaires de Liège.

11

Caprine husbandry at the Iron Age settlement of A Lanzada (Pontevedra, Spain)

Marta Moreno-García

Bad preservation has greatly hampered archaeozoological research in the north-western region of the Iberian Peninsula. This has limited the size of samples and the specific identification of the bone remains, preventing a better understanding of livestock production and husbandry strategies. With the aim to partially fill this gap, the present paper focuses on the study of the Late Iron Age caprine remains recovered at the coastal site of A Lanzada (Pontevedra, Galicia, Spain). The location of this settlement on seashore sandy sediments has favoured the conservation of a large faunal assemblage dominated by the main domesticates. Morphological and metrical determination of caprine teeth and bones has enabled to estimate sheep to goat ratios, to build age-at-death profiles for each species and in the case of goats, to recognise female and male specimens. The large proportion of immature and sub-adult animals suggests a balanced relationship between optimisation of production and herd security, as it would be expected in a self-sufficient economy. Most interesting are the osteometric trends observed for sheep and goats. While sheep evidence a large decrease in size with respect to other northern Iberian sites dated from the Neolithic to medieval times, goats appear to have undergone little variation in the course of time. Therefore, caprines biometric change cannot be related exclusively to environmental factors but rather to different feeding and husbandry strategies that need to be further researched.

Keywords: Sheep, goats, management, metrical data, Iron Age, north-western Iberia

Introduction

The archaeological site of Campo de A Lanzada occupies the area known as Punta Lanzada. This is located on a small cape that penetrates into the Atlantic Ocean, between the beaches of Foxos and Areagorda, in the parish of Noalla, Sanxenxo (Pontevedra, Galicia), on the north-west coast of Spain (Fig. 11.1). The site stands 15 m above sea level, at latitude 42º25'45'' N and longitude 08º52'40'' W, on sandy sediments and fossilised dunes which were formed by the action of the wind on top of a granitic cliff.

The first documentary references findings of historical interest from the site of A Lanzada date to the 18th century. But it was not until the archaeological campaigns carried out in the 1950s and 1960s that a number of structures belonging to an Iron Age settlement and also a Late Roman necropolis were discovered. New excavations in the 1970s revealed that the occupation sequence actually spanned from the Late Bronze Age to the 15th century AD. Unfortunately, the profuse digging activity was not mirrored by a corresponding number of research publications. At the beginning of the 21st century a complete overview of the site was still missing. With the aim to make up for these scientific shortcomings a team of archaeologists from the Diputación of Pontevedra and the INCIPIT-CSIC (Spanish National Research Council) set up a project in the summer of 2010 to re-excavate the sectors that were investigated in the 1950s and 1970s. Soon, the relevance of the new findings prompted the works to stop in the southern area to focus the attention on the 680 m^2 of the northern area. Here, the discovered Late Iron Age structures appeared to be organised in two separate zones, one dedicated to the transformation and conservation of marine resources and another identified with the living area and, maybe, also some storage spaces. Associated with the latter there were a number of large shell middens which were filled with debris consisting of pottery, metal (bronze and iron) and worked bone objects, stoneware

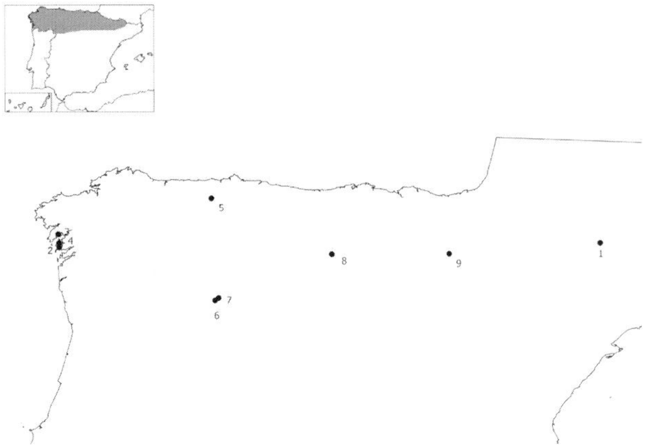

Figure 11.1. Location of sites mentioned in the text. 1. Cova de Els Trocs (Bisaurri, Huesca); 2. A Lanzada (Sanxenxo, Pontevedra); 3. O Achadizo (A Coruña); 4. Cantodorxo (Pontevedra); 5. El Castru (Vigaña, Asturias); 6. Peña Redonda (Zamora); 7. El Picón (Zamora); 8. Castrillo Mota de Judíos (Castrojeriz, Burgos); 9. La Noguera (La Rioja).

and the largest collection of faunal (molluscs, fish, mammal and a few birds) remains recovered so far in Galicia. The size of the structures and the abundant archaeological material evidence the intensive activity carried out by the human community settled in A Lanzada between the 2nd and 1st centuries BC. Indeed, with the arrival of the Romans at the change of era, this productive and residential area was partly modified and finally abandoned, suggesting perhaps the preferential interest of the Romans in the exploitation of the nearby salt marshes (Rodríguez Martínez *et al.* 2011).

This paper will focus on the archaeozoological study of the caprine remains. They are important because they comprise nearly 60% of the identified mammal assemblage that was recovered from both the Late Iron Age middens and the change of era layers (Table 11.1). Further, previous work on contemporaneous faunal collections from seashore sites in the north-western region of Iberia suggests that caprines, followed by cattle and pig, usually dominate the assemblages while equids and game are only occasionally present (Ramil Rego and Fernández Rodríguez 1999; Fernández Rodríguez 2000; Fernández Rodríguez and Pérez Ortiz 2007). However, bad preservation has frequently hampered the specific identification of sheep and goat remains and small samples' size have hindered a more comprehensive understanding of the role played by both species within the economy of these pre-Roman settlements. Little is known about management systems and husbandry production strategies as well as about the physical conformation of the animals (*i.e.* their size and shape). In this context, the large number of caprine bones and teeth from A Lanzada provides a unique opportunity to collect and analyse significant data and consequently, begin exploring fundamental questions surrounding livestock keeping in the Late Iron Age communities located on the Atlantic coast of this Iberian region.

Materials and methods

The A Lanzada mammal bones were sent to the Laboratorio de Arqueozoología, Instituto de Historia (IH), CCHS-CSIC, Madrid in 2011. Although at the time of the excavation a sieving program was designed to recover fish and molluscs remains (González Gómez de Agüero *et al.* 2011; 2014) the samples here analysed were recovered by hand. I did not have access to the sieved material. Bone, teeth and horncore/antler fragments were identified preferably to species level with the help of the IH modern reference collection. Taxonomically unidentifiable skeletal elements were recorded in two categories according to their size: large (LSM) and medium-sized (MSM). Given the scarcity of equids and red deer, LSM includes long-bone shafts, skull, vertebra and rib fragments most likely belonging to cattle; MSM comprises the same sort of fragments but of small ruminants (sheep and goat) and, occasionally, pig. Remains that could not be related to any body part were recorded as undetermined. Sheep bones and teeth were separated from those of goat following the morphological and metrical criteria described by Boessneck, Müller and Teichert (1964), Payne (1969), Halstead and Collins (2002), Zeder and Lapham (2010) and Zeder and Pilaar (2010).

All fragments were counted individually but in order to avoid the overrepresentation of the most fragmented species, taxonomic frequencies were recalculated on the basis of the minimum number of elements (MNE) identified for each taxon. MNE were estimated from the most frequent diagnostic zone in each skeletal element. By definition, a zone occurs only once on each bone so that the same zone of a single bone cannot be counted twice. Zones recorded follow Dobney and Rielly (1988).

Mandibular tooth eruption and wear stages of individual teeth and of mandibular tooth rows (Payne 1973; 1987) and epiphyseal fusion sequences of post-cranial bones (Zeder 2006) are the methods applied to estimate age-at-death. An attempt to distinguish the sex of sheep pelvises was made through the morphological characters described by Hatting (1995). Measurements were recorded for all fused and well-preserved bone elements after von den Driesch (1976). Additional measurements in sheep and goats follow Davis (1996). Data were fed into a database using IBM SPSS 25.

The composition of the mammal assemblage

As Table 11.1 shows 90 % of the remains derive from the Late Iron Age (2nd–1st centuries BC), 9 % from the change of era (1st *c.* BC–1st *c.* AD) and 1% are from the Roman period (2nd–4th centuries AD). The proportion of identified material exceeds 60%, increasing slightly in the course of time (61%, 65% and 66%). This result indicates the general good preservation of the faunal remains. Taking into account that the more recent assemblages represent only 10% of the

*Table 11.1. Number (N) and frequencies of analyzed remains (%N) and identified taxa (%ID). Minimum number of elements (MNE) of each taxon and their frequency (%MNE). [OVA+CAH+O/C] include all the remains belonging to the Caprinae subfamily. *Antler fragment.*

Taxa/ period	*Late Iron Age*					*Change of era*					*Roman*			*Total*		
Identified	*N*	*%N*	*%ID*	*MNE*	*%MNE*	*N*	*%*	*%ID*	*MNE*	*%MNE*	*N*	*%*	*%ID*	*N*	*%*	*%ID*
Equid (*Equus* sp.)	–	–	–	–	–	–	–	–	–	–	2	3	4	2	<1	<1
Cattle (*Bos taurus*)	1691	19	30	978	25	172	17	27	122	26	31	43	66	1894	19	30
Sheep (*Ovis aries*)	1108	12	20	993	26	96	10	15	89	19	–	–	–	1204	12	19
Goat (*Capra hircus*)	614	7	11	546	14	80	8	13	76	16	1	1	2	695	7	11
Caprines (*Ovis/Capra*)	1275	14	23	734	19	183	19	29	105	22	5	7	11	1463	14	24
[OVA+CAH +O/C]	[2997]	[33]	[54]	[2273]	[59]	[359]	[37]	[56]	[270]	[57]	[6]	[8]	[13]	[3362]	[33]	[54]
Suid (*Sus* sp.)	832	9	15	574	15	99	10	16	74	15	6	8	13	937	9	15
Red deer (*Cervus elaphus*)	1*	<1	<1	1*	<1	–	–	–	–	–	–	–	–	1*	<1	<1
Wolf (*Canis lupus*)	2	<1	<1	2	<1	–	–	–	–	–	–	–	–	2	<1	<1
Fox (*Vulpes vulpes*)	–	–	–	–	–	5	<1	<1	5	1	–	–	–	5	<1	<1
Rabbit (*Oryctolagus cuniculus*)	1	<1	<1	1	<1	1	<1	<1	1	<1	–	–	–	2	<1	<1
Dolphin	1	<1	<1	1	<1	–	–	–	–	–	–	–	–	1	<1	<1
Cetacean	13	<1	<1	13	<1	4	<1	<1	4	<1	2	3	4	19	<1	<1
Total	5538	61	100	3843	100	640	65	100	476	100	47	65	100	6225	62	100
Unidentified																
Large-sized mammal	2189	24				196	20				21	29		2406	24	
Medium-sized mammal	1110	12				118	12				3	4		1231	12	
Undetermined	219	2				24	3				1	1		244	2	
Total	9056	100				978	100				72	100		10106	100	
% Total		90					9					1			100	

analysed total the present work will primarily deal with the data from the Late Iron Age sample.

The faunal record of A Lanzada is dominated by the main domestic livestock: caprines, cattle and pig. Although the data based on MNE reflect equal relative frequencies of sheep and cattle followed at a certain distance by pig and goat, the total sum of caprine elements, including the generic sheep/goat category, clearly evidences the quantitative dominance of the small ruminants over cattle and pig. The presence of all skeletal parts and the observed butchery marks reveal that the bulk of the material derives from food debris. From this perspective, given the large biomass of cattle it may be assumed that more beef than mutton or goat meat and above all pork, was consumed at the site. Such results agree with the pattern recognised in other pre-Roman settlements located on the Atlantic coast of present-day Galicia (Fernández Rodríguez and Pérez Ortiz 2007). The single red deer antler fragment and rabbit bone, probably a later intrusion, point to the very marginal role that wild resources had in the subsistence economy of this human community, a circumstance already noted in previous studies (Fernández Rodríguez and Fuertes Prieto 2007). The recovery of the right scapula and humerus of a wolf among the domestic debris deposited in the shell middens suggests that the scarce hunting activity may have focussed on species potentially representing a threat to the survival of the local herds. Incidentally, this is the first archaeological record for wolf dated to the Iron Age in the region. Finally, in view of the seashore location of A Lanzada the occurrence of some cetacean remains is not surprising. One is a thoracic vertebra of a small dolphin and the others comprise butchered chunks of whale-sized unidentified bones. A cetacean single vertebral disc had already been recorded among the faunal remains from the 1983 excavations carried out at A Lanzada (Fernández Rodríguez 2000, 370). Findings of this kind have also been reported in the nearby province of Asturias (Morales Muñiz *et al.* 1992; Nores Quesada and Pis Millán 2001) hinting at the occasional use of stranded marine mammals (Clark 1947) which would have contributed not only to diversify the type of meat available for consumption but also to provide additional resources such as fat, oil or even bone for tool manufacturing (Pétillon 2008).

The faunal spectrum for the smaller assemblage dated towards the new era does not differ much from the preceding period. There appears to be a slight decrease in the frequency of identified sheep remains that is mirrored by an increase in goats and the mixed caprine category. Since the percentage of medium-sized fragments does not change, it is difficult to explain this shift in terms of differential preservation. Perhaps it may hypothetically be related to a modification of the husbandry strategies or to new dietary habits as the Romanisation process took place, certainly a question worth researching further with larger samples. The recovery of a complete fox mandible constitutes the second local pre-Roman record for this carnivore species. The first one comes from Castro de Punta de Cantodorxo (O Grove, Pontevedra) where a partially articulated lower part of a front leg and some vertebrae were interpreted as possible evidence of the use of the skin of this animal (Fernández Rodríguez *et al.* 1998).

The material dated to Roman times is too small to work with.

Identification of sheep and goats and their ratios

The specific identification of sheep and goat remains was a primary goal of the analysis because it would help to characterise the husbandry and production strategies followed with both species. Recognition of the morphological characters described in the references above enabled the separation of more than half of the total caprine remains

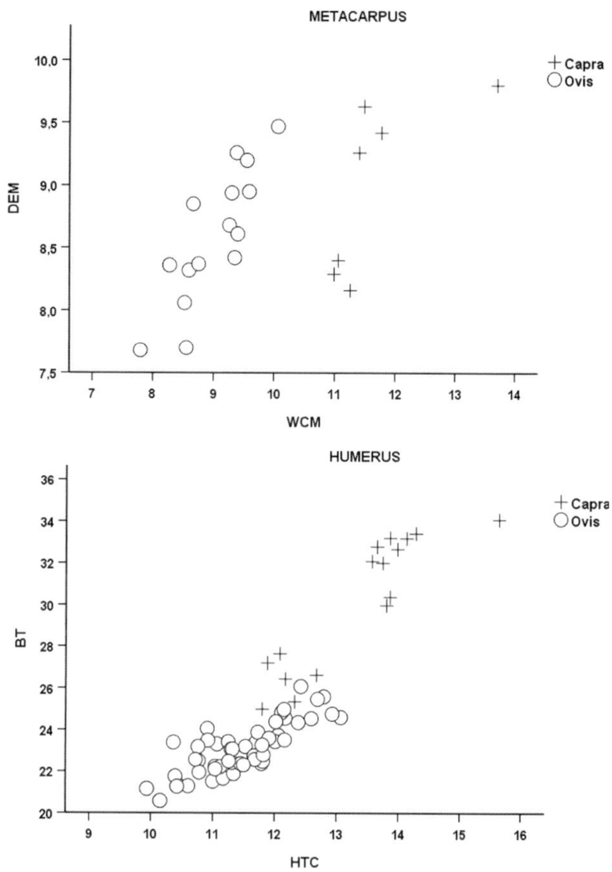

Figure 11.2. Identification of sheep and goats at A Lanzada (Late Iron Age). Above: scattergram of medial condyle width (WCM) against trochlea depth (DEM) measurements for the caprine metacarpals (after Payne 1969). Below: scattergram of humerus medio-lateral width of the trochlea (BT) against the vertical diameter of the trochlea at its central constriction (HTC) (after Davis 1996). All measurements are in mm. Sheep are represented as circles and goats as crosses.

(Table 11.1). Then, the analysis of metrical data of adult and well-preserved specimens helped to corroborate these identifications. For example, Fig. 11.2 plots (above) the Iron Age sheep and goat metacarpal DEM (antero-posterior diameter of the external trochlea of the medial condyle) against WCM (medio-lateral width of the medial condyle) measurements, taken after Payne (1969). No measurements were available for the assemblage dated to the change of era. Two clear clusters separate the sheep (represented as circles) on the left side from the goats (represented as crosses) on the right. Note that while the sheep spread upwards in a single group the goat metacarpals appear to be divided in two discrete clusters. It might be argued that the lack of intermediate goat specimens was due to their low number, however, taking into account that sex is one of the main factors influencing size in goats (Zeder 2001) it is more plausible that the two clusters represent different sex-groups.

The three specimens on the lower part of the graph may be interpreted as does while those spreading towards the upper right corner are probably billy goats, among which there is a very big individual.

A similar picture is observed with the largest sample of distal humeri (Fig. 11.2, below). The scatter plot of the medio-lateral width of the trochlea (BT) against the vertical diameter of the trochlea at its central constriction (HTC) (Davis 1996) provides a fair separation between the two species, with goats showing a greater width of the trochlea in relation to its height. Yet again, while sheep are relatively tightly clustered goats have a wide and divided spread. Six of them plot on the inner top part of the sheep cluster while the rest form a distinct group towards the upper right corner, with one outlying specimen of very large size.

These results are consistent with those reported both by Davis (2000) and Zeder (2001) in their respective metrical studies of modern known-sex unimproved Shetland sheep and wild and domestic goats from Iran and Iraq. Sheep distal humerus exhibits low sexual dimorphism whereas there is almost no overlap between the measurements of female and male goats. Further, Zeder (2001, 66) concludes that the measurements of male goats tend to show a consistently greater range and degree of variability than females. On these grounds, it may be assumed that the metrical data here analysed allowed not only the separation of both caprine species but also the sexual determination of the goat samples. The tentatively identified male goat specimens from A Lanzada are clearly larger than the females and show a wider distribution. Interpretation of the goat sex-ratios will be discussed in the next section.

Based on the specimens plotted in Fig. 11.2 the estimate ratio of sheep and goats would be 2:1 for the metacarpals and 3:1 for the humeri. That is 68% and 75% respectively, belong to sheep. These proportions of sheep to goats are within the range and close to the average value resulting from the quantification of the most easily identified skeletal elements (Table 11.2). The average contribution of sheep accounts to 75% in the Iron Age and 69% at the change of era. The calcaneus and astragalus ratios are higher than these values but they are probably biased by their scarce number. In fact, sample size may be the cause for the slight difference observed between the present results and the limited data available from contemporaneous assemblages, such as O Achadizo (A Coruña) (Phase II and III), Cantodorxo (Pontevedra) or even the old excavations in A Lanzada, where the sheep to goat ratios were estimated as 4:1 (Fernández Rodríguez 2000, 578).

In sum, although sheep clearly dominates the caprine samples at A Lanzada, goats in particular billies, may have played a relevant role in the local economy that due to previous lack of evidence has been overlooked.

Production and management strategies

The proportions of unfused limb bone epiphyses compared to fused limb bones indicate that nearly 40% of the caprines were slaughtered before 3 ½ years of age (Table 11.3) while according to the dental wear the frequency of immature (stages A–C) and sub-adult (stages D–E) individuals totals over 58% (Table 11.4). The discrepancy between both results may be explained by recovery bias and perhaps also differential preservation. Indeed, the loss of small bones during excavation is demonstrated by comparing the scarcity of sheep and goat astragali and calcanea to the large number of distal tibiae (Table 11.2) or by the few phalanges present in the assemblage (Table 11.3). Therefore, the data from the mandibular wear stages presumably reflect best the culling strategies practised by the human community settled at A Lanzada during the Late Iron Age.

Considering the total number of caprine mandibles, the peak of mortality occurs among sub-adult animals followed by those in stages F (3–4 years) and G (4–6 years). Such age-at-death distribution suggests production strategies primarily orientated to meat production but not exclusively, as killing animals in their second or third year would allow ensuring the replacement of the flocks and meeting the local demand for secondary products (*i.e.* milk, wool

Table 11.2. Numbers and ratio of sheep (O) and goats (C) and percentages of sheep. The numbers of deciduous lower fourth premolars (dLP4), scapulae (SC), distal humeri (dHU), proximal radii (pRA), distal metacarpals, (MTC), distal tibiae (dTI), distal metatarsals (MTT), astragali (AST) and calcanea (CAL) are given as x : y, where x = number of sheep teeth or bones and y = number of goat teeth or bones.

Period		*dLP4 O:C*	*SC O:C*	*dHU O:C*	*pRA O:C*	*dMTC O:C*	*dTI O:C*	*dMTT O:C*	*AST O:C*	*CAL O:C*	*Total*
Iron Age	N	36:12	13:8	83:24	77:19	29:28	103:32	48:12	18:4	18:4	425:143
	ratio	3:1	1,6:1	3,5:1	4:1	1:1	3,2:1	4:1	4,5:1	4,5:1	3:1
	% sheep	75%	62%	78%	80%	50%	76%	80%	82%	81%	75%
Change of era	N	1:1	2:1	13:5	8:6	4:3	9:3	6:2	1:0	2:0	46:21
	ratio	1:1	2:1	2,6:1	1,3:1	1,3:1	3:1	3:1	1:0	2:0	2,2:1
	% sheep	50%	66%	72%	57%	57%	75%	75%	100%	100%	69%

Table 11.3. Estimates of the percentages of juvenile caprines calculated from the proportion of unfused limb bone epiphyses. U= unfused; V= visible fusion line; F= fused. O/C includes the bones identified as sheep (OVA) and goat (CAH). Numbers of FV have been equally divided by the other two categories to calculate the percentage of juveniles.

		OVA			*CAH*			*O/C*		
Fusion		U	FV	F	U	FV	F	U	FV	F
Early	P RA	–	2	78	–	1	18	9	4	99
(0–1 ½ y)	D HU	13	6	58	1	3	17	17	11	78
	D SC	–	–	12	–	–	8	–	–	27
	PPH2	–	–	–	–	–	1	–	–	1
	PPH1	1	1	6	2	–	–	7	1	13
	% JUV		10.5%			10%			15%	
Medium	D TI	22	1	68	3	5	19	50	6	90
(1 ½ –2 ½ y)	DMTC	7	–	16	7	–	8	24	–	25
	DMTT	9	1	21	3	–	2	18	1	23
	% JUV		27%			33%			42%	
Late	CAL	4	1	12	3	1	–	7	2	12
(2 ½ –3 ½ y)	P FE	–	4	14	1	–	3	13	5	17
	DFE	1	–	5	–	1	2	19	4	14
	P UL	2		5	1	–	–	5	–	7
	D RA	9	–	22	6	–	4	33	–	26
	PTI	12	5	12	3	–	–	27	6	14
	P HU	11	3	2	2	1	–	15	4	4
	% JUV		37%			62,5%			55%	
% Juveniles			23%			30%			36.5%	
N			446			126			738	

Table 11.4. Age of slaughter of the caprines. Number and percentages of mandibles assigned to Payne's (1973) dental eruption and wear stages (MWS). O/C includes the identified sheep (OVA) and goat (CAH) mandibles.

			OVIS		*CAPRA*		*O/C*	
	MWS	*Age*	*N*	*%*	*N*	*%*	*N*	*%*
Immature	A	0–2m	1	1	–	–	1	<1
	B	2–6m	11	10	–	–	11	5
	C	6–12m	20	18	10	14	31	14
	Subtotal		32	**29**	10	**14**	43	**19**
Sub–adult	D	1–2y	11	10	11	15	30	13
	E	2–3y	27	24	17	24	57	26
	Subtotal		38	**34**	28	**39**	87	**39**
Adult	F	3–4y	13	12	9	12	25	11
	G	4–6y	22	20	17	24	49	22
	Subtotal		35	**32**	26	**36**	74	**33**
Senile	H	6–8y	5	5	7	10	14	6
	I	+8y	1	<1	1	1	4	2
	Subtotal		6	**5**	8	**11**	18	**8**
Total			111	100	72	100	222	100

and dung). The results for coastal Iron Age sites in north-western Iberia published by Fernández Rodríguez (2000, 579) conform to this trend, thus providing support for the observed pattern.

But, were flocks of sheep and goats managed in the same way? Exploring the data collected at A Lanzada may throw some light on this issue. Since both species have similar bone structure and are of analogous size, it may be assumed that any variation in their age-at-death would relate to other factors than the biases mentioned above. For example, the value placed on the products each of them provide, the physical characteristics of the animals and also the availability of grazing and feed (Payne 1973). At A Lanzada the greatest difference observed between sheep and goats relates to immature and senile individuals (Table 11.4). While the proportion of lambs killed before the first year of age doubles that of kids (29% *vs*. 14%), the opposite trend occurs with the very old animals. Senile goats were apparently kept in higher number than mature sheep (11% *vs*. 5%). At this point, one call for caution has to be made since rates of tooth wear might have been influenced by sheep and goat differences in feeding ecology, degree of hypsodonty and chewing mechanics

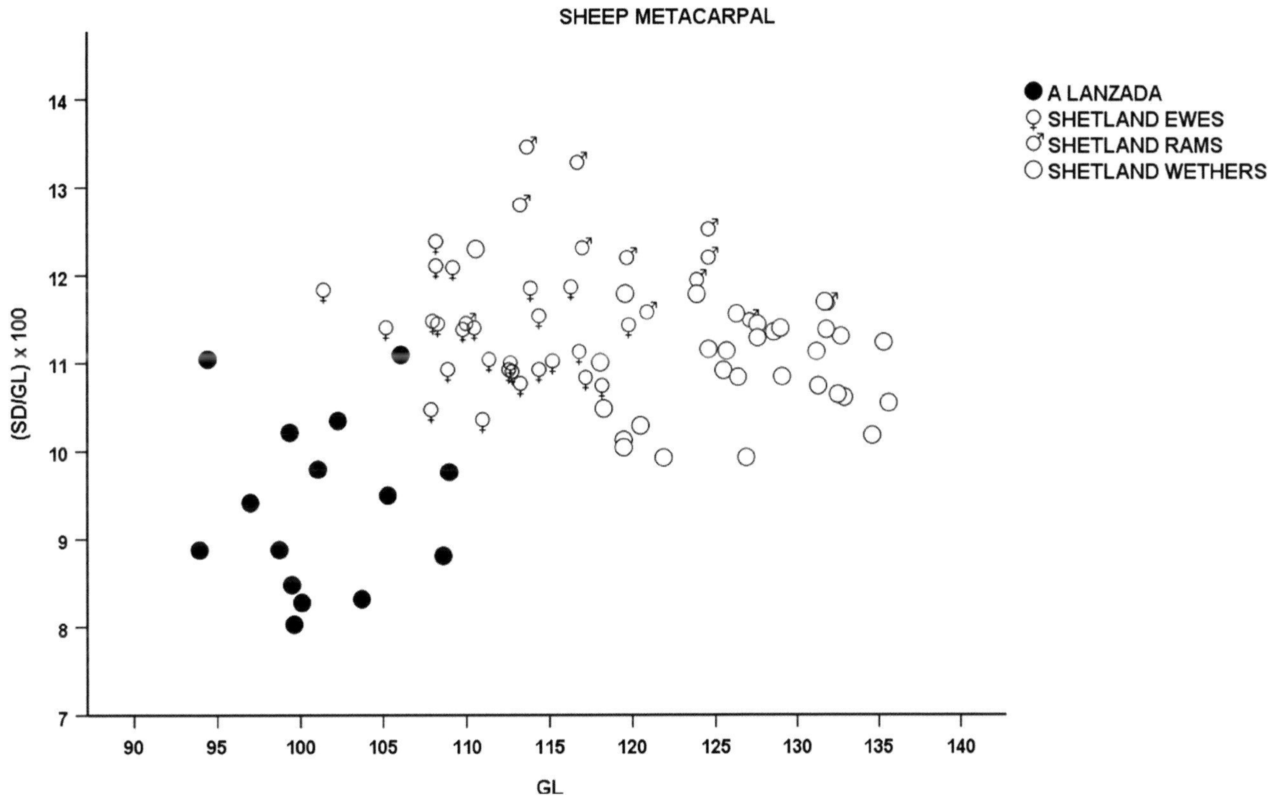

Figure 11.3. Sheep sex discrimination at A Lanzada. Plot of metacarpal length (GL) against a slenderness index of the shaft ((SD/GL) x 100). Modern unimproved Shetland ewes, rams and wethers (Davis 2000) are plotted together with the specimens from A Lanzada. All measurements are in mm.

(Zeder 2006, 98), causing sheep teeth to wear more slowly than goat teeth. As a result, it is plausible that more goat than sheep mandibles were assigned to the most advanced dental stages.

In all probability, a good number of the early slaughtered animals were the surplus males while the senile individuals would comprise mainly females. Taking into account that we are dealing with a subsistence economy, the success of livestock keeping laid on a balanced relationship between optimisation of production and herd security. The sheep to goat ratios revealed that goats represented only 25% of the animals in the flocks. From this perspective, it is likely that in order to minimise fluctuations in the goat population which may result in decreased annual yield from the flock (in particular meat), more kids and maybe old, but still reproductive, does were kept alive. Also, since goats are hardier than sheep and their feeding requirements are less, delaying the age-at-death of the goat flock most likely did not represent an expensive investment for the local community. The humerus metrical data suggest that 10 out of the 16 goat specimens represented in Fig. 11.2 are probably males. Bearing in mind that the distal fusion of the humerus occurs before the end of the first year of life these results demonstrate the regular survival of male kids beyond that age and consequently, the important role they played as meat suppliers. In view of their larger size, one carcass would provide more meat than a female goat, a ewe or even a ram.

Sex discrimination among the sheep bones is more difficult. The pubic bone of most pelvises was fragmented, preventing it to be measured. As an alternative, Fig. 11.3 shows the metacarpal length (GL) against the slenderness of the shaft ((SD/GL) × 100) for modern unimproved Shetland ewes, rams and wethers (Davis 2000) together with the specimens from A Lanzada. Davis (2000) noted that there was no clear-cut separation of the three sex groups in the Shetland sheep but at least they fall in different regions of the graph. While rams tend to be short and robust, ewes are short and slender and wethers show long and slender diaphyses. The clustering of most sheep metacarpals of A Lanzada in the inner low corner of the graph hypothetically indicates that ewes were more abundant than rams and that few wethers, if any, occurred. But perhaps the most outstanding result is the very small size of the Iberian Iron Age sheep compared to the unimproved Shetland breed, a characteristic that was observed during the identification process and which is another factor reinforcing the argument previously

12

Cattle from the East, cattle from the West: diversity of *Bos* morphotypes in the Iberian Peninsula during late prehistoric and Roman times

Ariadna Nieto-Espinet, Angela Trentacoste, Sílvia Guimarães and Silvia Valenzuela-Lamas

*This paper explores changes in cattle size in the North-East (NE) and South-West (SW) of the Iberian Peninsula from Chalcolithic to Roman times. Zooarchaeological data are contextualised with the ecological characteristics and settlement pattern of both areas. Results reveal significant differences in cattle size and morphology between the two areas, as well as changes through time. Integration of zooarchaeological, ecological and settlement data suggest that orography played a significant role both in the changes in settlement pattern and in the evolution of cattle morphology in the two regions. In this sense, the flatter orography and higher solar irradiation in the SW would have provided better ecological conditions for cattle husbandry. In addition, exchanges of animals were probably easier in the SW compared with the hillier orography of the NE. The Roman conquest had a different impact on cattle size as well: a clear size change was visible in the NE (where Iron Age cattle were very small), whereas size increase was more subtle in the SW. In both areas, cattle remained smaller compared to the large animals attested in North-Western Europe in Roman times (*e.g. *the Netherlands and France). These results provide a diachronic and wide territorial perspective to better assess what were the factors (ecological and/ or social) that contributed to the changes observed on cattle size at two opposite sides of the Iberian Peninsula.*

Keywords: *Bos taurus*, animal size, Chalcolithic, Iron Age, Romanisation

Introduction

Cattle have been crucial to animal production since their domestication approximately 10,000 years ago. Over the subsequent millennia, cattle size and morphology differed through time and geographically across Europe, reflecting variations and adaptations to environmental conditions and availability of resources (*e.g.*, Valenzuela-Lamas and Albarella 2017; Trentacoste *et al.* 2018). However, the underlying cause(s) of these developments in size and morphology remain unclear. Currently, a range of hypotheses coexist, ranging from desire for greater herd manageability, to responses to climate change, intensification, malnutrition, and the size of the socio-economic system (*e.g.* Clutton-Brock 1981; Davis 1981; Widdowson and Lister 1991; Manning *et al.* 2015; Valenzuela-Lamas and Albarella 2017).

The aim of this paper is to characterise cattle size and morphology in two distant areas of the Iberian Peninsula – the South-West (SW) and the North-East (NE) – across late prehistoric and Roman times, in order to better understand the developmental trajectories of cattle in these areas. Through evaluation of the height and robustness of cattle populations during late prehistory, we can more precisely assess the impact of the Roman conquest on cattle morphology at a regional level in these two parts of the Iberian Peninsula. Better understanding of regional patterns of biometric change, coupled with consideration of broader archaeological developments over the same period, sheds new light on the influence of ecological and socio-political factors on animal husbandry, especially in areas that exhibit greater continuity in the transition from the Iron Age to the Roman period. This is the first time that such comparison between the SW and the NE has been undertaken.

Materials and methods

Figure 12.1 shows the archaeological sites considered in the present analysis. Measurements derive from published data, grey literature produced before 2012, and the authors' own research (Table 12.1). Overall, the data derive from habitation/settlement sites, and any potential ritual deposits were excluded. Only measurements from well dated contexts were taken into account, so that the dataset for each site would be as representative as possible of animal husbandry production in that region and period of occupation.

The Iberian Peninsula is very diverse from a topographic and climatic perspective (Fig. 12.1). Nevertheless, the NE and the SW have some ecological similarities. Both have a similar amount of annual rainfall, with values between 300 mm and 700 mm per annum across the majority of the territory. The NE and SW also have a similar number of cloudy days per year – between 30 and 100 – and comparable potential vegetation cover, which broadly corresponds to thermo-mediterranean and meso-mediterranean bioclimatic classifications (Rivas-Martínez *et al.* 2002). The main ecological differences are that the SW has a much flatter topography, and solar irradiance (solar radiation per unit area) is much higher there compared to the NE (Aemet 2011).

From an archaeological perspective, the two areas experienced similar macro-processes during late prehistory, although differences exist in the chronology and the rhythm of changes, both between the two areas and within them (*e.g.* Ruiz Zapatero 2001; López-Cachero 2007; Albizuri 2011; Valente and Carvalho 2014; Valera 2015, for Chalcolithic and Bronze Age). Points in common between the SW and NE include the spread of metallurgy during the 3rd millennium BC, an Iron Age settlement pattern dominated by fortified settlements built using stone, mudbricks, rammed earth and vegetal matter (*e.g.* Arruda 1993; Cardoso 1996; Fabião 1998; Estrela 2013 and references in Table 12.1), an early conquest by Romans during Republican times (3rd–2nd century BC) (*e.g.* Gorges 1990; Fabião 1994; Olesti 1997; Bugalhão 1998; Asensio *et al.* 1998; Járrega 2000; Revilla 2004; Sanmartí 2004), and an increase in specialised production during Imperial times, in which both areas produced a surplus of specific items that were widely distributed across the Empire (fish sauces in the SW; cereal and wine in the NE) (*e.g.* Pena and Barreda 1997; Fabião 2008; Rodríguez-Martín 2011; Carreras and González 2012; Berni and Miró 2013). Differences between the two areas are found in the different material culture and early appearance of fortified settlements in the SW, which are documented from the 3rd millennium BC; fortified settlements are

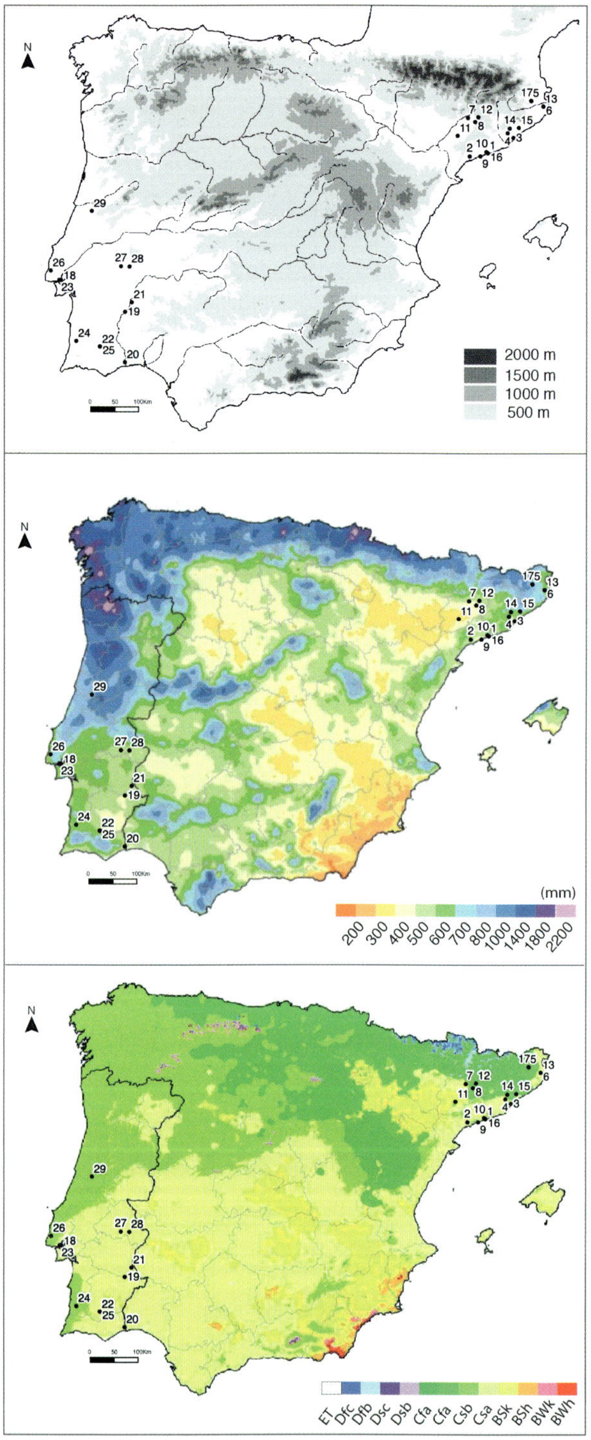

Figure 12.1. Location of the sites analysed, with the topography (top), mean annual rainfall (middle) and Köppen-Geiger Climate Classification (bottom) for the Iberian Peninsula. Numbers refer to sites as follows: 1 Alorda Park; 2 Antigons; 3 Baetulo; 4 Can Roqueta; 5 Draga; 6 Gou Batlle; 7 Hereuet; 8 Iesso; 9 La Llosa; 10 Mas d'en Gual; 11 Minferri; 12 Puig Castellar; 13 Puig de Sant Andreu; 14 Torre Roja; 15 Turó del Vent; 16 Vilarenc; 17 Vilauba; 18 Casa do Governador; 19 Castro dos Ratinhos; 20 Castro Marim; 21 Mercador; 22 Mesas do Castelinho; 23 NARC-BCP; 24 Odemira; 25 Santarem; 26 São Miguel de Odrinhas; 27 São Pedro da Fronteira; 28 Torre de Palma; 29 Zambujal.

Table 12.1. Sites, location area and chronologies considered in the present study.

Site	*Area*	*Chronology considered*	*Site type*	*Bibliographical references*
Zambujal	SW	Copper–Early Bronze Age	fortification	Kunst *et al.* 2002; Uerpmann and Uerpmann 2003
Castro dos Ratinhos	SW	Copper Age	fortification	Liesau and García-García 2010
Mercador	SW	Copper Age	fortification	Valera 2001
Castro Marim	SW	Iron	fortification, rural settl.	Arruda *et al.* 2006; Davis 2007
Mesas do Castelinho	SW	Iron Age and Roman	fortification, rural settl.	Fabião and Guerra 1994; Valenzuela-Lamas and Fabião 2012
Odemira	SW	Iron Age and Roman	fortification, rural settl.	Vilhena and Rodrigues 2009; Davis and Vilhena 2017
Santarem	SW	Iron Age and Roman	fortification, urban settl.	Arruda and Viegas 2002; Davis 2006
Torre de Palma	SW	Late Roman	villa	Heleno 1962; Maloney and Hale 1996; MacKinnon 1999–2000
Casa do Governador	SW	Roman	factory	Filipe and Fabião 2007; Valenzuela-Lamas 2014
NARC-BCP	SW	Roman	factory	Bugalhão 2001; Valenzuela-Lamas 2014
São Miguel de Odrinhas	SW	Roman	villa, well	Davis and Gonçalves 2017
São Pedro da Fronteira	SW	Roman	settlement	Carneiro and Sepúlveda 2004; Davis 2005
Can Roqueta	NE	Bronze	rural settlement, silos	Carlús *et al.* 2002; Albizuri 2011
Minferri	NE	Bronze	rural settlement, silos	Minferri 1997; Nieto-Espinet *et al.* 2014
Alorda Park	NE	Iron	fortification	Sanmartí and Santacana 1992; Valenzuela-Lamas 2008
Gou Batlle	NE	Iron	peri-urban neighbourhood	Martín *et al.* 2007; Valenzuela-Lamas *et al.* 2017
Mas d'en Gual	NE	Iron	rural settlement	Asensio *et al.* 2005; Valenzuela-Lamas *et al.* 2010
Puig de Sant Andreu	NE	Iron	fortification	Martín *et al.* 2005; Valenzuela-Lamas unpub.
Turó del Vent	NE	Iron	fortification	Rovira and Sanmartí 1982; Oliva 1998
Hereuet	NE	Iron Age and Roman	rural settlement	Garcia 2007; Valenzuela-Lamas 2013
Torre Roja	NE	Iron Age and Roman	fortification, rural settl.	Fortó *et al.* 2004; Valenzuela-Lamas 2011
Antigons	NE	Roman	villa	Járrega & Prevosti 2010; Valenzuela-Lamas 2010
Baetulo	NE	Roman	urban	Guitart and Padrós 1991; Colominas 2009
Iesso	NE	Roman	urban	Guitart and Pera 1994; Padrós unpubl.
La Llosa	NE	Roman	villa	Macias and Ramon 1993; Padrós 2010
Puig Castellar (Biosca)	NE	Roman		Pera *et al.* 2016; Padrós unpubl.
Vilarenc	NE	Roman	villa	Pou and Revilla 1996; Valenzuela-Lamas 2012
Vilauba	NE	Roman	villa	Castanyer and Tremoleda 1999; Colominas *et al.* 2017

not documented in the NE before the 1st millennium BC (Pellicer 1989; Sanmartí and Santacana 1994; Pons 1994; Chapman 1995; López-Cachero 2007).

In all cases, we used the chronology provided by the excavators of each site. Due to the limited data available for Chalcolithic settlements in the NE and Bronze Age contexts in the SW, we grouped these two periods into the same phase for analysis. In order to organise the data and analyse long-term changes, we used the following chronological phasing:

- Chalcolithic–Bronze Age: 3000–750 BC. This phase includes the Chalcolithic through Early Bronze Age (3000–2300 BC), Middle Bronze Age (2300–1200 BC) and Late Bronze Age (1200–750 BC).
- Iron Age: 750–200 BC. This phase includes the Early Iron Age (700–550 BC), Middle Iron Age (550–300 BC) and Late Iron Age (300–200 BC).
- Roman Republican: 200–27 BC.
- Roman Empire: 27 BC–AD 476. We considered two phases: Early Roman Empire (27 BC–270 AD) and Late Roman Empire (AD 270–476).

In order to assess the long-term dynamics of size variation, we included measurements from Mesolithic aurochs and Neolithic cattle (Saña 1998; Davis 2002) where available as, despite their low numbers, they show that a notable cattle size decrease occurred both in the SW and the NE through late prehistory, as attested in many other areas in Europe (*e.g.* Matolcsi 1970; Bökönyi 1974; Altuna 1980; Ijzereef 1981; Méniel 1984; Vigne 1988).

Changes in animal size were assessed using Log Standard Index values (LSI) and withers heights. LSI values were calculated using osteometric data from Nieto-Espinet (2018) as a baseline. Log ratios allow measurements from different elements to be pooled and thus provide a useful method for investigating animal size and changes through time. In order to avoid overrepresentation of articulating remains, our LSI analyses included only one measurement per bone, and only one bone from an articulating group of elements, following the method described in Trentacoste *et al.* (2018). Bone widths were considered separately from lengths in order to assess two-dimensional change. Bone depth measurements were scarce and were not considered here. Wither heights were calculated from metacarpals and metatarsals following the standard indices detailed in von den Driesch and Boessneck (1974) (6.15 for metacarpals and 5.45 for metatarsals). Differences between the populations, both in time and space, were statistically checked using the Wilcoxon test (Wilcoxon 1945) computed in R (R Core Team 2018).

In order to elucidate the potential impact of changing sex ratios (*e.g.* spread of castrates, or predominance of bulls) on cattle size, metacarpal shape (SD/GL *versus* Bd/GL indices) was compared with individuals of known sex: 11 individuals from Eketorp ringfort (Sweden, phase II: 5th–8th century AD and phase III: 12th–13th century AD; Telldahl *et al.* 2012) and 16 individuals from Late medieval urban contexts from Beja (Portugal, 15th *c.* AD; Davis *et al.* 2012).

Results

Cattle from the North-East

Figure 12.2 shows LSI results from cattle length measurements for the sites located in present-day Catalonia. During the

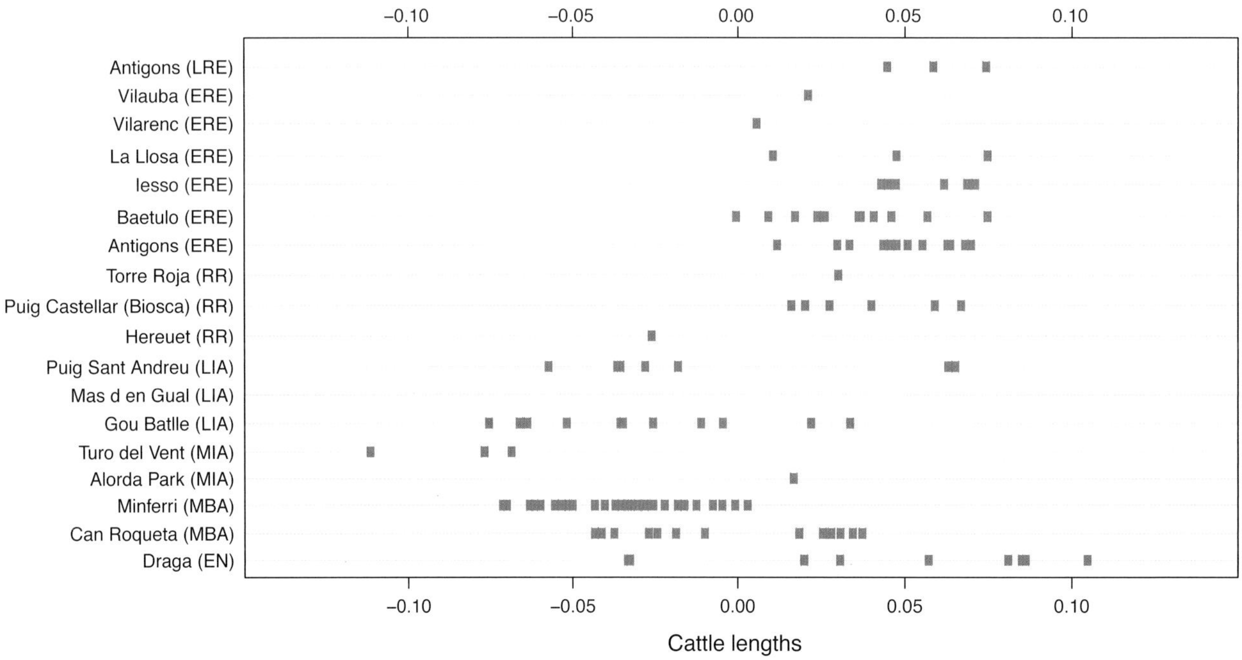

Figure 12.2. LSI values obtained from length measurements (GL and HTC). EN = Early Neolithic; MBA = Middle Bronze Age; MIA = Middle Iron Age; LIA = Late Iron Age; RR = Roman Republic; ERE = Early Roman Empire; LRE = Late Roman Empire.

Table 12.2. Results of the Wilcoxon test on cattle length and width LSI values from the NE in the different chronological periods.

	Lengths		*Widths*	
	p value	*code*	*p value*	*code*
Neolithic vs. Chalcolithic	0.0000	***	0.0000	***
Chalcolithic vs. Iron Age	0.4431	N	0.003357	**
Iron Age vs. Roman Rep.	0.01906	*	0.09247	N
Roman Rep. vs. Roman Emp.	0.01596	*	0.0003934	***

Table 12.3. Withers height calculations for cattle metacarpals and metatarsals from the NE.

Element	*Present-day Catalonia (NE-Ib. Peninsula)*	*Min*	*Max*	*Average*	*n*
Metacarpal	Middle Bronze Age (2100–1300 BC)	98.6	118.7	107.1	4
	Late Iron Age (3 BC)	100.1	126.5	109.8	6
	Roman Republican (2 BC)	102.6	116.0	110.5	3
	Early Roman (AD 1–3)	111.2	129.3	120.9	20
	Late Roman (AD 4–5)	120.7	129.2	124.8	3
	Late Antiquity (AD 5–7)	108.2	121.7	115.9	8
Metatarsal	Middle Bronze Age (2100–1300 BC)	96.7	114.7	106.9	7
	Mid Iron Age (5–4 BC)	104.8	104.8	104.8	1
	Late Iron Age (3 BC)	95.7	123.1	106.4	4
	Early Roman (AD 1–3)	113.8	127.1	121.2	8
	Late Antiquity (AD 5–7)	106.7	113.2	110.4	6

Bronze Age, cattle from the inland settlement of Minferri are significantly smaller than those at Can Roqueta, a site *c.* 25 km distant from the coast (Nieto-Espinet *et al.* 2014). When considered together, Bronze Age values were found not to be statistically different from Iron Age results (Table 12.2). Cattle size significantly increased in Roman times. This is observable immediately after the conquest (2nd century BC) during the Republican period, even where the site was occupied in pre-Roman times (*e.g.* Torre Roja). The largest cattle date to the Imperial period, including those from the Late Roman Empire. Interestingly, the palatial building at the urban site of Puig de Sant Andreu also produced some large cattle dated to the 3rd century BC, while contemporary cattle from a neighbouring extramural area, named 'Gou Batlle', were of smaller size.

Table 12.3 displays the results of the wither heights calculations for cattle metacarpals and metatarsals. The results from metacarpals suggest a constant height increase from the Middle Bronze Age to Republican times, followed by a sudden increase during the Imperial period, which was sustained until the late Imperial period (4th–5th century AD). Metatarsals show more continuity during late prehistory, and also reflect the sudden height increase in Imperial times.

Width values did not change as dramatically (Fig. 12.3), but the size change was significant between all periods with the exception of the Iron Age–Roman Republican transition (Table 12.2). An increase was observable in Roman times, although some individuals remained slender, even during the Imperial period. Again, the largest values correspond to cattle dated to Imperial times, including Late Roman individuals (4th century AD). Among Iron Age assemblages, the individuals from Late Iron Age levels of the palatial building at Puig de Sant Andreu (3rd century BC) have the largest Iron Age values, which fall within the range of Roman results. Again, this contrasts with the contemporary cattle from the extramural quarter of Gou Batlle.

In investigation of metacarpal shape (Fig. 12.4), Bronze Age specimens were comparable in slenderness to female cattle of known sex, although with relatively more robust distal breadths. Iron Age cattle from present-day Catalonia fall within the range of male comparative specimens, while Roman metacarpals occupy a larger range that spans male and female groups in the comparative data.

Cattle from the South-West

Figure 12.5 shows the results from cattle length measurements for the sites located in present-day central and southern Portugal. Overall, there was progressive increase in LSI values through time, although – as previous work has pointed out – cattle size seems to be slightly smaller in the Roman period compared to the Iron Age at some sites with long-term occupation (*e.g.* Santarém, Davis 2006), or change seems not to occurred at all (*e.g.* Castro Marim, Davis 2002). Conversely, at other settlements (*e.g.* Mesas do Castelinho) cattle LSI values from Republican times are notably higher than the Iron Age values from the site (Valenzuela-Lamas and Fabião 2012). The Wilcoxon test confirms that, when considering the sites per chronological phase, differences are significant in each chronological step from the Chalcolithic Age to Roman period (Table 12.4).

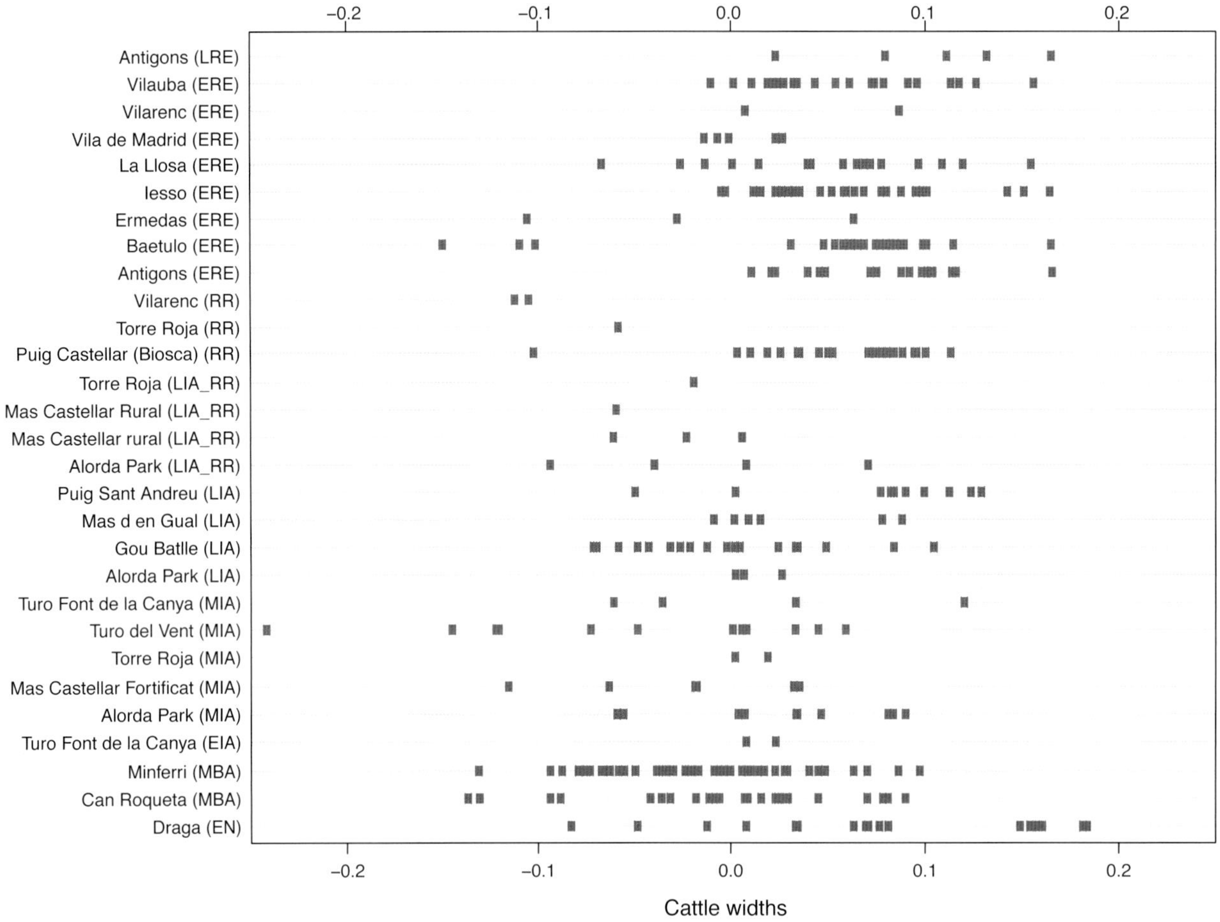

Figure 12.3. LSI values obtained from width measurements (Bd, BT and SD). MBA = Middle Bronze Age; EIA = Early Iron Age; MIA = Middle Iron Age; LIA = Late Iron Age; RR = Roman Republic; ERE = Early Roman Empire; LRE = Late Roman Empire.

Cattle widths (Fig. 12.6) display a different tendency, and values are higher at Iron Age Santarem than at Chalcolithic Zambujal and Mercador. Data for other sites are too scarce to draw conclusions. In Roman times, the cattle recovered from sites close to urban Olisippo (*e.g.* assemblages from NARC-BCP and Casa do Governador in present-day Lisbon) have values comparable with Iron Age ones. Consequently, the data suggest that the Roman conquest did not produce a significant change in cattle morphology in this area. The Wilcoxon test (Table 12.4) confirms the significant width increase between the Chalcolithic and the Iron Age, and suggests continuity in Roman times.

Table 12.5 shows the results of the withers heights calculations in the different time periods considered. Based on metacarpals, a notable height increase occurred between the Chalcolithic to the Iron Age, and metatarsals confirm this result. Withers height then display a slight increase in Republican times followed by a slight decrease in the Imperial period. More data, especially from sites with no Iron Age occupation, are necessary to confirm whether this was a trend across all sites or, rather, a feature typical of indigenous sites that preserved their livestock into Roman times.

Comparison of cattle morphology in the SW and the NE of the Iberian Peninsula

Based on LSI values, cattle from the SW were notably larger than cattle from the NE during both the Chalcolithic and the Iron Ages, and SW cattle did not change significantly during the Roman period (Fig. 12.7). Conversely, cattle from the NE were smaller in size than those from the SW from Chalcolithic to Republican times, and only display higher values during the Roman Imperial period, both in lengths and widths. In both cattle populations, lengths and widths appear to be correlated, although a closer look suggests that some variation in cattle morphology may have occurred: Iron Age cattle in the NE were shorter in length but had larger width values compared to Bronze Age cattle, and Republican cattle in the SW were significantly larger in length but with lower median width values compared to Iron Age cattle in the

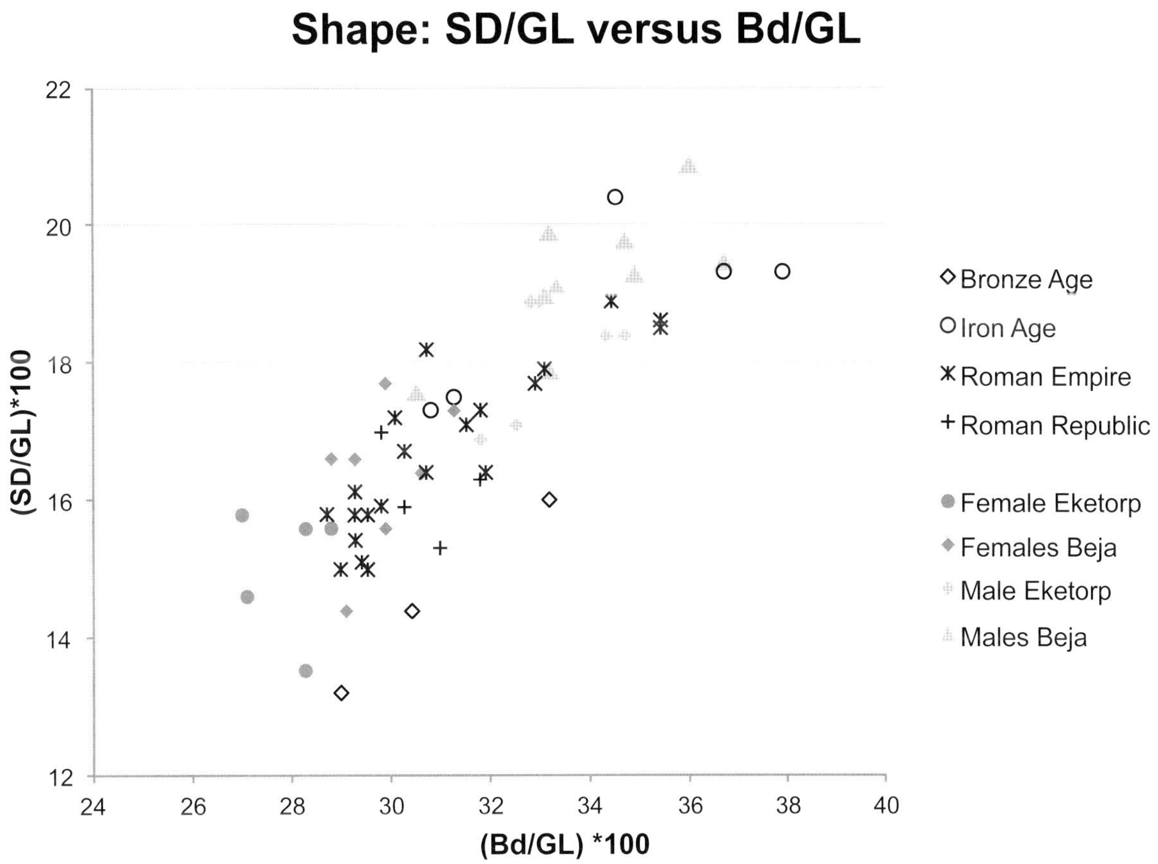

Figure 12.4. Metacarpal ratios (SD/GL and Bd/GL) from cattle from the NE and cattle of known sex from Beja (15th c. *AD, Portugal) and Eketorp ringfort (Sweden, 12th–13th* c. *AD).*

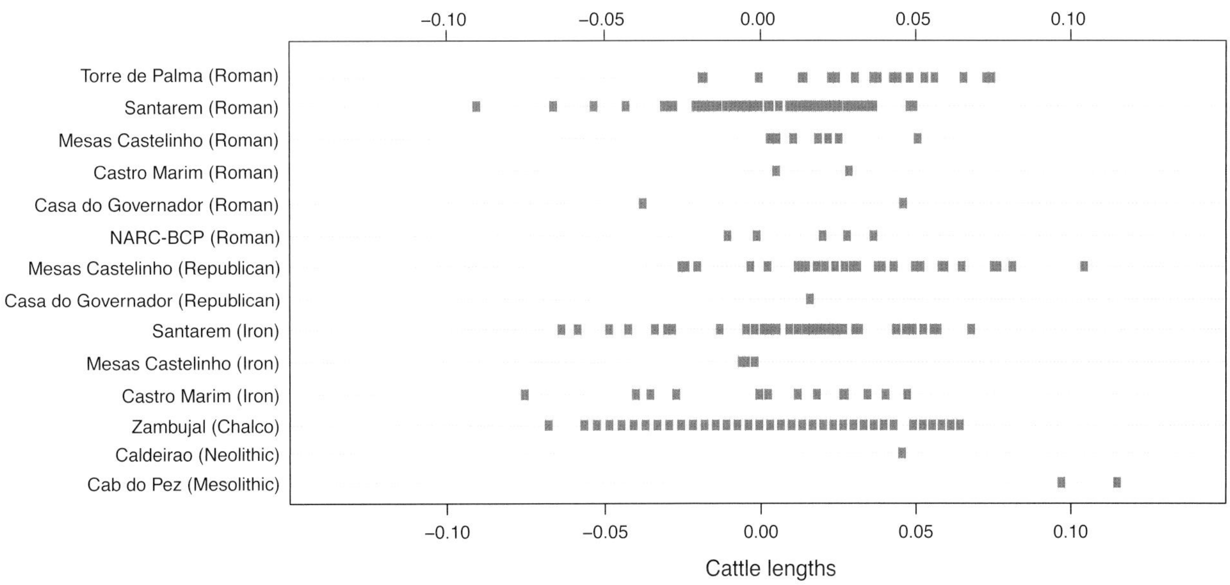

Figure 12.5. LSI values obtained from length measurements (GL and HTC).

Table 12.4. Results of the Wilcoxon test on cattle length and width LSI values from the SW in the different chronological periods.

	Lengths		*Widths*	
SW	*p value*	*code*	*p value*	*code*
Neolithic vs. Chalcolithic	0.1381	N	0.1483	N
Chalcolithic vs. Iron Age	0.04559	*	0.0000	***
Iron Age vs. Roman Rep.	0.00241	*	0.9735	N
Roman Rep. vs. Roman Emp.	0.001622	*	0.994	N

area. Consequently, some variation in cattle shape may have happened, but more data are needed to see the extent of these changes on a more precise geographical and temporal scale.

Discussion

Since their domestication, cattle have been crucial to animal husbandry and agricultural production. Cattle provided the power in traction, as well as meat and other products – a value which was reflected in their symbolic value. Our results demonstrate different patterns of cattle size and shape change in two regions of the Iberian Peninsula, with implications for animal husbandry strategies, the introduction of new morphotypes, and socio-economic changes in these areas. In the NE, cattle increased significantly in size during Roman times, as observed in previous works (*e.g.* Colominas *et al.* 2014; Colominas *et al.* 2017). A similar change is attested in many other areas of the Empire (*e.g.* Teichert 1984; Lauwerier 1988; Audoin-Rouzeau 1991; Lepetz 1996; Peters 1998; Breuer *et al.* 1999; Forest and Rodet-Belarbi 2002; Frémondeau *et al.* 2017; Groot 2017; Pigière 2017). Results from palatial buildings at the urban site of Puig de Sant Andreu suggest that large animals were also available to restricted groups during the 3rd century BC; contemporary cattle from Gou Batlle, a neighbouring extramural area, were of smaller size. Chemical analyses are needed to elucidate whether local animals reached these large sizes in the pre-Roman period, or whether the two specimens from Puig de Sant Andreu reflect the importation of animals from elsewhere.

In contrast to the NE, LSI values from the SW display greater continuity. Some sites with long-term occupation (*e.g.* Santarem, Davis 2006) produced LSI length values where cattle size seems to be slightly smaller in the Roman period compared to the Iron Age. Consequently, the data suggest that the Roman conquest did not produce a significant change in cattle morphology in this area. Cattle in the SW were also notably larger than those in the NE during later prehistory and the Republican period, until an increase in the size of NE cattle during Imperial times. It is difficult to imagine what these cattle looked like based exclusively on bone measurements, but the results displayed in Fig. 12.7 demonstrate that cattle populations were of different size and morphology in the SW and NE. More importantly, cattle morphology followed different trajectories in later prehistory, and the Roman conquest had a different impact in these two regions.

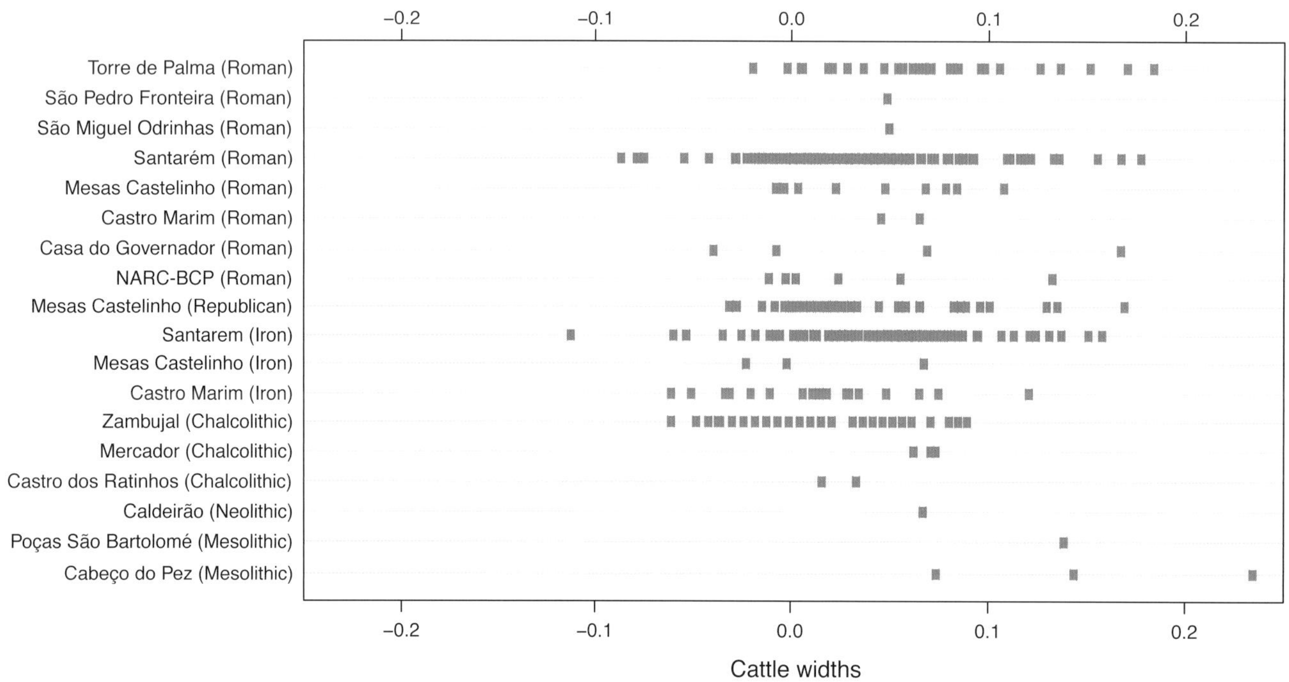

Figure 12.6. LSI values obtained from width measurements (Bd, BT and SD).

Table 12.5. Withers height calculation on cattle metacarpals and metatarsals from the SW.

Element	*SW (Southern Portugal)*	*Min*	*Max*	*Average*	*Difference cm (max–min)*	*n*
Metacarpals	Chalcolithic–Bronze (3000–700 BC)	104.7	104.7	104.7	0.0	1
	Iron Age (700–200 BC)	112.6	122.9	117.7	10.3	10
	Roman Republican (200–27 BC)	122.6	122.6	122.6	0.0	1
	Roman (27 BC– AD 575)	109.5	123.9	116.1	14.4	11
Metatarsals	Chalcolithic–Bronze (3000–700 BC)	107.9	118.9	113.4	11.0	2
	Iron Age (700–200 BC)	114.6	127.5	119.7	12.9	5
	Roman Republican (200–27 BC)	114.5	119.5	117.0	5.0	2
	Roman (27 BC–AD 575)	108.6	124.3	116.5	15.7	14

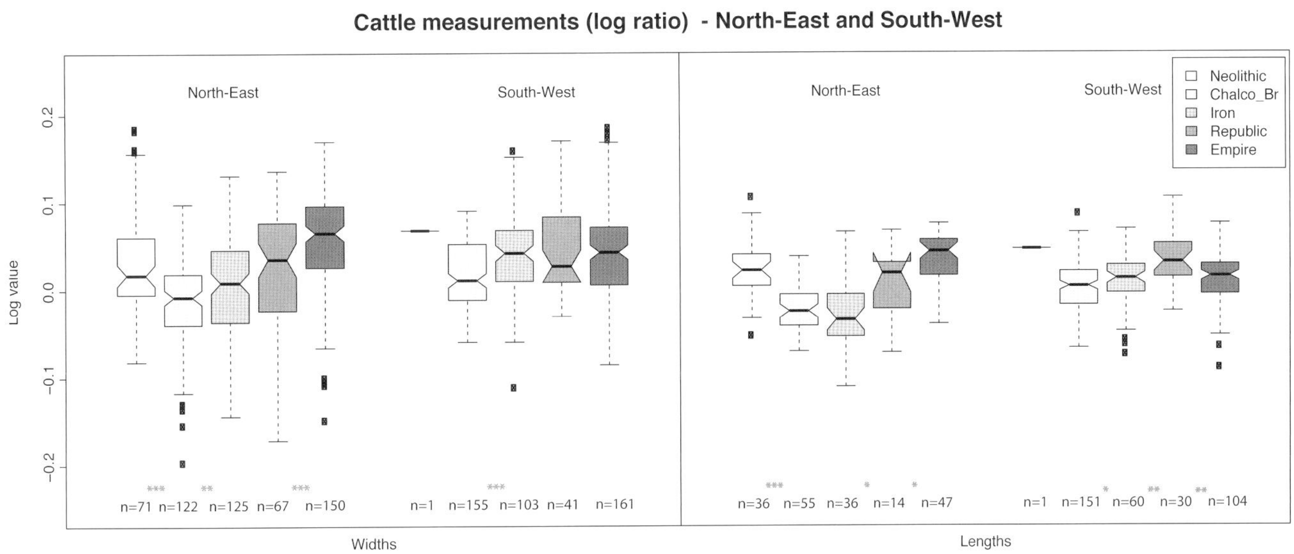

Figure 12.7. Comparison of cattle LSI values from the NE and the SW. The notch indicates two times the standard variation of the median.

What were the factors that influenced cattle morphology and induced different trends through time? Was there a major ecological determinant? Or do our results primarily demonstrate human decisions? Cattle production is favoured in flat biotopes where water and herbaceous fodder is abundant (Pérez-Ripoll 1999). The two areas of the Iberian Peninsula analysed here have a similar amount of annual rainfall and potential vegetation cover, as well as a similar number of cloudy days (Aemet 2011). The main ecological differences are that the orography of the SW is significantly flatter and solar irradiance is much higher there compared to the NE. Environmental factors, and notably heat stress, are key determinants in cattle production (*e.g.* Roman-Ponce *et al.* 1978; Hafez 2000; De Rensis and Scaramuzzi 2003; García-Ispierto *et al.* 2007). Factors contributing to the degree of heat stress experienced by animals include environmental temperature, radiant energy, relative humidity and wind speed (De Rensis and Scaramuzzi 2003). Although it is difficult to explain the differences on cattle morphology based on these ecological differences alone, higher solar irradiance, and hence greater potential for heat stress, may have contributed to the stability of cattle size in SW. For instance, higher solar irradiance may have encouraged continued exploitation of local heat-resistant morphotypes (as compared to imported varieties), or influenced the size of new morphotypes, through the negative impact of heat stress on fertility and development.

Some previous works on animal husbandry in Western Europe have focused on the role of socio-cultural and economic factors, suggesting that social hierarchisation and, later, Roman conquest, significantly influenced animal husbandry and especially cattle morphology (Valenzuela-Lamas and Albarella 2017 and references within; Trentacoste *et al.* 2018). On one hand, our results challenge this interpretation, as, from an archaeological point of view, both areas underwent similar macro-processes of socio-economic development in post-Neolithic times (*i.e.* trends toward social hierarchisation and territoriality – especially visible through the appearance of fortified sites – followed by Roman conquest). However, on the other hand, these social processes – which also occurred at a European scale – had different local manifestations in the two regions considered here. Metallurgy began in both areas during the 3rd millennium BC. During this period, several large fortified settlements, protected by complex defensive structures (*e.g.* Perdigões and Castro de Chibanes), were constructed in the SW (Tavares da Silva and Soares 2014; Valera 2015). In contrast, no permanent settlements can be found in the NE during the 3rd millennium BC, although local metalwork is attested from 2400 BC (Martín 2003). Significant population consolidation did not occur in the NE until the 1st millennium BC, when stone-built fortified settlements were erected on hilltops (*e.g.* López-Cachero 2007).

Therefore, the introduction of metallurgy had a different impact on the settlement patterns in the two areas, leading to an early concentration of population and fortification in the SW, while significant population consolidation in the NE did not occur before the end of the Bronze Age. The Roman conquest also had distinct regional impacts on settlement patterns, as most fortified settlements were abandoned in the NE after conquest (*e.g.* Alorda Park, Puig de Sant Andreu, Mas Castellar, Turó del Vent), whereas a number of them were still inhabited in Roman times in the SW (*e.g.* Castro Marim, Mesas do Castelinho, Santarém). Thus, like animal husbandry, settlement patterns exhibited greater continuity from late prehistory into Roman times in the SW of the peninsula compared to the NE.

Landscape morphology may also provide clues to understanding differences between the two cattle populations, and also perhaps the different rhythms seen in settlement patterns. The SW is significantly flatter than the NE (see Fig. 12.1). This topography facilitates movement, even in phases of significant social territoriality. Consequently, the rise of social complexity and territoriality in the SW may have entailed a reduction of the degree of mobility of animal populations, without precluding mixing and genetic flow between different cattle flocks in the area. The single Chalcolithic cattle sample in Žalaitė *et al.*'s (2018) preliminary investigation of strontium isotopes in Portuguese fauna corresponded with a local signature, while to the Neolithic samples were more diverse; further similar studies will better elucidate trends. Conversely, increased territoriality occurring from the 2nd and, especially, the 1st millennium BC in the NE may have increased the difficulty to move across a landscape where the topography limits potential routes and complicates movement over long distances. This may explain the reduced livestock mobility observed in the NE during the Iron Age based on strontium isotopes (Valenzuela-Lamas *et al.* 2016; Valenzuela-Lamas *et al.* 2018) and may have entailed a further reduction in the genetic flow of cattle populations, thus favouring a significant reduction of cattle size in the NE. Mobility probably expanded again with the greater connectivity facilitated by Roman conquest, for instance as demonstrated in strontium isotopic data from southern England and southern France (Minniti *et al.* 2014; Nieto *et al.* 2020).

Another aspect regarding the continuity of cattle size in the SW is that livestock at Roman sites with Iron Age occupation did not develop morphologically as much as animals in newly built Roman settlements, probably because indigenous populations preferred to keep their local morphotypes for economic and/or cultural reasons (*e.g.* Fernández Rodríguez 2003; Colominas *et al.* 2017). In this respect, and as mentioned earlier, a number of the published osteometric data from the SW come from sites previously inhabited during the Iron Age (*e.g.* Alcáçova de Santarém, Monte Molião, Castro Marim, Mesas do Castelinho). Secondly, the change in cattle size is very apparent in the NE because there was a significant size reduction during the 1st millennium BC and then a sudden increase in Roman times. Conversely, no size reduction is attested in the SW during the Iron Age, and cattle were already as big as Roman ones (Tables 12.2 and 12.3, Fig. 12.7).

Interestingly, cattle size both in the SW and the NE was similar in Roman times, but measurements do not seem to reach the values attested in France and the Netherlands (areas where flat and wet areas with abundant pasture probably existed) during the same period (Forest and Rodet Belarbi 2002; Duval *et al.* 2012; Lauwerier 1988). In these two areas, the mean withers height values obtained on metatarsals (calculated using the same indices as above) are around 131 cm, and the tallest animals ones reach 141 cm, to a maximum of 163 cm in north-western France (Duval *et al.* 2012). With the current data available from the Iberian Peninsula, it seems that only a few specimens from newly created settlements in the NW reached withers heights above 140 cm (see Colominas *et al.* 2017, fig. 3), but more data from newly created settlements in the SW are needed to evaluate the extent of these findings.

Conclusion

The present study provides a synthesis of cattle measurements from the South-West and the North-East of the Iberian Peninsula, and discusses the impact that regional ecology and socio-economic changes may have had on animal husbandry in these two areas during late prehistory and the Roman period. The study reveals that cattle size was more stable in the SW compared to the NE, where Roman conquest had a clear impact on cattle biometry. Factors accounting for different trajectories of cattle populations from the SW may include the region's flat topography and continuity in the settlement pattern over the period considered. Probably – for reasons that still require further investigation – pressure over the resources in the SW may have been less compelling compared to other areas. On the contrary, the NE follows a similar trend as the observed across Western Europe in response to the Roman conquest. The appearance of similar patterns in diverse environmental situations across Western Europe, suggests that territoriality during late prehistory and greater economic integration during the Late Iron Age and Roman times influenced animal husbandry production as much as ecology. Nevertheless, this study also reveals that ecologically similar areas can follow different trajectories even when subject to similar macro socio-political processes. These conclusions demonstrate the need to assess individual regions (and sub-regions) at different geographical and temporal scales, especially where we aim to understand factors driving biometric change.

Acknowledgements

This work first began as part of the FCT postdoctoral fellowship SFRH/BPD/46583/2008 (SVL) and could not be done without the guidance and help from S.J.M. Davis and C. Fabião, who provided data, access to materials and scientific discussion, and C. Detry who helped enormously with bibliographical references and invaluable friendship. The work has been subsequently expanded as part of the ERC-StG project ZooMWest (ERC-StG 716298), and is still in progress. This work presents the current state of our research and provides the first comparison of cattle morphology at these two extremes of the Iberian Peninsula.

References

Aemet (2011) *Atlas climático ibérico*, Agencia Estatal de Meteorología (Aemet).

Albizuri, S. (2011) *La ofrenda animal durante el Bronce Inicial en Can Roqueta II (Sabadell, Vallès Occidental). Arqueozoología del ritual funerario*. Unpublished PhD Thesis, Univrersity of Girona.

Altuna, J. (1980) Historia de la domesticación en el País Vasco desde sus orígenes hasta la romanización, *Munibe* 32, 9–163.

Arruda, A.M. and Viegas, C. (2002) A Alcáçova. In *De Scallabis a Santarém*. 73–81. Lisboa, Museu Nacional de Arqueologia.

Arruda, A.M., Viegas, C., Bargão, P. and Pereira, R. (2006) A importação de preparados de peixe em Castro Marim: da Idade do Ferro à Época Romana. *Setúbal Arqueológica 13*, 153–176.

Asensio, D., Belarte, M.C., Sanmartí, J. and Santacana, J. (1998) Paisatges ibèrics: Tipus d'assentaments i formes d'ocupació del territori a la costa central de Catalunya durant el període ibèric ple. *Saguntum* Extra 1, 373–385.

Asensio, D., Jornet, R., López, D. and Morer, J. (2005) La troballa d'una màscara grotesca de terra cuita en el jaciment ibèric de Mas d'en Gual (el Vendrell, Baix Penedès), *Fonaments* 12, 223–233.

Arruda, A. (1993) A Idade do Ferro no Centro/Sul. *História de Portugal* 2, 45–68.

Audoin-Rouzeau, F. (1991) La taille du boeuf domestique en Europe de l'Antiquité aux temps modernes, *Fiches d'ostéologie animale pour l'archéologie*, Série B, Mammifères, 2, 40.

Berni Millet, P. and Miró Canals, J. (2013) Dinámica socioeconómica en la Tarraconense Oriental a finales de la República y comienzos del Imperio. El comercio del vino a través de la epigrafía anfórica. In J. López Vilar (ed.) *1er Congrés Internacional d'Arqueologia i Món Antic: Govern i Societat a la Hispània Romana: Novetats epigràfiques: Homenatge a Géza Alföldy*, 63–83. Tarragona.

Bökönyi, S. (1974) *History of Domestic Animals in Central and Eastern Europe*. Budapest, Akadémiai Kiadó.

Breuer, G., Rehazek, A. and Stopp, B. (1999) Grossenveränderungen des Hausrindes. Osteometrische undersuchungen grosser fundserien aus der Nordschweiz von der Spätlatènezeit bis ins Frühmittelalter am biespiel von Basel, Augst (Augusta Raurica) und Schleitheim-Brüel. *Jahtesberichte aus August und Kaiseraugst* 20, 207–228.

Bugalhão, J. (1998) O povoamento rural romano no Alentejo: contribuição da arqueologia preventiva. *Revista portuguesa de Arqueologia* 1(2), 123–136.

Bugalhão, J. (2001) *A indústria romana de transformação e conserva de peixe em Olisipo: Núcleo Arqueológico da Rua dos Correeiros*. Lisboa, Instituto Português de Arqueologia.

Cardoso, J.L. (1996) O povoamento no Bronze Final e na Idade do Ferro na região de Lisboa. In J. de Alarcão (ed.) *De Ulisses a Viriato: O primeiro milénio aC*, 73–81. Lisboa, Ministério de Estado da Cultura.

Carlús, X., Lara, C., López, J., Oliva, M., Palomo, A., Rodríguez, A., Terrats, N. and Villena, N. (2002) El paraje arqueológico de Can Roqueta (Sabadell, Vallès Occidental): Diacronía y tipología de las ocupaciones. *Bolskan* 19, 121–139.

Carneiro, A. and Sepúlveda, E. (2004) Terra sigillata hispânica tardia do concelho de Fronteira: exemplares recolhidos entre 1999 e 2003. Lisboa, *Revista Portuguesa de Arqueologia* 7(2), 435–458.

Carreras, C. and González, H. (2012) Ánforas tarraconenses para el limes germano: una nueva visión de las Oberaden 74. In D. Bernal and A. Ribera (eds) *Cerámicas hispanorromanas II: producciones regionales*, 207–230. Monografías. Historia y Arte, Universidad de Cádiz.

Castanyer, P. and Tremoleda, J. (1999) *La vil.la romana de Vilauba. Un exemple de l'ocupació i explotació romana del territori a la comarca del Pla de l'Estany.* Girona, Ajuntament de Banyoles.

Chapman, R. (1995) Urbanism in Copper and Bronze Age Iberia? In B. Cunliffe and S. Keay (eds) *Social Complexity and the Development of Towns in Iberia*, 29–46. Proceedings of the British Academy 86, Oxford, Oxford University Press.

Clutton-Brock, J. (1981) *Domesticated Animals from Early Times*. London, British Museum Natural History.

Colominas, L. (2009) *La gestió dels animals al nord-est de la península ibèrica entre el segle v ane. I el segle v dne. Proposta metodològica d'integració de les anàlisis arqueozoològiques als estudis de cronologies històriques*. Unpublished PhD thesis, Universitat Autònoma de Barcelona.

Colominas, L., Antolín, F., Ferrer, M., Castanyer, P. and Tremoleda, J. (2019) From Vilauba to Vila Alba: Changes and continuities in animal and crop husbandry practices from the Early Roman to the beginning of the Middle Ages in the north-east of the Iberian Peninsula. *Quaternary International* 499, 67–79.

Colominas, L., Fernández Rodríguez, C. and Iborra, P. (2017) Animal husbandry and hunting practices in Hispania

Tarraconensis: an overview. *European Journal of Archaeology* 20(3), 510–534.

Colominas, L., Schlumbaum, A. and Saña, M. (2014) The impact of the Roman Empire on animal husbandry practices: study of the changes in cattle morphology in the north-east of the Iberian Peninsula through osteometric and ancient DNA analyses. *Archaeological and anthropological sciences* 6(1), 1–16.

Davis, S.J.M. (1981) The effects of temperature change and domestication on the body size of Late Pleistocene to Holocene mammals of Israel. *Paleobiology* 7(1), 101–114.

Davis, S.J.M. (2002) The mammals and birds from the Gruta do Caldeirão, Portugal. *Revista Portuguesa de Arqueologia* 5(2), 29–98.

Davis, S.J.M. (2005) Animal bones from Roman São Pedro, Fronteira, Alentejo. *Trabalhos do CIPA 88*, 10.

Davis, S.J.M. (2006) *Faunal Remains from Alcáçova de Santarém, Portugal*. Lisbon, Instituto Português de Arqueologia.

Davis, S. (2007) The mammals and birds from the Iron Age and Roman periods of Castro Marim, Algarve, Portugal. *Trabalhos do CIPA* 107, 61.

Davis, S.J.M. and Gonçalves, A. (2017) Animal remains from the 4th–5th century AD well at São Miguel de Odrinhas, Sintra, Portugal. *Revista Portuguesa de Arqueologia* 20, 139–156.

Davis, S.J.M. and Vilhena, J. (2017) Animal remains from Iron Age and Roman Odemira, Portugal. *Archaeofauna* 26, 199–217.

Davis, S.J., Svensson, E.M., Albarella, U., Detry, C., Götherström, A., Pires, A.E. and Ginja, C. (2012) Molecular and osteometric sexing of cattle metacarpals: a case study from 15th century AD Beja, Portugal. *Journal of Archaeological Science* 39(5), 1445–1454.

De Rensis, F. and Scaramuzzi, R.J. (2003) Heat stress and seasonal effects on reproduction in the dairy cow–a review. *Theriogenology* 60, 1139–1151.

Díaz-del-Río, P. (2008) El contexto social de las agregaciones de población durante el Calcolítico peninsular. *Era Arqueologia* 8, 128–137.

Driesch, A. Von den and Boessneck, J. (1974) Kritische Anmerkungen zur Widerristhöhenberechnung aus Längenmassen vor-und frühgeschichtlicher Tierknochen. *Säugetierkundliche Mitteilungen* 22(4), 325–348.

Duval, C., Lepetz, S. and Horard-Herbin, M.P. (2012) Diversité des cheptels et diversification des morphotypes bovins dans le tiers nord-ouest des Gaules entre la fin de l'âge du Fer et la période romaine. *Gallia* 69, 79–114.

Estrela, S. (2013) Mesas do Castelinho (Almodôvar): uma aldeia amuralhada na paisagem da Idade do Ferro do Baixo Alentejo. *Arqueologia & História* 62–63, 101–115.

Fabião, C. (1994) Garum na Lusitânia rural? Alguns comentários sobre o povoamento romano do Algarve. In J.G. Gorges and M. Salinas (eds) *Les Campagnes de la Lusitanie Romaine*, 227–252. Madrid, Casa de Velázquez.

Fabião, C. (1998) *O Mundo Indígena e a sua Romanização na área Céltica do território hoje Português*. Unpublished PhD thesis, Faculdade de Letras da Universidade de Lisboa (Portugal).

Fabião, C. (2008) *Las ánforas romanas de Lusitania*. Servicio de Publicaciones de la Universidad de Cádiz. In D. Bernal and A. Ribera (eds) *Cerámicas hispanorromanas II: producciones regionales* 725–745. Monografías. Historia y Arte, Universidad de Cádiz.

Fabião, C. and Guerra, A. (1994) As ocupações antigas de Mesas do Castelinho (Almodôvar). Resultados preliminares das campanhas de 1990–1992. *V Jornadas Arqueológicas* 2, 275–289.

Fernández-Rodríguez, C. (2003) Ganadería, caza y animales de compañía en la Galicia romana: estudio arqueozoológico. *Brigantium: Boletín do Museu Arqueolóxico e Histórico da Coruña* 15, 1–238.

Filipe, I. and Fabião, C. (2007) Uma unidade de produção de preparados de peixe de época romana na Casa do Governador da Torre de Belém (Lisboa): uma primeira apresentação, *Arqueologia & História* 58/59, 103–118.

Forest, V. and Rodet-Belarbi, I. (2002) À propos de la corpulence des bovins en France durant les périodes historiques. *Gallia* 59, 273–306.

Fortó, A., Fidalgo, X.M., Pelegero, B., Pisa, J. and Vidal, À. (2004) El poblat ibèric de la Torre Roja (Caldes de Montbui-Sentmenat). *Lauro: revista del Museu de Granollers* 26, 5.

Frémondeau, D., Nuviala, P. and Duval, C. (2017) Pigs and cattle in Gaul: the role of Gallic societies in the evolution of husbandry practices. *European Journal of Archaeology* 20(3), 494–509.

García, C. (2007) Informe de l'intervenció arqueològica preventiva al jaciment iberoromà de l'Hereuet (Seró, Artesa de Segre, La Noguera).

García-Ispierto, I., López-Gatius, F., Bech-Sabat, G., Santolaria, P., Yániz, J.L., Nogareda, C., De Rensis, F. and López-Béjar, M. (2007) Climate factors affecting conception rate of high producing dairy cows in northeastern Spain. *Theriogenology* 67(8), 1379–1385.

Gorges, J.-G. (1990) Villes et *uillas* de Lusitanie (interaction – échanges – autonomies). *Les villes de Lusitanie romaine*, 91–113. Paris, CNRS.

Groot, M. (2017) Developments in animal husbandry and food supply in Roman Germania Inferior. *European Journal of Archaeology* 20(3), 451–471.

Guitart i Duran, J. and Padrós i Martí, P. (1991) La ciutat romana de Baetulo (Badalona): història i urbanisme. *Espais: revista del Departament de Política Territorial i Obres Públiques* 31, 50–56.

Guitart i Duran, J. and Pera i Isern, J. (1994) La ciutat romana de 'Iesso' (Guissona, la Segarra). In X. Dupré (ed.) *La ciutat en el món romà= La ciudad en el mundo romano: XIV* Congreso Internacional de Arqueología Clásica *Tarragona, 5 al 11-9-1993: Actas*, 186–187. Tarragona, Consejo Superior de Investigaciones Científicas.

Hafez, E.S.E. and Hafez, B. (2000) *Reproduction in Farm Animals*. Lea & Febiger, Wiley-Blackwell.

Heleno, M. (1962) A 'villa' lusitano-romana de Torre de Palma (Monforte). O Arqueólogo Português, Nova Série, IV. *Separata de O Arqueólogo Portugués*, 313–318.

Ihaka, R. and Gentleman, R. (1996) R: a language for data analysis and graphics. *Journal of Computational and Graphical Statistics* 5(3), 299–314.

Ijzereef, G.F. (1981) Bronze Age animal bones from Bovenkarspel: the excavations at Het Valkje. *Nederlandse Oudheden 10*, Project Noord- Holland 1, 228.

Járrega, R. and Prevosti, M. (2010) *Els Antigons (Reus, Baix Camp). Una vil·la romana senyorial de l'ager Tarraconensis*, Collecció Documenta 4. Tarragona, Institut Català d'Arqueologia Clàssica.

Járrega, R. (2000) El poblament rural i l'origen de les villae al nord-est d'Hispania durant l'època romana republicana (segles II–I aC). *Quaderns de prehistòria i arqueologia de Castelló* 21, 271–302.

Kunst, M., Morán, E. and Parreira, R., (2002) Zambujal (Torres Vedras, Lisboa): relatório das escavações. *Revista Portuguesa de Arqueologia* 16, 103–131.

Lauwerier, R.C.G.M. (1988) *Animals in Roman Times in the Dutch Eastern River Area.* Rob, University of Michigan.

Lepetz, S. (1996) Effets de la romanisation sur l'élevage et la diete carnee dans les etablissements ruraux du nord de la Gaule. L'exemple de l'augmentation de la stature des animaux domestiques. In D. Bayard and J.-L. Collart (eds) *De la ferme indigène à la villa gallo-romaine*, 317–324. Amiens, Revue Archéologique de Picardie.

Liesau, C. and García, J. (2004) La fauna de Ratinhos: Estudio de la fauna y de la industria óssea procedente de la Tercera línea de muralla. O Castro dos Ratinhos (Barragem do Alqueva, Moura). Escavações num povoado proto-histórico do Guadiana. *O Arqueólogo Português,* sup. 6, 329–348.

López-Cachero, J. (2007) Sociedad y economía durante el Bronce final y la Primera edad del Hierro en el noreste peninsular: una aproximación a partir de las evidencias arqueológicas. *Trabajos de Prehistoria* 64(1), 99–120.

Macias, J.M. and Ramon, E. (1993) La vil· la romana de la Llosa, Cambrils (Baix Camp). *Tribuna d'Arqueologia* 1993, 125–135.

MacKinnon, M. (1999) O papel dos animais na economia rural da Lusitânia romana: zooarqueologia de Torre de Palma. *A Cidade. Revista Cultural de Portalegre* 13, 129–40.

Maloney, S.J. and Hale, J.R. (1996) The villa of Torre de Palma (Alto Alentejo). *Journal of Roman Archaeology* 9, 275–294.

Manning, K., Timpson, A., Shennan, S. and Crema, E. (2015) Size reduction in early European domestic cattle relates to intensification of Neolithic herding strategies. *PloS one 10*(12), e0141873.

Martín, A. (2003) Els grups del neolític final, calcolític i bronze antic. Els inicis de la metal·lúrgia. *Cota Zero* 18, 76–105.

Martín, A., Casas, S., i Falgàs, F.C., i Sastre, J.M. and de Prado, G. (2005) La zona 14 de l'oppidum del Puig de Sant Andreu d'Ullastret. Un conjunt arquitectònic dels segles IV i III aC. *Cypsela: revista de prehistòria i protohistòria* 15, 265–284.

Martín, A., Mallart, R.P., Falgàs, F.C. and Gay, C. (2007) El jaciment camp d'en Gou-gorg d'en Batlle, un barri periurbà de l'Oppidum d'Ullastret (Baix Empordà). *Cypsela* 17, 161–183.

Matolcsi, J. (1970) Historische Erforschung der Körpergrösse des Rindes auf Grund von ungarischem Knochenmaterial. *Zeitschrift für Tierzüchtung und Züchtungsbiologie* 63, 155–194.

Méniel, P. (1984) Contribution à l'histoire de l'élevage en Picardie du Néolithique à la fin de l'Âge du Fer. *Revue archéologique de Picardie,* numéro spécial 3, 1–56.

Méniel, P. (1996) Importation de grands animaux romains et amélioration du cheptel à la fin de l'Âge du Fer en Gaule Belgique. *Revue archéologique de Picardie* 3–4, 113–122.

Minferri, Equip (1997) Noves dades per a la caracterització dels assentaments a l'aire lliure durant la primera meitat del II milleni cal. BC: primers resultats de les excavacions en el jaciment de Minferi (Juneda, les Garrigues). *Revista d'arqueologia de Ponent* 7, 161–211.

Minniti, C., Valenzuela-Lamas, S., Evans, J. and Albarella, U. (2014) Widening the market. Strontium isotope analysis on cattle teeth from Owslebury (Hampshire, UK) highlights changes in livestock supply between the Iron Age and the Roman period. *Journal of Archaeological Science* 42, 305–314.

Nieto-Espinet, A. (2018) Referencial Element Measure Standard Biometrical data from a cow dated to the Early Bronze Age (Minferri, Catalonia), DOI: 10.13140/RG.2.2.13512.78081.

Nieto, A., Moya, A., López, J.B. and Agustí, B. (2014) Ofrenes o deixalles? El cas dels bovins (Bos taurus) en context funerari del jaciment del bronze ple de Minferri (Lleida, Catalunya). In A. Gardeisen and C. Chandezon (eds) *Équidés et bovidés de la Méditerranée antique. Rites et combats. Jeux et savoirs*, 53–112. Montpellier, Monographies d'Archéologie Méditerranéenne, Publications CNRS-UMR 5140.

Nieto-Espinet, A., Valenzuela-Lamas, S., Bosch, D., and Gardeisen, A. (2020) Livestock production, politics and trade: A glimpse from Iron Age and Roman Languedoc. *Journal of Archaeological Science: Reports* 30, 102077.

Olesti Vila, O. (1997) El origen de las villae romanas en Cataluña. *Archivo Español de Arqueología* 70(175–176), 71–90.

Oliva, M.A., (1998) *Anàlisi arqueozoològica de l'assentamenl ibèric del Turó del Vent Llinars del Vallès.* Treball de recerca inedit. Universitat Autònoma de Barcelona, Bellaterra.

Padrós, N. (2010) Les restes faunístiques de la vil· la romana de la Llosa: gestió ramadera, consum alimentari i paisatge. In *Ager Tarraconensis*, 192–199. Tarragona, Institut Català d'Arqueologia Clàssica (ICAC).

Pellicer Catalán, M. (1989) Observaciones sobre la problemática tartesia. *Habis* 20, 205–216.

Pena Gimeno, M.J. and Barreda Pascual, A. (1997) Productores de vino del nordeste de la Tarraconense. Estudio de algunos nomina sobre ánforas Laietana 1 (= Tarraconense 1). *Faventia* 19(2), 51–73.

Pera, J., Carreras, C., Rodrigo, E., Romaní, N., Padrós, N., De Solà, G. and Guitart, J. (2016) Puig Castellar de Biosca (La Segarra). Una fortificació de segle II aC a l'interior de Catalunya.(Campanyes 2013–2015). *Tribuna d'Arqueologia* 2015–2016, 31–56.

Pérez-Ripoll, M. (1999) La explotación ganadera durante el III milenio a.C. en la Península Ibérica. *Sagvntvm Extra* 2, 95–103.

Peters, J. (1998) *Römische Tierhaltung und Tierzucht: eine Synthese aus archäozoologischer Untersuchung und schriftlich-bildlicher Überlieferung.* Passaver Universitäts Scriften zur Archäologie Band 5. Rahden/Westf, Marie Leidorf GmbH.

Pigière, F. (2017) The evolution of cattle husbandry practices in the Roman period in Gallia Belgica and western Germania Inferior. *European Journal of Archaeology* 20(3), 472–493.

Pons, E. (1994) L'hàbitat a Catalunya durant el primer mil·leni aC: els precedents de l'habitació consolidada. *Cota zero: revista d'arqueologia i ciència* 10, 9–18.

Pou, J. and Revilla, V. (1996) La vil· la romana del Vilarenc (Calafell, Baix Penedès): campanyes 1988–1994, *Tribuna d'Arqueologia* 1995–1996, 105–112.

R Core Team (2018) R: A Language and Environment for Statistical Computing. Vienna, R Foundation for Statistical Computing.

Revilla, V. (2004) El poblamiento rural en el noreste de Hispania entre los siglos II aC y I dC: organización y dinámicas culturales y socioeconómicas. In P. Moret and M.T. Chapa (eds) *Torres, atalayas y casas fortificadas: explotación y control del territorio en Hispania (s. III a. de C.-s. I d. de C.),* 175–204. Universidad de Jaén.

Rivas-Martínez, S., Rivas-Saenz, S. and Penas, A. (2002) *Worldwide Bioclimatic Classification System*. Backhuys Pub, Universidad de León.

Rodríguez-Martín, F.G. (2011) La producción de vino y aceite entre Augusta Emérita y el occidente atlántico *Anales de Prehistoria y Arqueología* 27–28, 451–469.

Roman-Ponce, H., Thatcher, W.W., Canton, D., Barron, D.H. and Wilcox, C.J. (1978) Thermal stress effects on uterine blood flow in dairy cows. *Journal of Animal Science* 46, 175–180.

Rovira i Port, J. and Sanmartí i Grego, E. (1982) *Excavaciones en el poblado layetano del Turó del Vent, Llinars del Vallés: campañas de 1980 y 1981*. Barcelona, Diputació de Barcelona, Institut de Prehistòria i Arqueologia.

Ruiz Zapatero, G. (2001) Las comunidades del Bronce Final: enterramiento y sociedad en los Campos de Urnas. In M. Ruiz-Gálvez Priego (ed.) *La Edad del Bronce, ¿Primera edad de oro de España? Sociedad, economía e ideología*, 257–288. Barcelona, ed. Crítica.

Sanmartí, J. (2004) From local groups to early states: the development of complexity in protohistoric Catalonia. *Pyrenae* 35, 7–41.

Sanmartí, J. and Santacana, J. (1992) *El poblat ibèric d'Alorda Park: Calafell, Baix Penedès: campanyes 1983–1988*. Barcelona, Generalitat de Catalunya, Departament de Cultura, Direcció General del Patrimoni Cultural, Servei d'Arqueologia.

Sanmartí, J. and Santacana, J. (1994) L'urbanisme protohistòric a la costa de Catalunya. *Cota zero: revista d'arqueologia i ciència* 10, 27–37.

Saña M. (1998) Arqueozoologia i faunes neolítiques a Catalunya. Problemàtica plantejada entorn a la dinàmica del procés de domesticació animal. *Cypsela: revista de prehistòria i protohistòria* 12, 99–110.

Tavares da Silva, C.T. and Soares, J. (2014) O Castro de Chibanes (Palmela) e o tempo social do III milénio BC na Estremadura. *Setúbal Arqueológica-II Encontro de Arqueologia da Arrábida. Homenagem a AI Marques da Costa* 15, 105–172.

Telldahl, Y., Svensson, E.M., Götherström, A. and Storå, J. (2012) Osteometric and molecular sexing of cattle metapodia. *Journal of Archaeological Science* 39(1), 121–127.

Teichert, M. (1984) Size variation in cattle from Germania Romana and Germania Libera. In C. Grigsonand J. Clutton-Brock (eds) *Animals and Archaeology 4: Husbandry in Europe*, 93–104. British Archaeological Reports International Series 227. Oxford.

Trentacoste, A., Nieto-Espinet, A.and Valenzuela-Lamas, S. (2018) Pre-Roman improvements to agricultural production: evidence from livestock husbandry in late prehistoric Italy. *PLOS ONE* 13(12): e0208109.

Uerpmann, H. and Uerpmann, M., 2003. Zambujal: Die stein und beinartefakte aus den grabungen 1964 bis (1973) *Madrider Beiträge Band 5*, Zambujal Teil 4. Mainz, Philip von Zabern.

Valente, M.J. and Carvalho, A.F. (2014) Zooarchaeology in the Neolithic and Chalcolithic of southern Portugal. *Environmental Archaeology* 19(3), 226–240.

Valenzuela-Lamas, S. (2008) *Alimentació i ramaderia al Penedès durant la Protohistòria, segles VII-III aC*. Barcelona, Societat Catalana d'Arqueologia.

Valenzuela-Lamas, S., (2010) Paisatge, alimentació i gestió dels ramats als Antigons a partir de les restes de fauna (vertebrats i mol· luscos). In *Ager Tarraconensis*, 181–191. Tarragona, Institut Català d'Arqueologia Clàssica (ICAC).

Valenzuela-Lamas, S., (2011) Estudi arqueozoològic del jaciment de la Torre Roja (Caldes de Montbui, Vallès Oriental), campanyes 2008 i 2009. Unpublished report.

Valenzuela-Lamas, S., (2012) Informe de les restes de fauna recuperades al jaciment del Vilarenc (Calafell, Baix Penedès).

Valenzuela-Lamas, S., (2013) Informe de les restes de fauna recuperades al jaciment de l'Hereuet (Serós, La Noguera).

Valenzuela-Lamas, S. (2014) Mammal remains from the Governor's House (Belém Tower, Lisbon) and Rua dos Correeiros (Baixa, Lisbon) in the context of fish processing factories in Lusitania. In C. Detry and R, Dias (eds) *Proceedings of the First Zooarchaeology Conference in Portugal*, 57–68. Oxford, BAR International series 2662.

Valenzuela-Lamas, S. and Albarella, U. (2017) Animal husbandry across the Western Roman Empire: changes and continuities. *European Journal of Archaeology* 20(3), 402–415.

Valenzuela-Lamas, S. and Fabião, C. (2012) Ciervos, ovejas y vacas: el registro faunístico de Mesas do Castelinho. In M. de Deus (ed.) *Actas do V Congresso de Arqueologia do Sudoeste Peninsular*, 413–432. UNIARQ-Universidade Lisboa.

Valenzuela-Lamas, S., Jiménez-Manchón, S., Evans, J., López, D., Jornet, R. and Albarella, U. (2016) Analysis of seasonal mobility of sheep in Iron Age Catalonia (north-eastern Spain) based on strontium and oxygen isotope analysis from tooth enamel: First results. *Journal of Archaeological Science: Reports* 6, 828–836.

Valenzuela-Lamas, S., Nadal, J. and López, D. (2010) La fauna del jaciment de Mas d'en Gual (el Vendrell, Baix Penedès) en el context dels assentaments especialitzats d'època ibèrica. *Revista d'arqueologia de Ponent* 20, 215–224.

Valenzuela-Lamas, S., Orengo, H.A., Bosch, D., Pellegrini, M., Halstead, P., Nieto-Espinet, A., Trentacoste, A., Jiménez-Manchón, S., López-Reyes, D. and Jornet-Niella, R. (2018) Shipping amphorae and shipping sheep? Livestock mobility in the north-east Iberian peninsula during the Iron Age based on strontium isotopic analyses of sheep and goat tooth enamel. *PloS one* 13(10), e0205283.

Valenzuela-Lamas, S., Valenzuela-Suau, L., Jimenez-Manchón, S., Codina-Falgas, F., Plana-Mallart, R. and Martín, A. (2017) Consum carni i artesanat en os als afores del Puig de Sant Andreu: el jaciment ibèric del Camp d'en Gou, Gorg d'en Batlle (Ullastret, Baix Empordà). *Empúries* 57, 7–26.

Valenzuela-Oliver, A., Alcover, J.A. and Cau, M.A. (2013) Tracing changes in animal husbandry in Mallorca (Balearic Islands, Western Mediterranean) from the Iron Age to the Roman Period. In M. Groot, D. Lentjes and J. Zeiler (eds) *Barely Surviving or More than Enough?*, 201–223. Leiden, Sidestone Press.

Valera, A.C. (2001) A ocupação pré-histórica do sítio do Mercador (Mourão). *Era Arqueologia* 3, 42–58.

Valera, A.C. (2015) Social Change in the Late 3rd Millennium BC in Portugal: The Twilight of Enclosures. In H. Harald Meller, H. Wolfgang Arz, R. Jung and R. Risch (eds) *Social Change in the Late 3rd Millennium BC in Portugal: The Twilight of*

Enclosures, 7th Archaeological Conference of Central Germany October 23–26, 2014 in Halle (Saale), 409–427. Barcelona, Universitat Autonoma de Barcelona.

Vigne, J.-D. (1988) Les Mammifères post-glaciaires de Corse, étude archéozoologique. *XXVIe suppl. à Gallia Préhistoire*. Paris, CNRS éd.

Vilhena, J. and Rodrigues, J. (2009) O grande fosso: a escavação arqueológica no cineteatro Camacho Costa e o Cerro do Castelo de Odemira na Idade do Ferro tardia. In I. Silva, J. Madeira and S. Ferreira (eds) *1o Encontro de Historia do Alentejo Litoral*, 204–214. PSA – Sines, Câmara Municipal de Grândola.

Widdowson, E.M. and Lister, D. (1991) Nutritional control of growth. In A.M. Pearson and T.R. Dutson (eds) *Growth Regulation in Farm Animals. Advances in Meat Research* 7, 67–101. Springer.

Wilcoxon, F. (1945) Individual comparisons by ranking methods, *Biometrics Bulletin* 1, 80–83.

Žalaitė, I., Maurer, A.F., Grimes, V., Silva, A.M., Ribeiro, S., Santos, J.F., Barrocas Dias, C. and Valera, A.C. (2018) Diet and mobility of fauna from Late Neolithic–Chalcolithic site of Perdigões, Portugal. *Journal of Archaeological Science: Reports* 19, 674–685.

13

Animal remains from 17th-century Carnide, Lisbon, Portugal

Cleia Detry, Ana Beatriz Santos, Tânia Casimiro, Ana Caessa and Nuno Mota

Excavations undertaken in 2012 by the Centro de Arqueologia de Lisboa (CAL), in Largo do Coreto in Carnide (Lisbon, Portugal), uncovered over 7,000 faunal remains. These came from 60 underground pits previously used for storage, especially cereal, and subsequently, between 1550 and 1660 AD, filled with domestic rubbish. Most remains belonged to mammals and birds, with a significant number of molluscs. They are presumably food waste which therefore tell us something about the way of life of the inhabitants of 17th-century Carnide. These people clearly depended primarily on domestic animals such as cattle, pig, sheep and goat as well as chicken and goose. Large wild animal remains were strikingly absent though there were some bones of small game like rabbit and partridge. For the rabbit it is unclear if it was the domestic or wild variety. A few remains of ferret and raptors point to their possible uses for hunting. A number of whole skeletons of cats and dogs, with no trace of butchery, were probably deposited as garbage.

Keywords: Zooarchaeology, Carnide-Portugal, 17th century, storage pits

Introduction

The archaeological intervention in Carnide (Lisbon) took place between May 2012 and April 2013 and was supervised by archaeologists of the Lisbon Archaeological Centre (CAL). It took the form of archaeological monitoring of floor removal in order to replace pavements and the excavation of 62 archaeological trenches, responding to a need to open ditches to install infrastructures in the Largo do Coreto and adjacent streets.

The archaeological work recorded the considerable urban transformation that Carnide underwent in the middle of the 19th century – an extensive change of the landscape (Caessa and Mota 2013; 2014). Despite this destructive episode, it was possible to identify, in the western area of the Largo, the foundation of a medieval church dedicated to the Holy Spirit and some of the burials that date from the 17th century and continued until the 19th century, made in and around this building, (Garcia, Caessa and Mota in press). Various archaeological features were identified, and 136 storage pits, excavated in the bed-rock, were found by Ana Caessa and Nuno Mota (a total of 172 of these structures have so far been identified in Carnide since the 19th century). They seem to correspond to the 'covas de pão' (bread/cereal pits) mentioned in the medieval documentation available for Carnide and dated to the end of the 12th century. The complete or partial archaeological excavation of 71 of these silos brought to light a large amount of material culture indicating that throughout the Middle Ages they were used as underground pits to store cereal. Subsequently from the mid-16th to mid-17th century they served as garbage dumps (Caessa and Mota 2013; 2014; 2016). Of the 71 silos excavated only 60 provided faunal remains.

Our aim here is to describe the faunal remains recovered from these pits, associating them, as far as possible, to what other studies concerning the ceramics can tell us about the consumption of the animals' meat. Most of the remains belonged to domestic animals, as well as molluscs. It is a large collection and therefore provides useful quantitative information about post-medieval Portuguese domestic animals and seafood. The numerous cat and dog skeletons are the largest known so far from this period in Portugal.

The abundance of domestic animals indicates a level of dependence upon these – little difference from other contemporary sites in Portugal (Davis 2009; Detry 2014; Detry and Pimenta 2017).

The Helicid (*Cornu aspersum*) a large land snail, with a globular shell is very common in Portugal and is still eaten today. This species lives in humid places under stones or vegetation in calcareous regions (Matos 2014). These animals although relatively numerous (a total of 27 shells) appeared scattered among the silos, indicating that they are probably intrusive and not part of the original diet of the inhabitants of Carnide. Only three shells of a smaller land snail were found, they are similar to the genus *Candidula* (family Hygromiidae), also a common taxon in Portugal (Matos 2014). Terrestrial gastropods are usually intrusive in archaeological contexts since they like humid and organic rich environments such as pits filled with garbage.

The most common species of bivalve is the common cockle (*Cerastoderma edule*) which comprised 30% of the invertebrate remains (Table 13.2). This inhabitant of sandy bottoms can endure low salinity (20–25‰) and nowadays is typically found in estuaries such that of the Tagus in the Alcochete region (Rodrigues *et al.* 2006; Verdelhos *et al.* 2015). The second most common bivalve is the grooved carpet shell (*Ruditapes decussatus*) a species still commonly eaten today, although replaced in present-day Lisbon by an invasive oriental but similar species – *R. philippinarium* (López Garaulet 2011). The autochthonous species of carpet shell (*R. decussatus*) is present in sub-littoral sandy environments (Sterry and Cleave 2012).

Almost 400 shells of the edible mussel (*Mytilus edulis*) were counted. This is a species that also endures brackish water but needs a fixed substrate, such as rocks, and occurs on rocky beaches, like the aforementioned gastropods (Macedo 1996; Sterry and Cleave 2012).

The family Ostreidae is a particularly interesting group. Most of the shells (*c.* 200), belong to the autochthonous species *Ostrea edulis*, an oyster with a rounded shell. A few shells (six) may have belonged to *Crassostrea angulata* and eight could only be identified to family level. The shells of these species are longer and are not striated in the hinge area of the shell indicating that they may have belonged to *C. angulata*. Mitochondrial and nuclear DNA analysis have shown that the Portuguese oyster was probably introduced to Europe from Asia in historical times and is closely related to *C. gigas,* a very common species in Japan (Foighil *et al.* 1998; Huvet *et al.* 2000). The presence of Chinese ceramics in Carnide (Casimiro *et al.* 2017) shows how commercial exchanges from the orient could explain the presence of this exotic mollusc. Are these specimens the first representatives of this species in Europe? Were 17th-century Portuguese vessels responsible or did they arrive subsequently attached to ships? In other archaeological sites in the Lisbon area we have not found *C. angulata* in the Iron Age (Detry *et al.* 2016) and in the 13th and 15th centuries in Vila Franca de Xira (Detry and Pimenta 2017) indicating that this taxon probably arrived here after the 16th century.

The last of the most frequent species is the peppery furrow shell (*Scrobicularia plana*), a species that lives in muddy sediments and tolerates low levels of salinity in the far interior of estuaries.

Mimachlamys varia is one of the taxa present with 16 shells. It has a small shell – around 50 mm – needs a fixed substrate and lives as deep as 400 m in the sea (Macedo 1996). Three shells of *Acanthocardia* sp. and eight of *Pecten maximus* were recorded. Both are marine sandy coast species, and both have beautiful strong shells that can have other uses besides consumption.

Only nine elements belonging to *Mactra stultorum* and seven of *Venus verrucosa* (warty venus) were found. Both species prefer sand at depths no greater than 100 m (Macedo 1996).

Anomia ephippium with only four exemplars also appears in tidal areas and needs rocky substrates to fix its shell.

The most common bivalves in this assemblage were all edible species, still very popular today in Portugal. According to contemporary cookery books the common cockle, grooved carpet shell, oyster and mussel were all commonly eaten during the 16th and 17th centuries (Casimiro *et al.* 2017).

The molluscs found in Carnide came from sandy beaches with rocks. Another group came from other areas such as the estuaries. Since Lisbon is located on the Tagus Estuary and near the beaches of Cascais and Estoril with their rocky shores, it is possible that these resources could come from all of these localities.

Other invertebrates

A few other species of invertebrates, namely cephalopods and arthropods, were also found in this assemblage. Five fragments of the internal shell of the common cuttlefish (*Sepia officinalis*) were found. The internal shell of cuttlefish is very fragile, so they are rare in archaeological assemblages. In Carnide a few fragments were registered confirming its capture and consumption although no recipes for these animals were found in contemporary cookery books (Casimiro *et al.* 2017). This species inhabits the infralittoral – 200 m depth but can be caught closer to shore (Neves *et al.* 2009) – is commonly eaten today by Portuguese people.

Crustaceans are represented by only three elements attributed to *Carcinus maenas* (the green crab, a species present in estuaries) and another seven elements from *Maja squinado* (European spider crab). Both are common on the Portuguese coast and are still economically very important. These may have come from local markets and from nearby habitats. The old cookery books confirm the consumption of European spider crab (Casimiro *et al.* 2017).

The Balanidae are sessile crustaceans that attach themselves to rocks located in the mediolittoral and infralittoral. More than 35 shells were found, sometimes attached to the shells of other invertebrates. Although this species is usually consumed, most of the shells were very small indicating that they may have been introduced attached to other animals.

Amphibians and Testudines

Amphibians are scarce and most probably intrusive. They can easily become integrated in humid sediments like silos where they may hibernate and eventually die. Only one tibia of *Bufo bufo*, the common toad, was found. This species occurs throughout Portugal (Loureiro *et al.* 2008). Therefore, its presence in this assemblage is not surprising. A pelvis, femur and one tarsal bone of frog were also found. Since *Rana iberica* is not present in the Portuguese Estremadura region and is more common north of the Tagus River and in higher altitudes, it is more probable that these bones belong to *Rana perezi*, the Iberian water frog, common throughout the Iberian Peninsula (Loureiro *et al.* 2008).

Chelonians were also present, but only represented by a single fragment of the top plates/plastron. The fragment belongs to *Mauremys leprosa*, the Spanish pond turtle, an animal very dependent upon water sources. The reason for its presence in this assemblage is difficult to understand. Since no cutmarks or fire marks were observed, it may have belonged to a pet. Casimiro *et al.* (2017) refer to turtle soup and broth recipes, popular at that time, especially for sick people, indicating the possibility that this specimen was consumed.

Mammals

Mammals comprise more than half of the identified specimens (NISP = 58%) but only 12% when the minimum number of individuals (MNI) is considered. The MNI reverses the proportions as molluscs constitute 82% of the MNI sample and very little usable biomass. This is because each individual mollusc represented consists of only one or two shells and provides a very small quantity of meat compared to mammals. Each mammal comprises over 200 bones and represents a much larger amount of meat. Hence mammals constituted a very important part of the diet of the inhabitants of Carnide.

Approximately 87% of the mammals in Carnide belonged to the four domestic species most commonly exploited by humans in Europe since the Neolithic: cattle (*Bos taurus*), sheep (*Ovis aries*), goat (*Capra hircus*) and pig (*Sus domesticus*) (Table 13.3). The other 11% belong to animals that are generally considered as pets and not usually consumed, such as the cat (*Felis catus*) and dog (*Canis familiaris*). The remaining bones belong to equids, lagomorphs and rodents and two interesting bones of European polecat (*Mustela putorius*). Lagomorphs are the only hunted mammals in the Carnide assemblage,

Table 13.3. Number of Identified Specimens (NISP) and Minimum Number of Individuals (MNI) of the different species of vertebrates.

	NISP		*MNI*	
	N	*%*	*N*	*%*
Amphibia + Testudines				
Bufo bufo	1	20%	1	33%
Rana sp.	3	60%	1	33%
Mauremys leprosa	1	20%	1	33%
Total Amphibia + Testudines	5		3	
Mammalia				
Bos taurus	1128.5	28%	50	24%
Ovis/Capra	1361.5	33%	68	33%
Ovis aries	253	6%	21	10%
Capra hircus	123	3%	12	6%
Sus sp.	679.5	17%	28	14%
Equus sp.	10	0.2%	2	1%
Equus asinus	7	0.2%	1	0.5%
Equus caballus	9	0.2%	1	0.5%
Lepus sp.	8.2	0.2%	2	1%
Oryctolagus cuniculus	8	0.2%	1	0.5%
Rattus rattus	4	0.1%	1	0.5%
Rattus norvegicus	3	0.1%	1	0.5%
Canis lupus familiaris	275	7%	9	4%
Felis catus	202.4	5%	8	4%
Mustela putorius	2	0.05%	1	0.5%
TOTAL MAMMALS	4074.1		206	
AVES	5	1%	1	1%
Burhinus oedicnemus	6	1%	2	2%
cf. Athene noctua	2	0%	1	1%
Columba sp.	3	0%	1	1%
Falco sp.	1	0%	1	1%
Buteo buteo	4	1%	1	1%
Anas sp.	2	0%	1	1%
Anser sp.	177	28%	22	26%
Alectoris rufa	2	0%	1	1%
Gallus domesticus	429	68%	54	64%
Total Aves	631		85	
Total Vertebrata	4705.1		291	

although some of the rabbits could have been domestic. The distribution of the seven most common species in the Carnide silos (Fig. 13.2) shows that the animals that were consumed appear in almost every pit, while the cats and dogs appear concentrated in a few areas. The silos with more cattle and caprines (silos nr. 24-XVI, 28-VII and 18-VIII) contained no cat remains and very few dogs. The silo with more dog bones (24-XII) had no birds and very few cattle, sheep, goat or pig. It is interesting that the contexts with more cats (29-IX and 24-XXII) had few or no dogs.

*Cattle (***Bos taurus***)*

The wild ancestor of cattle, the aurochs, has probably been extinct in Iberia since the Bronze or Iron Age (Castaños 1991, Cardoso 2002). Therefore, it is assumed that all remains of *Bos* at Carnide belonged to the domestic form – *B. taurus*.

In terms of numbers, cattle are second in the Carnide silos, indicating their economic importance in the 16th/17th century, although given the method of collection and urgent nature of the excavation, frequencies of different-sized taxa need to be viewed with caution. This effect might explain why bones such as metapodials, scapulae and humeri are so abundant (Fig. 13.3). Cattle were certainly consumed but also used for their power to transport goods and people and no doubt for ploughing. Evidence for their use for power is indicated by pathological deformations on some of the metapodials as well as their pattern of age-at-death (see below). A little more than a third of the cattle remains are teeth. This is hardly surprising given that teeth are more resistant to erosion than bones. Of the 403 teeth studied, only 14 were deciduous indicating the scarcity of very young animals.

We identified almost 100 metacarpals but only 70 metatarsals, around 50 scapulas, humerus and tibias, other long bones would be no more than 25 each. Metapodials are large and very dense, so post-mortem decay may, at least in part, be an important factor accounting for these very different numbers. Metapodials often serve as raw material in the production of utensils, since they have long robust circular or semi-circular (in cross-section) diaphysis. Some had served as anvils for the production of the serrations on scythes (Aguirre *et al.* 2004; Moreno-García *et al.* 2005). The measurements taken on these bones (Appendix online) show that the cattle at Carnide were more robust than those from earlier times, which corroborates previous findings (Davis 2008). The adult metacarpals from 17th-century Carnide were of similar shape and size as the ones from 15th-century Beja but were slightly larger and more robust than Iron Age to Moslem period Portuguese cattle (Davis *et al.* 2012; 2018). It could also be shown that there were more adult males than adult females. Davis *et al.* (2018) suggested that they could have belonged to fighting bulls or maybe males were more useful to carry heavy agricultural machinery and cereals. The same authors developed an

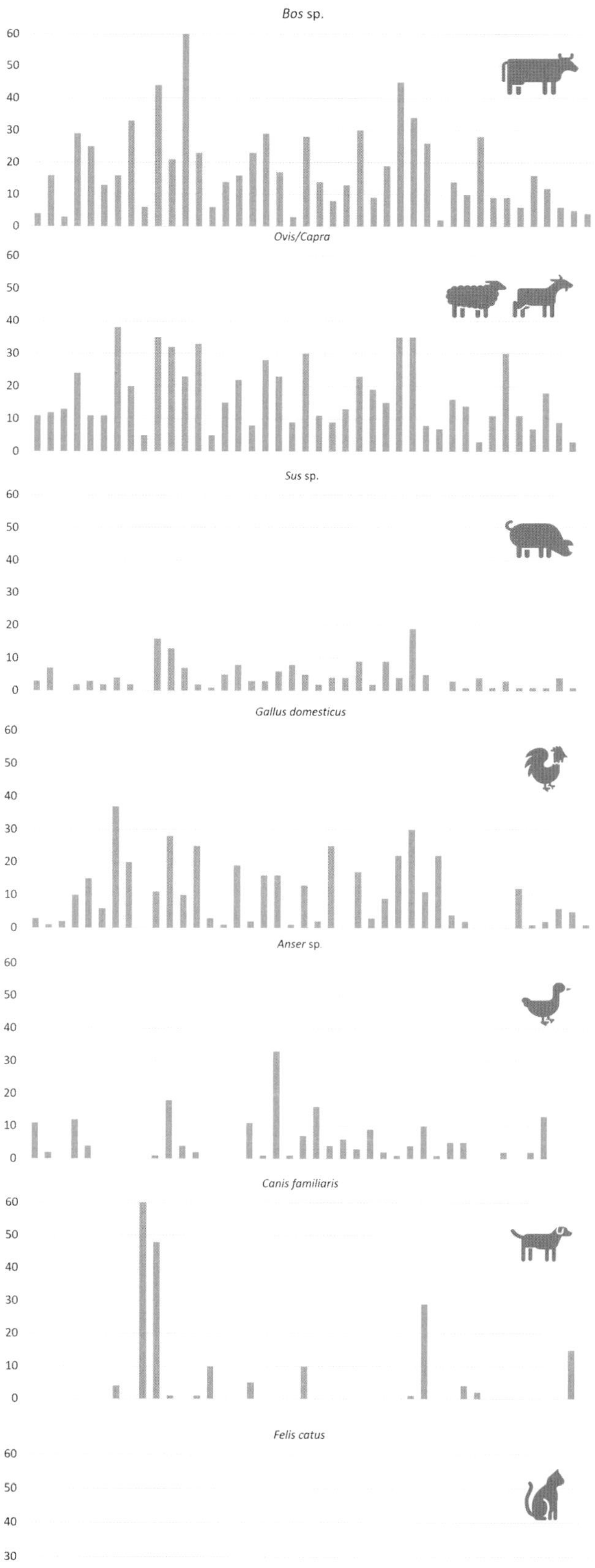

Figure 13.2. Distribution of the Number of Identified Specimens by taxa in selected silos with more abundant remains (>100 NISP).

osteometrical method to sex cattle metacarpals (Davis *et al.* 2012). These bones were also used to create a metrical baseline for comparing cattle metacarpals from other sites and periods in the Iberian Peninsula. For more details see Davis *et al.* (2018).

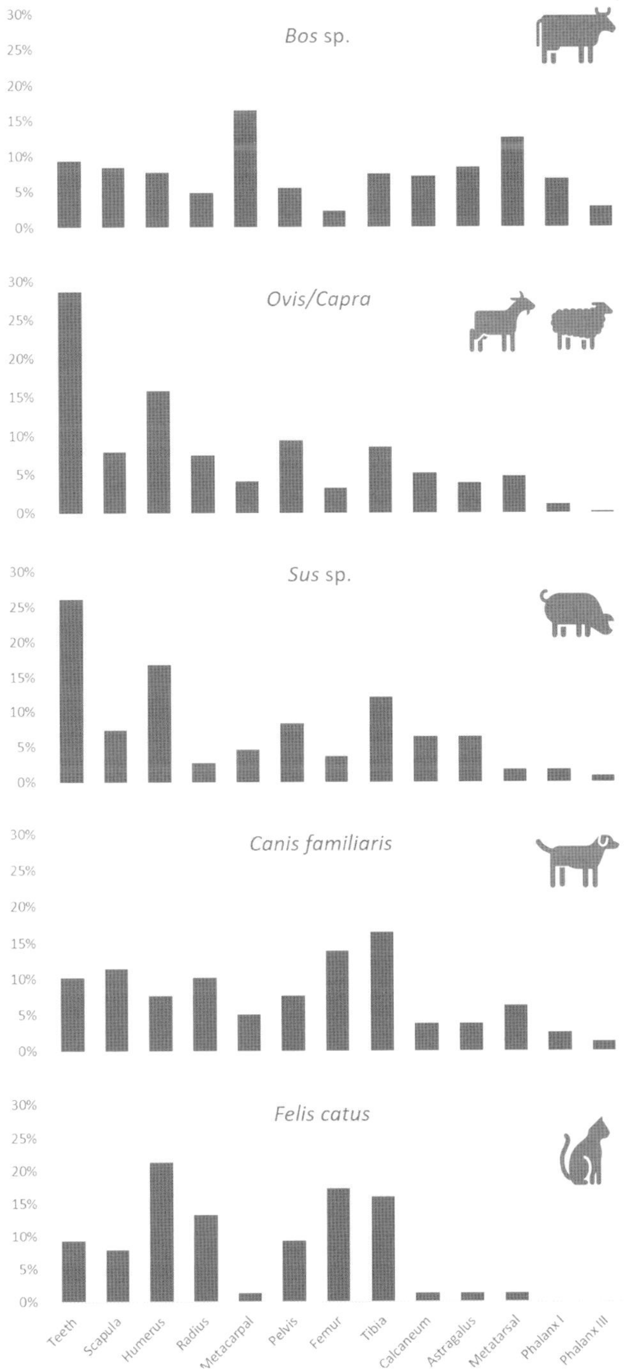

*Figure 13.3. Distribution (in %) of the Minimum Number of Individuals by parts of the skeleton for cattle (*Bos taurus*), sheep and goat (*Ovis aries *and* Capra hircus*), pig (*Sus *sp.), dog (*Canis familiaris*) and cat (*Felis catus*).*

*Sheep (***Ovis aries***) and goat (***Capra hircus***)*

These two small bovids are the most common at Carnide in terms of number of identified specimens (NISP) and minimum number of individuals (MNI). In this respect Carnide is little different from other sites from the Neolithic onwards in southern Portugal (Davis and Mataloto 2012; Valente and Carvalho 2014; Detry and Arruda 2013; Detry *et al.* 2014; Valente and Carvalho 2014; Detry and Pimenta 2017). Contemporary cookery books indicate that these species were also the most common in recipes in post-medieval Portugal (Casimiro *et al.* 2017).

Figure 13.4 is a plot of two indexes of measurements of sheep and goat astragali that may partially or completely separate sheep from goat (Davis 2016). In this figure, the astragali were first identified to species based on their morphological characters and subsequently their measurements were plotted. One astragalus identified as goat had measurements indicating sheep, showing there are some exceptions to this osteometric distinction. Davis (2016) also finds some wild and feral goats with measurements more similar to sheep. Something similar may be the case here at Carnide.

For those parts of the skeleton that could be identified to species sheep were clearly more common than goat by a factor of around two. The predominance of sheep over goat seems to be the rule in more humid regions of southern Portugal with goats being better adapted to arid areas.

The parts of the skeleton that are present in this assemblage (Fig. 13.3; Table 13.4) indicate a probable preference for bones with more meat surrounding them. Teeth are the most common in this profile probably because they preserve better. If we observe the number of fused bones and the high level of wear of the teeth and low number of deciduous teeth we can conclude that most of these sheep and goats were older and had been exploited for their so-called secondary products such as milk or wool. Since sheep are twice as

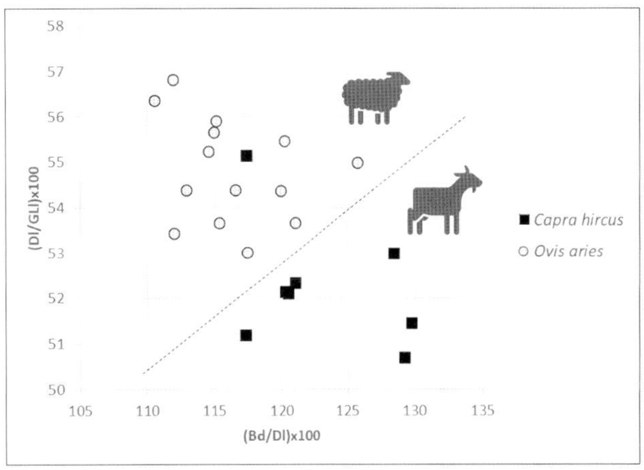

Figure 13.4. Measurements of the astragalus of Capra hircus *and* Ovis aries*, in millimeters. Measured by C. Detry.*

Table 13.4. Parts of the skeleton of each species of mammal found in the Carnide silos.

	Bone Fusion	*B*	*O/C*	*OVA*	*CAH*	*S*	*EQ*	*EQC*	*EQA*	*LEP*	*ORC*	*RAR*	*RAN*	*CAF*	*FEC*	*MUP*
Teeth																
Incisors		1		1		(6)10										
Canine						(4)										
Deciduous Pre-molar 2		(4)	(11)3	(36)	(18)	(11)								(3)		
Deciduous Pre-molar 3		(5)	(22)2	(41)3	(23)	(16)								(5)	(10)	
Deciduous Pre-molar 4		(4)	(32)4	(34)	(21)1	(15)								(3)	(10)	
Incisors		34	(8)29			(75)25	1							(6)	(3)	
Canine						(42)12								(12)	(11)	
Pre-molar 1														(2)		
Pre-molar 2		(33)2	(47)3		(3)	(30)1								(5)		
Pre-molar 3		(47)7	(74)			(51)	1		(1)					(7)	(13)	
Pre-molar 4		(43)2	(82)10			(52)			(1)					(12)	(13)	
Molar1		(52)	(130)1	(26)	(10)	(56)			(1)					(14)1	(14)	
Molar 2		(57)	(133)3	(5)	(4)	(50)			(1)					(10)		
Molar 3		(47)	(106)32	(1)		(28)2			(1)					(1)		
Molar 1/2		12	(7)61			(2)7										
Molar		(2)1	(2)			0			1		(4)					
Total teeth		403	834	167	88	510	2	5	6	0	4	0	0	81	74	0
Bones																
Scapula	F	49	61			13	2							11	8	
	UM		8			1								5	3	
	U	3	6													
Humerus	F	46	29	39	16	30	1			1	1	1		8	15	1
	UM	2	4			5								4	13	
	UE	1				1										
	FV		1												2	
Radius	F	25	26				1							8	10	
	UM	2	41											8	7	
	UE	2	5			6										
	FV	2	1												2	
Metacarpal	F	98.5	7	3	2	5.5		1	1					18	1.4	
	UM	2	14	2		3								13		

B – Bos taurus*; O/C* – Ovis/Capra*; OVA* – Ovis aries*; CAH* – Capra hircus*; S* – Sus *sp.; EQ* – Equus *sp.; EQC* – Equus caballus*; EQA* – Equus asinus*; LEP* – Lepus *sp.; ORC* – Oryctolagus cuniculus*; RAR* – Rattus rattus*; RAN* – Rattus norvegicus*; CAF* – Canis lupus familiaris*; FEC* – Felis catus*; MUP* – Mustela putorius. *Number of teeth in brackets are included in the mandible. Bone Fusion status: F – Fused; UE – Unfused epiphysis; UM – Unfused metaphysis; FV – Fusion visible; U – Undetermined.*

(Continued)

Table 13.4. (Continued)

	Bone Fusion	*B*	*O/C*	*OVA*	*CAH*	*S*	*EQ*	*EQC*	*EQA*	*LEP*	*ORC*	*RAR*	*RAN*	*CAF*	*FEC*	*MUP*
Pelvis	F	19	59			8	3			3	2		1	9	9	1
	UM		5			1								2	3	
	U	14	22			8									1	
Femur	F	10	7							2	1			8	7	
	UM	1	20			5				1		2		14	13	
	UE	3	7			7								1		
	FV		1										1		4	
Tibia	F	37	64			7	1					1		10	7	
	UM	5	41			15				1				17	14	
	UE	3	5			2										
	FV	1	4			2									2	
Calcaneum	F	17	4	8	5	1								2	2	
	UM	2	8	5	2	7								4		
	U	27	2	2	1	5										
Astragalus	F	51	2	22	9	11								5	1	
	U		1	1		1										
Metatarsal	F	70	13	3		1.5		1		0.2			1	24	2	
	UM	8	21	1		2								14		
	UE	2														
Metapodial	F	6.5						2							2	
	UE	3	3.5			1.5										
	UM					4								1		
	U	1														
Phalanx I	F	158	25			7								6		
	UM	5	7			7								1		
	UE	1														
	FV		2			1										
Phalanx III	F	67	1			1								1		
Total Bones		725.5	527.5	86	35	169.5	8	4	1	8.2	4	4	3	194	128.4	2
Total		1147	1361.5	253	123	679.5	10	9	7	8.2	8	4	3	275	202.4	2
MNI		50	68	21	12	28	2	1	1	2	1	1	1	9	8	1

common as goats it is likely that wool was much esteemed. The age profiles obtained via the fusion of long bones are similar to those observed in the contemporary assemblage at Santa Clara-a-Velha in Coimbra (Detry *et al.* 2014).

Suids (*Sus sp.*)

We recorded 680 fragments of suids, 75% of which are teeth. This probably reflects the considerable robustness of pig teeth as well as pig cranial bones. Remains of wild boar and its domestic relative, the pig, are difficult to distinguish. Generally, bones and teeth of wild boar are larger than those of the pig, but there is a substantial amount of overlap of their measurements, especially in the Iberian Peninsula where the wild boar is relatively small (Albarella *et al.* 2005). Figure 13.5 shows the measurements of lower third molar teeth of *Sus* from Carnide. Note they tend to be small and more

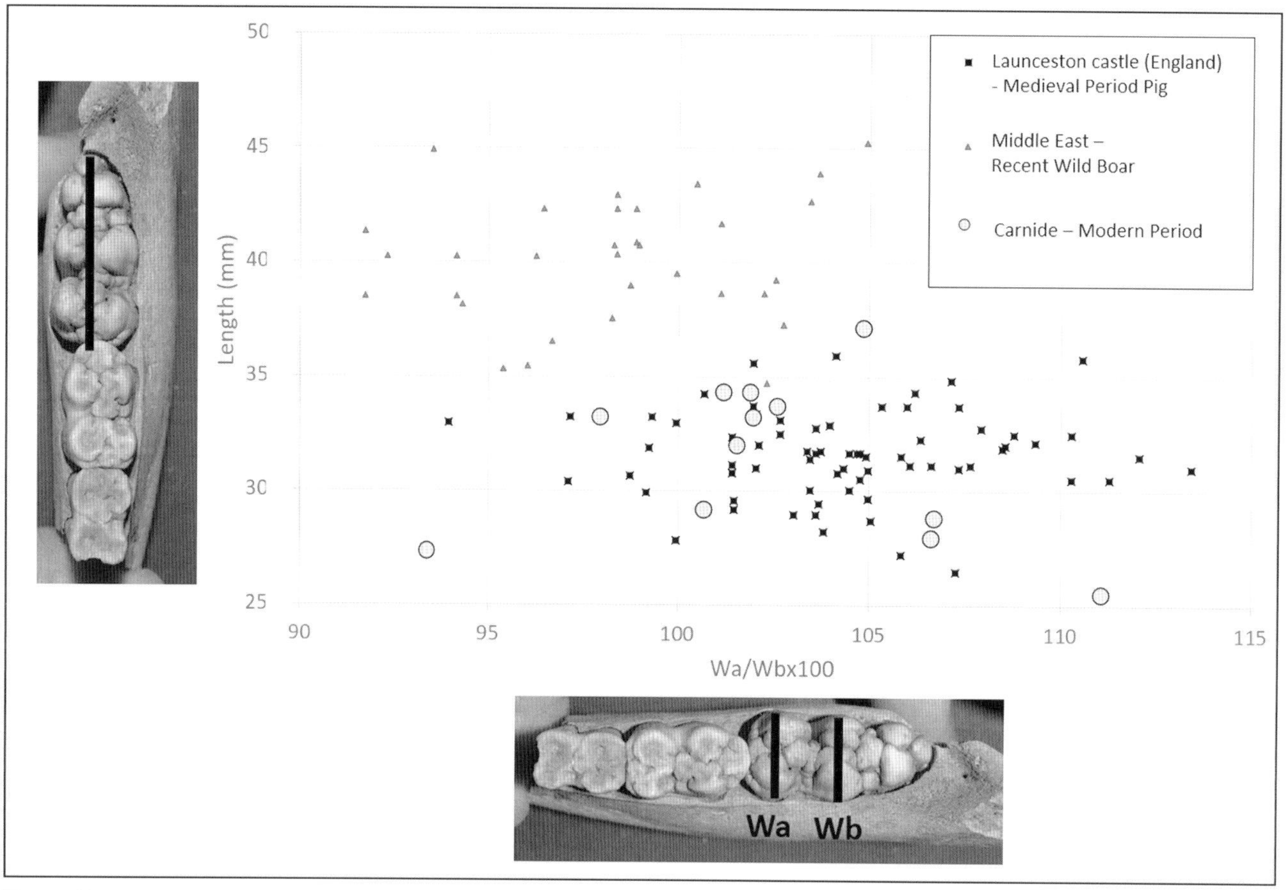

Figure 13.5. Measurements of the inferior third molar of suids recovered in Carnide measured by C. Detry. The pig measurements from Launceston Castle are from Albarela and Davis (1996) and the measurements of wild boar were taken by U. Albarella.

similar in shape to those of the domestic form. Moreover, none of the bone measurements were compatible with those of the larger wild boar. The fact that no big game hunting of animals like red deer was undertaken by the inhabitants of Carnide makes it seem unlikely that wild boar was hunted. It is therefore likely that all the *Sus* remains belonged to pig. Their remains derive from various parts of the carcass and most of their long bones have unfused epiphyses. Unlike sheep, goats and cattle, secondary products of the pig are not exploited by man. The pig is raised primarily for its meat and fat – and so is generally slaughtered young before many of its long-bone epiphyses fuse.

*Equids (*Equus *sp.)*

Equids are represented by 22 identified remains. Among these, ten could only be identified as equid; most bones and teeth of horse (*Equus caballus*) and donkey (*E. asinus*) are very similar and difficult to identify to species. However, a metacarpal and a mandible with five teeth could be identified as donkey (the internal enamel fold of the teeth form are 'V' shaped, see Fig. 13.6). The horse is represented by a metacarpal, a metatarsal and two distal metapodials.

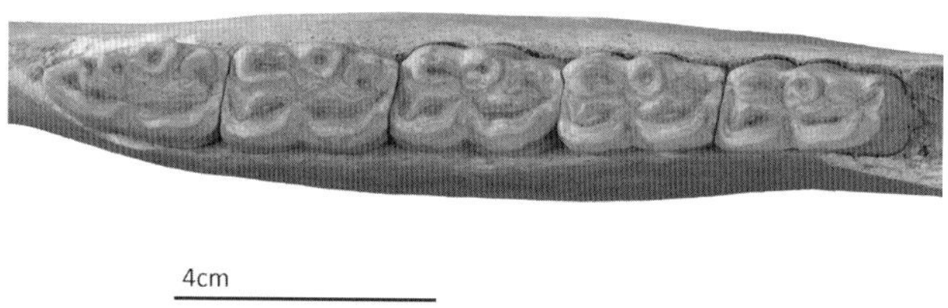

Figure 13.6. Equid mandible, probably Equus asinus *(donkey). Photo by Guilherme Cardoso.*

It is probable that both species played an important role transporting people and goods. The unidentified equid bones include a pelvis, radius and humerus. None of the remains have cut marks. Hippophagy was and still is uncommon so that equid bones, unlike those of animals like sheep, pig and cow, are scarce on archaeological sites.

Lagomorphs

Lagomorph bones are scarce in the Carnide silos. This could well reflect the problem of recovery rather than any dietary avoidance. The lagomorph bones identified include eight long bones of hare (*Lepus* sp.) and four long bones and two mandibles of rabbit (*Oryctolagus cuniculus*). The rabbit was probably domesticated in medieval times (Callou 2003) or even earlier so we cannot be sure if they belonged to the wild or domesticated form of this animal.

*Dogs (*Canis lupus familiaris*) and cats (*Felis catus*)*

A large number of skeletons of dogs and cats was found inside the Carnide silos. Most are well preserved, and the bones of each animal appeared in connection and with most parts of the skeleton represented. This indicates that they were deposited in the silos as complete carcasses. The smaller bones however are underrepresented, so most of the skeletons are represented mainly by large bones. The absence of cut and burn marks, as well as the relatively complete state of the skeletons indicates that these animals were not consumed.

The silos do not appear to have served as graves and these cat and dog skeletons were found among other animal bones that were clearly domestic refuse and suggests these animals were not a part of a ritual. The fact that dogs were of medium size, being around 45 cm high at the shoulder, indicates that they had not undergone controlled breeding or artificial selection. Normally as a result of human selection, dogs display a wide variety of shapes and sizes (*i.e.* morphotypes or breeds).

In relation to cats, measurements of the lower first molar tooth compared with measurements of both wild and domestic cats (as in Davis *et al.* 2008) indicate that the Carnide cats were small and therefore domestic (Fig. 13.7).

The 275 and 202 bones of canids and felids, respectively, belonged to at least 25 dogs and 28 cats. The complete skeletons provided some information about their ages at death. Given their epiphysial fusion and dental eruption states, nine dogs were juvenile (under six months), the other 16 were adults, and of those, one was probably some eight months and another 12 months old at death. Knowing how normally juveniles are underrepresented because of the fragility of their bones, it seems that juvenile dogs are quite frequent. Of the cats however, only five were under six months old and 18 older, but of these as many as nine are between six and 12 months; *i.e.* the majority of the cats died before attaining one year, perhaps indicating that this pet was even less cared for than the dogs.

They were probably stray cats. None of the cat or dog remains proved to be from older animals.

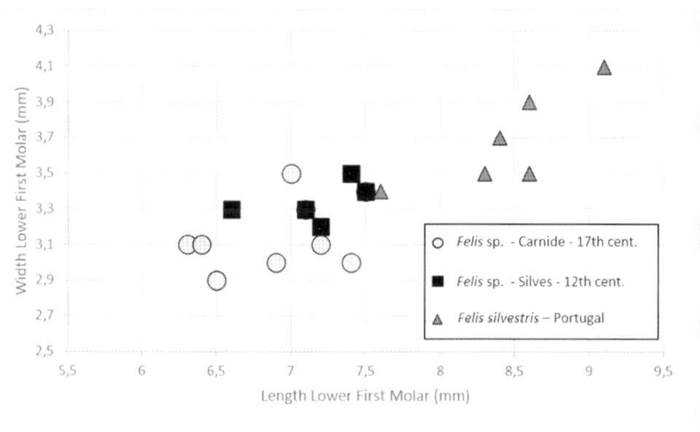

Figure 13.7. Measurements of the lower first molar of Felis *sp (in mm). Measurements from Silves archaeological site and from the reference collection of wild and domestic cats were taken by Simon Davis (Davis* et al. *2008). The measurements of the Carnide cats were taken by C. Detry.*

*Rodents (*Rattus *sp.)*

Four bones identified as *Rattus rattus* (black rat) were found in two different silos: one femur and one humerus in stratigraphic unit [2491] and one femur and one tibia in stratigraphic unit [2467]. Another three bones of its close relative, *Rattus norvegicus* (brown rat) were also found in Carnide but in different silos, one pelvis and one femur in stratigraphic unit [4313] (Fig. 13.8) and a small metatarsal from a different silo [2419]. The animals were most probably intrusive since they feed on human garbage and make burrows, probably using the silos to eat and hide. Black rats are supposed to have been introduced into Europe in Roman times (Audoin-Rouzeau and Vigne 1994) and the brown rat probably in medieval times, therefore it is not surprising to find them in 16th and 17th century Carnide.

*European polecat (*Mustela putorius*)*

Two bones of a small mustelid, a humerus and a pelvis, probably from the same individual, were also found (Fig. 13.9). This is the only occurrence of this species in

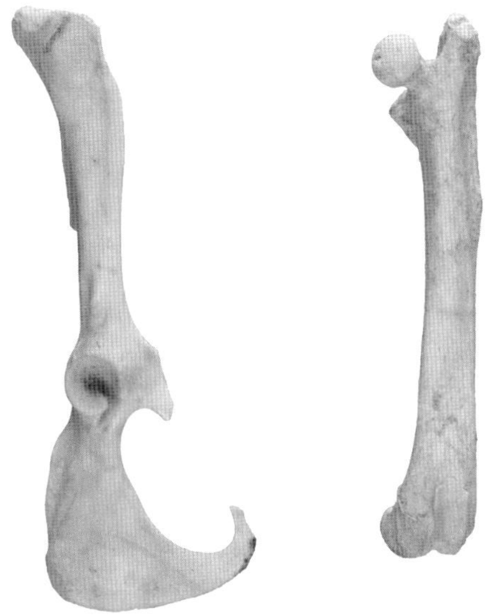

Figure 13.8. Pelvis (left) and complete femur (right) belonging to Rattus norvegicus *(brown rat) recovered in Carnide silos in stratigraphic unit [4313]. Photo by Guilherme Cardoso.*

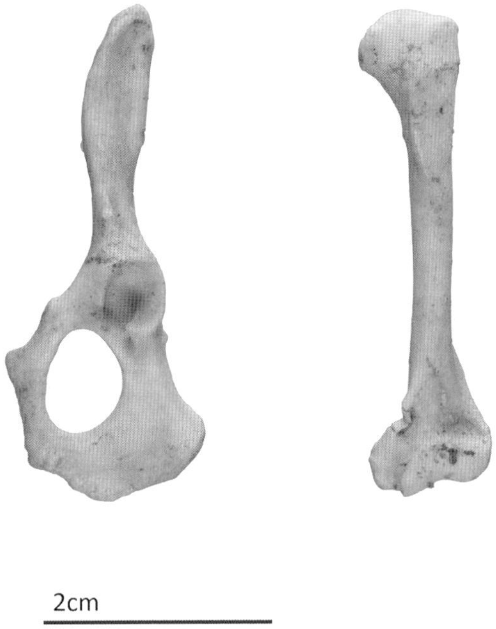

Figure 13.9. One complete pelvis (left) and one complete humerus (right) identified as Mustela putorius *(ferret) found in stratigraphic units [2855] and [2861] both in the same silo and possibly belonging to the same individual. Photo by Guilherme Cardoso.*

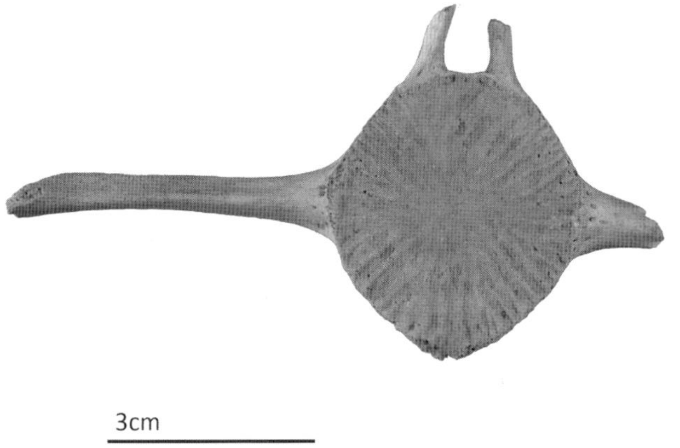

Figure 13.10. Vertebra of a small cetacean, probably a dolphin, recovered in stratigraphic unit [2873] of the Carnide silos. Photo by Guilherme Cardoso.

the whole assemblage. It was not possible to determine if they belonged to a wild (*Mustela putorius*) or domestic (*M. putorius furo*). This species can be tamed and is used to control rats or help hunting rabbits since they can easily fit into their small burrows. In the wild they are still welcome since they prey upon small rodent pests and their fur was also exploited (Thompson 1951; Vinke and Schoemaker 2012).

The origin of this particular animal in Carnide is difficult to ascertain, although appearing in this domestic context, it could have been a tamed animal used in hunting, rodent control or simply amusement.

*Dolphin (*Delphinus delphis*)*

One vertebra of a small cetacean (Fig. 13.10), probably a dolphin (*cf. Delphinus delphis*) was found in one of the silos, curiously the same silo where the polecat was found. The consumption of dolphin is attested in medieval and modern layers of the Castelo de Palmela, south of Lisbon and contemporary to Carnide (Fernandes *et al.* 2012), with the presence of vertebrae with cutmarks. In 15th-century Beja, a vertebra of a small cetacean was found inside the silos of the Avenida Miguel Fernandes, also with no cutmarks (Moreno-García, pers. comm.). In medieval England dolphin was a much-appreciated delicacy (Albarella and Davis 1996). The presence of dolphin in archaeological sites can also be found in the Pleistocene levels of Figueira Brava (Sesimbra) (Antunes 2000).

The Carnide dolphin bone has no cut marks but it is possible these animals were consumed, since it was appreciated in Portugal until not long ago.

Birds

More than half the bird remains belonged to chicken (Table 13.3) and no other domestic Galliformes were identified. Another 28% of the remains belonged to goose (*Anser* sp.), most probably the domesticated form. The rest of the birds were wild and appeared in very small numbers.

At contemporary Mosteiro de Santa Clara-a-Velha (Coimbra) and in Vila Franca de Xira (near Lisbon), the same species were preferred (Moreno-García and Detry 2010; Detry and Pimenta 2017).

Among the wild birds, two bones of red partridge (*Alectoris rufa*), a commonly hunted bird in Mediterranean habitats, were found. Other species such as pigeon (*Columba* sp.), owl (*Athene noctua*), common buzzard (*Buteo buteo*) and falcon (*Falco* sp.) are represented by very few bones. These probably bear witness to occasional additions, maybe of dead animals that occupied nearby abandoned constructions or that were living around the area and hunted. Birds of prey do eat rats which would have been common in a human environment especially in an area where food refuse and other types of garbage were found.

Six bones of the Eurasian stone curlew (*Burhinus oedicnemus)* were found. They belonged to at least two individuals. This species is not easy to hunt and has a rather low food value. Recipes for partridge, pigeon and chicken were common in 16th-century cookery books, as well as the recipes for turkey too (Casimiro *et al.* 2017). This last galliform, of American origin, was brought to Europe in the early 16th century and interestingly is not present in Carnide but was found at 17th-century Mosteiro de Santa Clara-a-Velha (Moreno-Garcia and Detry 2010).

Age at death

The age at death of animals can provide useful information concerning their management. Animals tend to be slaughtered at younger ages if they are reared primarily for their meat and at older ages if they are reared more for their secondary products such as milk, wool and traction power.

For cattle, the percentage of unfused bones (Table 13.4) is very low. In other words, few young animals are represented. The scarcity of very young cattle makes it unlikely that cows were exploited exclusively for their milk. The distribution of wear stages of the cattle teeth (Fig. 13.11) shows two or even three peaks. One represents the young adults and a large one of older adults, and even a smaller peak of very old animals. This pattern may represent the slaughter of some young adults for consumption and the majority of the animals were slaughtered when old or very old; *i.e.* retired from other functions – perhaps in the case of cattle – they were used for traction. Several metapodials displayed pathological alterations probably due to excess stress in life like carrying heavy weights (Davis *et al.* 2018). Carnide was surrounded by fields that were certainly cultivated and it was also a central trading point supporting the hypothesis that cattle would be have been used for these activities (Caessa and Mota 2016).

Figure 13.12 shows the distribution of caprine mandibular tooth wear stages. The peak seems to be around three to six years old. A smaller peak of younger animals (one to three years) indicates, as for the cattle, that some animals were slaughtered for consumption, but the majority was kept to older ages during which time they could have provided milk and wool. The fact that we have more sheep may indicate the production of wool.

Pigs on the contrary show a peak for the very young (Fig. 13.13) with a gradual decline in the dental wear stages of older teeth. Since the pig is an animal kept for its

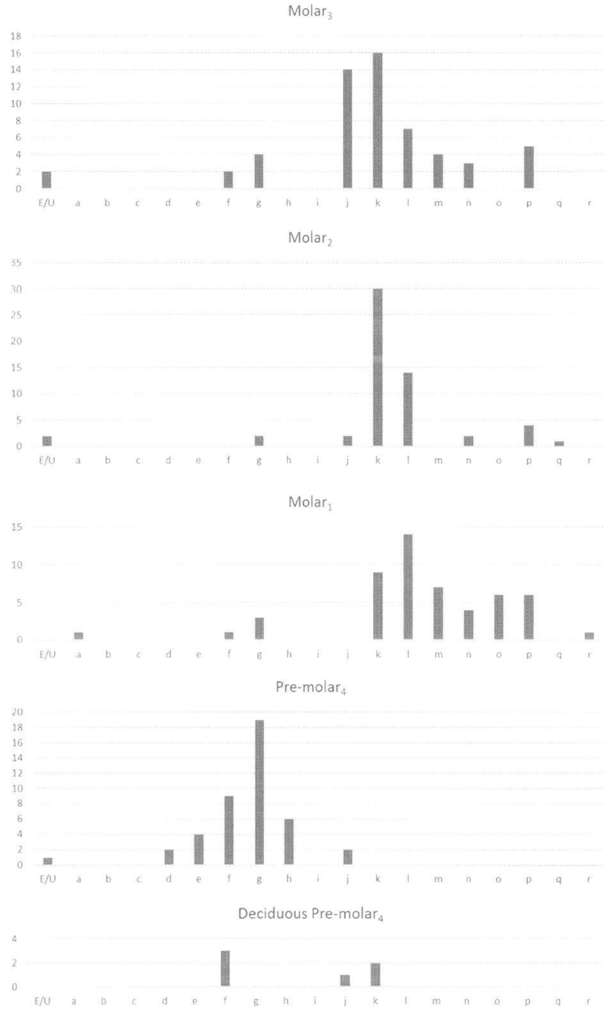

Figure 13.11. Tooth wear-stage frequencies for Bos taurus *(cattle) lower teeth found in Carnide. The wear stages follow Grant (1982).*

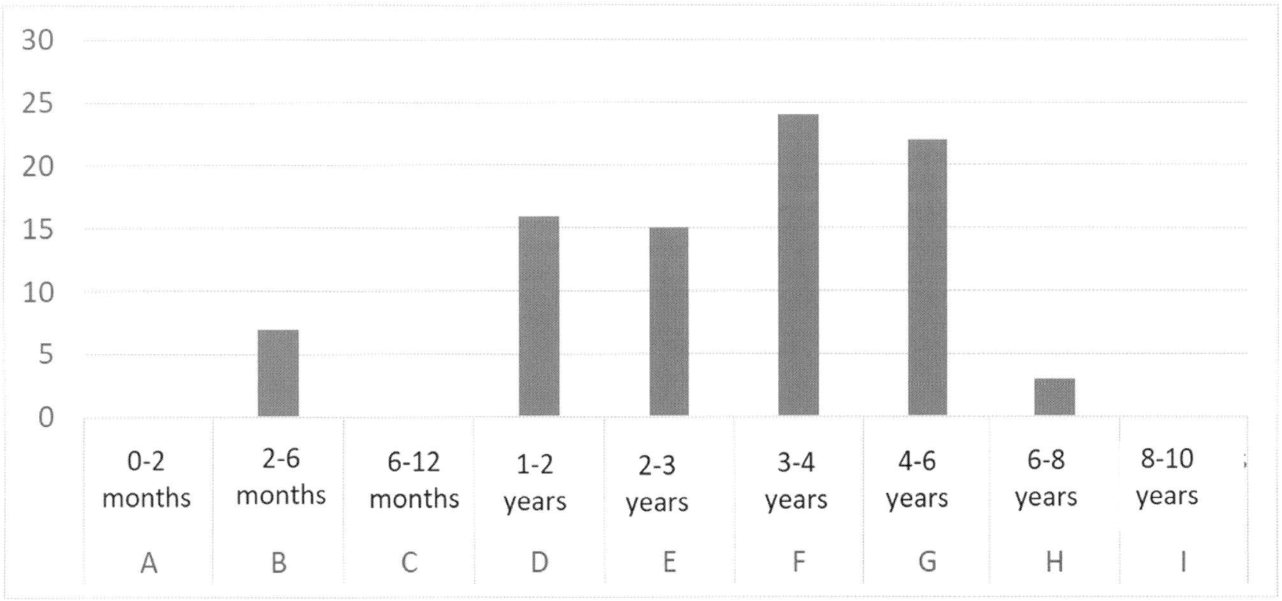

Figure 13.12. Carnide Capra/Ovis *mandible wear stages as defined by Payne (1973).*

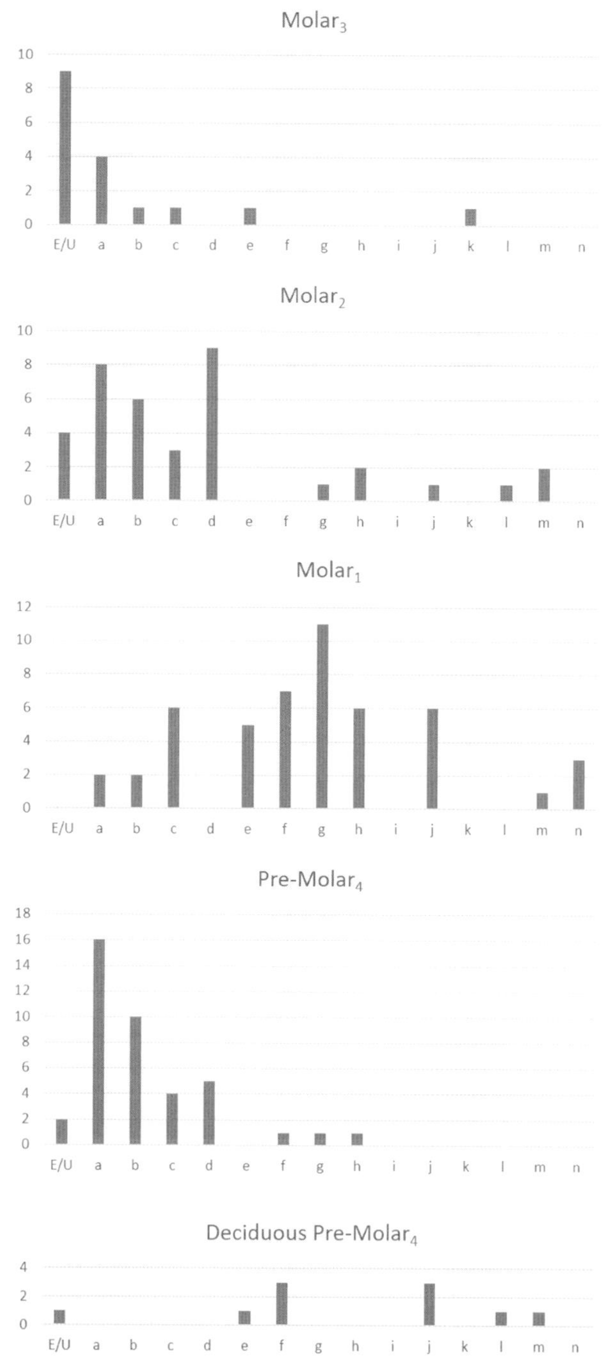

Figure 13.13. Tooth wear stage frequencies for Sus *sp. (pig) mandibular teeth found in Carnide. The teeth wear stages follow Grant (1982).*

slaughter products (meat and fat), it is general practice to slaughter pigs when still young, keeping only a few animals for breeding purposes.

Taphonomy

It is assumed that the animal remains that are described here are mainly the leftovers of meals and carcass processing. First, all parts of the skeleton, including those with less meat appear to be present. It is possible that the carcasses would have been prepared nearby and deposited in the same place as leftovers of domestic consumption. Butchering animals, removing the meat from the bones, cooking and eating could in fact be made in close spaces, especially in a rural environment such as Carnide where, unlike urban centres, people could deal with all the processing of animal consumption.

Table 13.5. Percentages of parts of the skeleton showing anthropic and animal marks.

	Bos taurus	*Ovis/Capra*	*Sus sp.*
Knife-cut	1%	1%	1%
Chopped	6%	6%	2%
Sawed	0%	1%	0%
Gnawed	2%	2%	7%
Punctures	0%	3%	1%
Burn	1%	0%	1%

While cut and burn marks (Table 13.5) are rare, there is abundant evidence for chop marks, made by coarser instruments like a cleaver. This kind of use seems more prevalent than slender cutmarks to remove meat. It is probable that carcasses were divided into smaller pieces and then cooked, since cooked meat would be very easy to detach from bones. Some bones show evidence of gnawing, probably by dogs. This is hardly surprising given the abundance dog remains in the Carnide faunal assemblage. The percentage of gnawed bones seems higher for pig, while it is strange that pig bones appear to carry fewer cut marks than the other species. Perhaps the pigs were processed or cooked differently.

Conclusion

The large number of faunal remains found in the Carnide silos is unique for this period and provides a new insight into the day-to-day life of this 16th/17th-century community. The great quantity of cattle metacarpals allowed Davis *et al.* (2018) to confirm the osteometric method for identifying the sex of the cattle (Davis *et al.* 2012). It also corroborates previous findings of a size increase of cattle after the 15th century in Portugal (Davis 2008; Davis *et al.* 2012; 2013; 2018). Clearly the human population of Carnide, depended mainly upon domesticated species for meat supply and only to a very small extent on venison. They did however consume a fair amount of marine resources such as molluscs and fish. Contemporary cookery books confirm the consumption of most of the species found on this site (Casimiro *et al.* 2017).

As evidenced in other early modern sites such as the Palmela Castle, the consumption of dolphin was probably common at that time. The presence of ferret represents a first occurrence in Portugal. The absence of any cutmarks in the bones of this animal does not allow to conclude if it was used as a food resource or a pet. The Carnide faunal assemblage also includes the largest collection of cat and dog remains from this period in Portugal, most likely stray animals that would share the streets. As mentioned, the storage pits with more cattle and caprine bones have fewer dogs and cats, indicating that these were not domestic garbage and possibly did not share the household with people.

Acknowledgements

We are most grateful to staff of the Centre for Archaeology of Lisbon (CAL) for providing us with the opportunity to study the faunal remains from Carnide and to the Laboratory of Archaeosciences (Direcção Geral do Património Cultural) for providing access to their reference collection of modern animal skeletons. We thank Guilherme Cardoso for taking the excellent photographs and André Pereira for editing them. We also thank the FCT grant from Ana Beatriz Santos (Ref.: 2020.06605.BD).

Finally, we are grateful to Simon Davis for editing the text and making suggestions.

Supplementary materials (Appendix with measurements):

Word: https://repositorio.ul.pt/bitstream/10451/43657/2/SupMat-Measurements.docx

Excel: https://repositorio.ul.pt/bitstream/10451/43657/3/SupMat-Measurements.xlsx

References

Aguirre, A., Etxeberria, F. and Herrasti, L. (2004) El yunque de hueso para afilar la hoz metálica dentada. *Munibe* 56, 113–121.

Albarella, U. and Davis, S.J. (1996) Mammals and birds from Launceston Castle, Cornwall: decline in status and the rise of agriculture. *Circaea* 12(1), 1–156.

Albarella, U., Davis, S., Detry, C. and Rowley-Conwy, P. (2005) Pigs of the 'Far West': a biometrical study of wild boar and pig remains from archaeological sites in Portugal. *Anthropozoologica* 40(2), 27–54.

Antunes, M.T. (2000) Gruta da Figueira Brava: Pleistocene marine mammals. *Memórias da Academia das Ciências de Lisboa. Classe de Ciências* 38, 245–558.

Audoin-Rouzeau, F. and Vigne, J.D. (1994) La colonisation de l'Europe par le rat noir (*Rattus rattus*). *Revue de paléobiologie* 13(1), 125–145.

Boavida, C. (2017) Dos objectos inúteis, perdidos ou esquecidos. Os artefactos metálicos do Largo do Coreto (Carnide, Lisboa). In J.M. Arnaud and A. Martins (eds) *Arqueologia em Portugal 2017: estado da questão*, 1821–1834. Lisboa, Associação dos arqueólogos Portugueses.

Boavida, C. and Medici, T. (2018) Da importação à inspiração. *Meios, vias e trajectos… Entrar e sair de Lisboa. Fragmentos de arqueologia* 2, 177–196.

Boessneck, J. (1969) Osteological differences bet-ween sheep (*Ovis aries*, Linné) and goat (*Capra hircus*, Linné). In D. Brothwell and E.S. Higgs (eds) *Science in Archaeology* 2nd ed., 331–358. London, Thames and Hudson.

Caessa, A. and Mota, N. (2013) Redescobrindo a história de Carnide: a intervenção arqueológica no largo do coreto e envolvente. *Arqueologia em Portugal* 150, 1025–1032.

Caessa, A. And Mota, N. (2014) O núcleo histórico de Carnide. O contributo da investigação arqueológica. In C.M. Veiga and M.F. Reis (eds) *Quadros da História: a freguesia de Carnide*, 83–104. Lisboa, Academia Portuguesa de História.

Caessa, A. and Mota, N. (2016) A Arqueologia em Carnide e a intervenção arqueológica no Largo do Coreto. *Rossio – Estudos de Lisboa* 6, 96–107. Lisboa, Câmara Municipal de Lisboa/Direcção Municipal de Cultura/Departamento de Património Cultural.

Callou, C. (2003) *De la garenne au clapier. Histoire et archéologie du lapin européen*. Paris, Publications Scientifiques du Muséum.

Cardoso, J.L. (2002) Arqueofaunas: balanço da sua investigação em Portugal. *Arqueologia e História* 54, 281–298.

Casimiro, T., Boavida, C. and Detry, C. (2017) Cozinhar e comer: cerâmicas e alimentação em Carnide (1550–1650). *Diz-me o que comes… A alimentação antes e depois da cidade. Fragmentos de arqueologia* 1, 110–121.

Casimiro, T.M., Boavida, C. and Moço, A. (2017) Louça 'de fora' em Carnide (1550–1562). *I Encontro de Arqueologia* 1, 50–67.

Castaños, P. (1991) Animales domésticos y salvajes en Extremadura. Origen y evolución. *Revista de Estudios Extremeños* 47, 9–67.

Davis, S.J.M. (1992) A rapid method for recording information about mammal bones from archaeological sites. *Ancient Monuments Laboratory Report* 19, 1–14. London, Ancient Monuments Laboratory.

Davis, S.J.M. (1996) Measurements of a group of adult female Shetland sheep skeletons from a single flock: a baseline for zoo-archaeologists. *Journal of Archaeological Science* 23, 593–612.

Davis, S.J.M. (2008) Zooarchaeological evidence for Moslem and Christian improvements of sheep and cattle in Portugal. *Journal of Archaeological Science* 35(4), 991–1010.

Davis, S.J.M. (2009) Animal remains from an 18th–19th century AD pit in the Palácio Centeno, Lisbon. *Revista Portuguesa de Arqueologia* 12(2), 239–250.

Davis, S.J.M. (2016) Hacia una distinción métrica entre los astrágalos de oveja y cabra. In Ll. Lloveras, C. Rissech, J. Nadal and J.M. Fullola (eds) *What bones tell us. El que ens expliquen els ossos, Monografies del SERP 12*, 35–58. Barcelona, Universitat de Barcelona.

Davis, S.J.M. and Mataloto, R. (2012) Animal remains from Chalcolithic São Pedro (Redondo, Alentejo): evidence for a crisis in the Mesolithic. *Revista Portuguesa de Arqueologia* 15, 47–85.

Davis, S.J.M., Albarella, U., Detry, C., Ginja, C., Götherström, A., Pires, A.E., Sendim, A. and Svensson, E.M. (2018) An osteometrical method for sexing cattle bones: the metacarpals from 17th century Carnide, Lisbon, Portugal. *Annalen des Naturhistorischen Museums in Wien, Serie A* 120, 367–387.

Davis, S.J.M., Gonçalves, M.J. and Gabriel, S. (2008) Animal remains from a Moslem period (12th/13th century AD) lixeira (garbage dump) in Silves, Algarve, Portugal. *Revista portuguesa de Arqueologia* 11(1), 183–258.

Davis, S.J.M., Svensson, E.M., Albarella, U., Detry, C., Götherström, A., Pires, A.E. and Ginja, C. (2012) Molecular and osteometric sexing of cattle metacarpals: a case study from 15th century AD Beja, Portugal. *Journal Archaeological Science* 39(5), 1445–1454.

Davis, S.J.M., Svensson, E., Albarella, U., Detry, C., Götherström, A., Pires, A., Ginja, C. (2013) Evidencia de mejoras de ovino y vacuno durante época andalusí y cristiana en Portugal a partir del análisis zooarqueológico y de ADN antíguo. *Debates de Arqueologia Medieval* 3, 241–287.

Detry, C. and Arruda, A.M. (2013) A fauna da Idade do Ferro e da Época Romana de Monte Molião (Lagos, Algarve): continuidades e rupturas na dieta alimentar. *Revista Portuguesa de Arqueologia* 16, 213–226.

Detry, C. and Pimenta, J. (2017) Animal remains from medieval and modern Vila Franca de Xira, Portugal: Excavations at the Neo-Realism Museum. *CIRA Arqueologia* 5, 238–259.

Detry, C., Cardoso, J.L. and Bugalhão, J. (2016) A alimentação em Lisboa no decurso da Idade do Ferro: resultados das escavações realizadas no núcleo arqueológico da rua dos Correeiros (Lisboa, Portugal). *SPAL* 25, 67–82.

Detry, C., Gambini, L. and Corte-Real, A. (2014) At table with the nuns: the mammals of 17th century Santa-Clara-a-Velha Monastery (Coimbra, Portugal). In C. Detry and R. Dias (eds) *Proceedings of the First Zooarchaeology Conference in Portugal. British Archaeological Reports* 2662, 117–128. Oxford, Archaeopress.

Driesch, A. von den. (1976) *A Guide to the Measurement of Animal Bones from Archaeological Sites*. Peabody Museum of Archaeology and Ethnology Bulletin 1. Cambridge, Harvard University.

Fernandes, I.C.F., Cardoso, J.L. and Detry, C. (2012) Cozinhar e comer no Castelo Medieval de Palmela. In S. Gómez Martínez (ed.) *Memória dos sabores do Mediterrâneo*, 113–127. Mértola, Campo Arqueológico de Mértola.

Foighil, D.Ó., Gaffney, P.M., Wilbur, A.E., and Hilbish, T.J. (1998) Mitochondrial cytochrome oxidase I gene sequences support an Asian origin for the Portuguese oyster *Crassostrea angulata*. *Marine biology* 131(3), 497–503.

Garcia, S., Caessa, A. and Mota N. (in press) Enterramentos no Largo do Coreto em Carnide: vestígios do cemitério da Ermida do Espírito Santo.

Grant, A. (1982) The use of tooth wear as a guide to the age of dome s tic ungulates. In B. Wilson, C. Grigson and S. Payne, *Ageing and Sexing Animal Bones from Archaeological Sites*, 91–108. British Archaeological Reports British series 109. Oxford, Archaeopress.

Huvet, A., Lapegue, S., Magoulas, A. and Boudry, P. (2000) Mitochondrial and nuclear DNA phylogeography of *Crassostrea angulata*, the Portuguese oyster endangered in Europe. *Conservation Genetics* 1(3), 251–262.

López Garaulet, L. (2011) *Estabelecimento do bivalve exótico Ruditapes philippinarum (Adams & Reeve, 1850) no estuário do Tejo: caracterização da população actual e análise comparativa com a congénere nativa Ruditapes decussatus (Linnaeus, 1758) e macrofauna bentónica acompanhante*. Unpublished PhD thesis, University of Lisbon.

Loureiro, A. Ferrand de Almeida,,N. Carretero, M.A. and Paulo, O.S. (eds) (2008) *Atlas dos Anfíbios e Répteis de Portugal*. Lisbon, Instituto da Conservação da Natureza e da Biodiversidade.

Macedo, M. (1996) *Rei Dom Carlos. Campanhas oceanográficas. Estudo das colecções malacológicas*. Lisboa, INAPA.

Matos, R. (2014) *Atlas dos caracóis de Portugal Continental*. Lisboa, Edição da Autora.

Moreno-Garcìa, M. and Detry, C. (2010) The dietary role of hens, chickens and eggs among a 17th-century monastic order: the Clarisse of Santa Clara-a-Velha, Coimbra (Portugal). *Birds in Archaeology, Proceedings of the 6th Meeting of the ICAZ Bird Working Group in Groningen* (23.8–27.8.2008), *Groningen Archaeological Studies* 12, 45–55. Groningen, Barkhuis Publishing.

Moreno-García, M., Pimenta, C. and Ruas, J.P. (2005) Safras em osso para picar foicinhas de gume serrilhado… a sua longa historia! *Revista Portuguesa de Arqueologia* 8(2), 571–627.

Neves, A., Cabral, H., Sequeira, V., Figueiredo, I., Moura, T. and Gordo, L.S. (2009) Distribution patterns and reproduction of the cuttlefish, Sepia officinalis in the Sado estuary (Portugal). *Journal of the Marine Biological Association of the United Kingdom* 89(3), 579–584.

Payne, S. (1969) A metrical distinction between sheep and goat metacarpals. In P.J. Ucko and G.W. Dimbleby (eds) *The Domestication and Exploitation of Plants and Animals*, 295–305. London, Duckworth.

Payne, S. (1987) Reference Codes for wear states in the mandibular cheek teeth of sheep and goats. *Journal of Archaeological Science* 14, 609–614.

Rodrigues, A.M., Meireles, S., Pereira, T., Gama, A. and Quintino, V. (2006) Spatial patterns of benthic macroinvertebrates in intertidal areas of a Southern European estuary: the Tagus, Portugal. *Marine Biodiversity* 183, 99–113.

Saldanha, L. (2003) *Fauna Submarina Atlântica*, 4th edition. Lisboa, Publicações Europa-América.

Sterry, P. and Cleave, A. (2012) *Collins Complete Guide to British Coastal Wildlife*. London, HarperCollins Publishers.

Thomson, A.P. (1951) A history of the ferret. *Journal of the History of Medicine and Allied Sciences* 6, 471–480.

Tiley, S. and Burger, E. (2002) Cowries in the archaeological and maritime record. *Strandloper, April* 267, 5–7.

Valente, M.J. and Carvalho, A.F. (2014) Zooarchaeology in the Neolithic and Chalcolithic of southern Portugal. *Environmental Archaeology* 19(3), 226–240.

Verdelhos, T., Marques, J.C., and Anastácio, P. (2015) The impact of estuarine salinity changes on the bivalves *Scrobicularia plana* and *Cerastoderma edule*, illustrated by behavioural and mortality responses on a laboratory assay. *Ecological indicators* 52, 96–104.

Vinke, C.M. and Schoemaker, N.J. (2012) The welfare of ferrets (*Mustela putorius furo* T): A review on the housing and management of pet ferrets. *Applied animal behaviour science* 139(3–4), 155–168.

Watson, J.P. (1979) The estimation of the relative frequencies of mammalian species: Khirokitia 1972. *Journal of Archaeological Science* 6(2), 127–137.

Zeder, M.A. and Lapham, H.A. (2010) Assessing the reliability of criteria used to identify postcranial bones in sheep, Ovis, and goats, Capra. *Journal of Archaeological Science* 37(11), 2887–2905.

Zeder, M.A. and Pilaar, S.E. (2010) Assessing the reliability of criteria used to identify mandibles and mandibular teeth in sheep, Ovis, and goats, Capra. *Journal of Archaeological Science* 37(2), 225–242.

14

The contribution of Islamic culture to the medieval faunistic redefinition of the Iberian Peninsula

Marco Masseti

From its first appearance in the 7th century AD, Islam generated a profound cultural renewal along the Mediterranean shores, from the Levant to the Iberian Peninsula. Together with the new religion, Islam brought with it new customs and cultural traditions. These included the utilisation of foreign biological elements which comprised domestic animals, such as the dromedary and special breeds of sheep, that revolutionised the traditional subsistence strategies. Arabs are considered responsible for the definitive introduction into Europe of a variety of foreign zoological species, including the Algerian hedgehog, several African and Near Eastern medium-sized carnivores, perhaps, the Mediterranean chameleon, and others. They inherited structures of the hunting park type from previous cultures and they passed them on to subsequent civilizations. The Islamic faunistic and ecological rebalancing continued in the periods immediately following those of Muslim influence on the ancient oecumene. Certain consequences may indeed have been one of the original sources of the by now global story of nature conservation, playing a fundamental role of cultural filter between the Late Antique and the modern world.

Keywords: Islam, Mediterranean region, Iberian Peninsula, exotic zoological species, domestic animals.

Introduction

Among the various human groups that migrated to the west in the Early Middle Ages, the Arabs were the bearers of a cultural complex destined to leave a particularly deep mark on the oecumene of the time. In 711 AD, the conquest of the Rock of Gibraltar (*Jebel el Tariq*), by the Muslim commander Tariq ibn Ziyad, allowed the Umayyad – the first Islamic caliphate – to extend their dominion over most of the Iberian Peninsula, in part of which the Islamic supremacy lasted up to the end of the 15th century. In the 9th century, the conquest of Sicily provided to this island the role of geographic centre of the Muslim empire, which extended from the Iberian Peninsula up to the western Indian subcontinent.

Together with the new religion, Islam brought with it new customs and cultural traditions. Since its first appearance, Muslim culture seemed to pay particular attention to the natural elements. The Syrian Desert was crowded with fortifications and castles decorated with frescoes documenting its faunistic richness. These artistic representations have been described as the last pure and untroubled efflorescence of Hellenistic grace (Brown 1971). Even in the Iberian Peninsula the new art brought by Islam was deeply influenced by the pre-existing Late Antique productions (Fig. 14.1).

In some of the westernmost parts of the Mediterranean, the introduction of new botanical species, such as the white mulberry, *Morus alba* L., the lemon, *Citrus lemon* (L.) Osbeck, or the sugarcane, *Saccharum officinarum* L. (Gabrieli 1966; Lo Valvo *et al.* 1993; Bresc 2003), revolutionised the traditional subsistence strategies. Not to mention the artificial diffusion of domestic animals, such as the dromedary, *Camelus dromedarius* L., 1758 (*cf.* Cardoso and Fernandes 2012), and some breeds of sheep characterised by larger dimensions than those previously recorded (Davis 2005; 2017; Masseti 2019). Due to the long duration of Islamic rule over the Iberian Peninsula, the latter has proved to be one of the territories most receptive to faunistic innovations and to the introduction of alien biological elements which were carried out by Muslim people over nearly eight centuries.

craniological differences between them (Pocock 1951; Heptner and Sludskii 1992). Various genetic investigations have confirmed that the two groups belong to the same polymorphic species, differentiating only at a sub-specific level (Ragni and Possenti 1991; Hemmer 1999; Ottoni *et al.* 2017). Apart from Sicily, which was colonised naturally by the European wild cat during an unspecified moment of the Upper Pleistocene, all the other extant populations of the Mediterranean islands are the result of human activity and derived from the *F.s. libyca* group (Masseti 2010a; 2018).

Already known in classical antiquity (*cf.* Herodotus, *The Stories* II, 66–67), and very likely tamed – but, possibly, not yet domesticated – in Neolithic Near East (*cf.* Vigne *et al.* 2004), the African cat made its first sporadic appearances in eastern Europe from at least the start of the 6th century BC or, possibly, even earlier (*cf.* Toynbee 1973; Malek 1993; Masseti 2002). Known in western Europe from the early centuries of our era, at the end of the Roman Empire the domestic cat is documented more or less everywhere by scattered osteological remains (see also Zeuner 1963). Bobis (2000) believes that its diffusion was favoured by the trade routes, in particular the tin road that linked the British Isles to the Mediterranean. The Roman army too must have represented a crucial carrier for the penetration of the animal in northern Europe, since several fortresses guarding the Rhine-Danube *limes* were home to cats. Nevertheless, according to the results of research carried out by MacKinnon (2004), cat bones were infrequent in most Roman sites of the Italian Peninsula between the Republican period and Late Antiquity (end of the 6th century BC–6th century AD), having been found in only 16 of the 146 rural and urban sites considered in the study. Furthermore, it should be understood whether they were all domestic cats or whether they could include even a number of wild specimens. In the Classical Age, however, another small-sized carnivore was used in domestic environments to control unwanted populations of rodents. This was the weasel, *Mustela nivalis* (L., 1766), as, among other things, widely documented also by Bon and Masseti (1995). In ancient Greek, this animal was indicated with the term gale (*gale*) (Masseti 2002). In the *Batracomiomachia*, for example, an encomiastic poem from the 5th century BC, the mouse says to the frog: 'There are two beings I fear most on earth: the hawk and the weasel'. In his comedy *The Wasps*, Aristophanes (about 445 – after 388 BC) indicates in the *gale* the domestic hunter of mice *par excellence*, not considering the cat. Still in Roman times, literary documents do not seem to assign to the cat the specialised role of domestic mice hunter. In the comedy *Stichus* (499–501) by the Latin author Plautus (*c.* 259/251–184 BC) is a weasel and not a cat to catch a mouse at the foot of an actor. Pliny the Elder (1st century AD), in his *Naturalis historia* (XXIX, 60–61), handed down to us that the weasel was allowed to walk inside the houses, where: '*according to Cicero, every day it transports its puppies and changes location*' (*cf.* Capitani and Garofano 1986). In antiquity, other carnivores of larger size, such as the stone marten, *Martes foina* (Erxleben, 1777, the pine marten, *Martes martes* (L., 1758), and the polecat, *Mustela putorius* L., 1758 – generically indicated by the ancient Greek term of iktis (*ictis*) – were also used to defend domestic larders from damage produced by rodents. (Masseti 2002). In this regard, there is the later witness of Pecci (1760) who explains, still in the 18th century, that on the island of Giglio in the Tuscan Archipelago (Italy) the hunt for martens: 'is forbidden at any time due to the fact that they have been properly introduced in ancient times because, together with the many non-poisonous snakes, they diminish the large number of mice that will otherwise devastate all the country'.

Although the presence of the domestic cat in Europe is testified by a range of documents dating to the Early Middle Ages, prior to any Islamic influence (*cf.* Bobis 2000), it may not have become widespread until the establishment of the Arab culture, in concomitance with which the felid finally became more extensively diffused, at least in the countries of the northern Mediterranean and western Europe. In this regard, it is interesting to note that in the towns of medieval Britain too there is some evidence that the frequency of domestic cats increased in the years following the Norman conquest (O'Connor 1982; 1992), that is, from the second half of the 11th century (*cf.* Rowley 1999; Crouch 2002). A noticeable diffusion in Sicily of the feline can be traced to shortly after the year 1000, vestiges having been found in the excavations at the 'Antonino Salinas' Regional Archaeological Museum of Palermo, in chronological contexts referred to the second half of the 10th–early 11th century (Sarà 1997). Prior to this discovery, the oldest Sicilian finds of the domestic cat dated to the 12th century, found in a pit of the castle of Fiumedinisi (Villari 1995) and at Brucato (Bossard-Beck 1984). Both sites yielded remains bearing butchering cut marks and traces of burns (Villari 1995), as has also been documented from other European sites of 14th–18th century (*cf.* Villari 1995; Corbino 2009). In Sicily, the domestic cat is also present in 13th-century layers in Calathamet (Sarà 2005), and in the 14th- and 15th-century contexts of Palazzo Steri, in Palermo (Lupo 2006–2007).

In the Iberian Peninsula too, the more consistent spread of the domestic cat appears to coincide with the full affirmation of Islamic culture (Masseti 2009a; 2019). Finds referred to the carnivore have in fact been yielded by the exploration of the Spanish sites of Granada (Califal period, 10th–11th century) (Riquelme 1992), Calatrava La Vieja (Almohad period) (Morales Muñiz *et al.* 1988), Saltés (Huelva) (12th–13th century) (Lentacker and Ervynck 1999), Motril (Granada) (16th–18th century) (Riquelme Cantal 1993), and the Portuguese site of Alcáçova de Mértola (second half of the 12th century–first third of the 13th century) (Telles Antunes 1996). Nevertheless, it cannot be ruled out that for the Iberian Peninsula – together with continental Italy and Sicily – the absence of finds for slightly earlier historic periods may very plausibly be attributable to the lack of excavations or the absence of specific archaeozoological studies. Unfortunately, it is not always possible to document the past presence of a certain zoological species in a specific territory and/or a particular cultural context, solely on the basis of the data offered by archaeozoological research (Masseti 2009a). Consequently, it would be advisable to adopt a different approach to this subject, taking into account a multidisciplinary view which includes also literature and iconography.

Figure 14.3. Adult male black francolin, Francolinus francolinus *(L., 1766). Photo by Giuliano Cappelli.*

Foreign wild game

Also the occurrence of the black francolin, *Francolinus francolinus* (L., 1766) (Fig. 14.3), a Galliform bird of the size of a partridge, is attested in Europe since, at least, Early medieval times (Masseti 2009a; 2016). With the adult males characterised by black underparts spotted in white, a chestnut collar and a white ear-patch, this is another biologic element native to the Near East, where its present distribution ranges from Cyprus, Asia Minor and the southern Caspian,south-eastward through Iraq, Iran and southern Afghanistan to Pakistan, and across India east to Assam and south to the Deccan plateau (Johnsgard 1988). Various authors, such as Amari (1937), Orlando (1958), Johnsgard (1988) and Jiménez (2013), consider that its first introduction in the western World has been effected by the Arabs, or alternatively may have occurred following the return of the first Crusades (Arrigoni Degli Oddi 1929; Baldacci 1964; Pratesi 1976; Perco 1981). Baccetti (1992), partly following Baldacci (1964), observes that in Spain, Italy and the respective islands, numerous introductions of this bird took place over the course of the centuries, and it has been possible to document for long periods of time the presence of various populations of uncertain origin. In Sicily, the oldest bones of black francolin have been provided by the archaeological exploration of the site of Calathamet, dating to the 13th century (Sarà 2005). Moreover, as observed by Maluquer and Travé (1961) and by Muntaner *et al.* (1983), the western artificial diffusion of this bird coincides with surprising precision with the historic territory of dominion of the Catalan-Aragonese Confederation, which comprised Sicily starting from 1282 and took in even more easterly regions during the following century (*cf.* Rodríguez-Picavea Matilla 2006), thus offering ample possibilities of contact between the various areas of the Mediterranean. During this period, black francolins were frequently the subject of lively interest on the part of kings and nobles, possibly due to their value as excellent meat and prized courtly game, as we can glean from a number of literary and legal documents. In order to protect the fowl, severe legal provisions prohibited the hunting and poaching of the species. One of the oldest documents indicating the value of the bird is a letter sent by the Spanish king Peter IV of Aragon, from Sicily to the governor of Mallorca dated 26 May 1368. This letter informs on the delivering to the latter island of several individuals of pheasants and francolins 'in order that they shall breed and multiply' (Maluquer and Travé 1961). Evidence of the former occurrence of black francolins in the Balearics is to date restricted to a single mounted specimen without data, still preserved in the museum at Mahon (Menorca): no further mention of the introduction of the species and no other trace of it are available (Bannerman and Bannerman 1983). Apart from Cyprus, where it is likely to have been artificially introduced in historical times too, the black francolin no longer belongs to the European fauna. The species disappeared, in fact, from the Iberian and Balkan peninsulas, France and Italy during the 19th century (Arrigoni Degli Oddi 1929; Ghigi 1968; Ferrer *et al.* 1986; Diaz *et al.* 1996; Jiménez 2013). Its extinction is mainly regarded as a consequence of the combined effect of hunting and land reclamation (Amari 1937; Pratesi 1976; Iapichino and Massa 1989; Masseti 2016).

Under Islamic rule, other allochthonous game animals were imported into the Iberian Peninsula. Literary references document also the importation of gazelles, possibly from North Africa. These ungulates do not occur at present along the northern Mediterranean mainland or on any island, but on some of the latter their former presence is testified not only by osteological evidence but also by literary and artistic documents (Masseti 2016). According to several Arab scholars, such as Al-Idrisi (12th century) and Sahâb ad dîn Abû al-Abbâ Ahmad ibn Yahyâ Al-Umarî (14th century), gazelles occurred on the small island of Marettimo, in the Egadi archipelago (off the north-western coast of Sicily) (Masseti 2009c; 2016; *cf.* Amari and Schapparelli 1883; Rizzitano 1994). On the not far distant Malta, some of these animals were let loose in large private grounds for hunting purposes in mediaeval times (Lanfranco 1969); and a set of gazelle bones of undetermined age has been uncovered from strata of soil and loose rocks on the nearby islet of Comino (Boffa 1966). Literary references document the introduction of gazelles also on Mallorca, in the Balearic Archipelago,

Figure 14.4. Tile decorated with the image of a gazelle tied to a collar inside of a stylised garden. Probable Iberian production, Manises (Valencia), 15th century (Florence, private collection).

between the 12th and the 15th centuries (Llabrès Ramis and Vallespir Soler 1983). It is practically impossible today to trace back to the taxonomic identification of these animals due to the absence of an accurate description of their morphology and a reliable artistic representation. We can only assume that, in the majority of cases, they could have been dorcas gazelles, *Gazella dorcas* (L., 1758), which were the most common gazelles wherever they occurred in the ancient oecumene (Masseti 2010b) (Fig. 14.4).

Domestic breeds

Hilzheimer (1909), Diffloth (1924), Epstein (1971) and Lombard (1971) are of the opinion that the ancestors of the famous modern *Merinos* sheep, generally regarded as a domestic breed that has been fixed in the Iberian Peninsula, have been previously selected in the North African highlands at the time of the Islamic rule. The world *merinos* seems, in fact, to derive from the name of the Berber tribe *Banū marīn* of western Maghreb, or from the Arab *marīn* which translates 'sweet, soft, fine', derived from the fleece of particularly fine hair of these animals. Metrical studies of sheep bones from archaeological sites in southern Portugal indicate that sheep became larger in Muslim times (Davis 2003; 2017). Such an increase in size could reflect an ethnic improvement of animals in the more complex set of the so-called 'Arabic agro-culture revolution' of *al-Andaluz* (Davis 2005; 2008), that is to say Muslim Spain. It could be an objective testimony of the moment in which the genetic pool of the Iberian sheep became substantially enriched, thus representing one of the fundamental stages in the history of modern breeding in Spain and Portugal (Pereira *et al.* 2002; Davis 2005; 2008). According to Davis (2017), however, whether new stock such as the *Merinos* was imported from abroad or whether local animals were improved must remain within the realm of speculation. While the absence of any observable change in the shape of the sheep bones between Roman and Muslim periods (*cf.* Davis 2008) tends to point in favour of a local improvement rather than import of stock, the genetic evidence based on modern sheep in the Iberian Peninsula does indicate some input from overseas, thought just when this occurred is unknown (Pereira *et al.* 2006).

Following these arguments, it could still be observed that Muslim breeders must have preferred fat tailed sheep because they could have not obtained fat from other domestic animals, such as pigs which are regarded as taboo in Islamic culture. Thus, the *Merinos* – which, apart from the development of the fleece, is a short-tailed animal, not very distant from the body morphology of its wild ancestor, the Asiatic mouflon, *Ovis orientalis* Gmelin, 1774 – very likely originated from a separate breed, mainly devoted to the production of wool; and possibly pre-existent to the Islamic spread. We know that a peculiar breed of sheep characterised by the medium-fine wool flake of the Iberian populations (that would later be destined to evolve into the *Merinos* breed) was already known in southern Italy in Roman times and even earlier. This was called as *gentile di Puglia*, being Apulia its homeland (Volpe *et al.* 2012). It should not be a coincidence that an archaeological complex called *Merino*, is today located 7 km from the town of Vieste, in Apulia, covering a large area known as the Roman Villa of *Merino*, which remained in activity between the 1st and the 10th centuries AD.

Although the Romans had already imported dromedaries in several parts of their empire, the Iberian remains of this species from the Caliphate period refute the long-held notion that these camelids arrived in Iberia with the Almoravids and Almohads, and support the statement that the Muslims took them to Spain from Northern Africa in 1019 AD (Morales Muñiz *et al.* 1995; *cf.* Zeuner 1963; Cardoso and Fernandes 2012) (Fig. 14.5). In times of peace, these animals were set free to roam in the steppes of south-east Spain, in Albacete and Murcia, but rather curiously, it seems that they never established permanent feral populations in the Iberian Peninsula.

Concluding remarks

More than other Early medieval cultures, Islam seems to have promoted a real faunistic renewal in the Iberian Peninsula, with the diffusion of zoological species that until then had been regarded as partially or completely extraneous. In simple terms, with the affirmation of Muslim culture, we are

Figure 14.5. Image of a dromedary in the frescoes of the hermitage of San Bauledio de Berlanga, at Casillas de Berlanga (province of Soria, Spain), built in the first half of the 11th century in what was then a border area, or no-man's-land, between the Christian kingdoms and the territories dominated by the Muslims.

witnessing a pilot invasion of alien zoological elements that become part of the western natural ecosystems, eventually altering the original equilibrium (Masseti 2019).

Since the time of the Muslim dominion of Spain and Portugal, the common genet has become part of the European fauna within which its distribution extends today to a large area of France, as well as small portions of north-western Italy (Fig. 14.6). After being able to adapt to the Iberian ecosystems over at least seven centuries, the black francolin became extinct in Spain only in the 19th century, and the extant diffusion of the Mediterranean chameleon in Andalusia seems to be due to repeated anthropic introductions. It cannot even be excluded that the Arabs introduced crocodiles in a few territories beyond their homeland (Masseti 2009a; 2016). Among the domestic varieties, sheep have evolved into a breed that generated the modern *Merinos*, while the African Guinea fowl, *Numida meleagris* L., 1766, has become so well acclimatised in Spain that Bru De Ramon defines it as a nesting bird as far as 1784.

The conquest of the Levant, Asia Minor, North Africa and south-western Europe brought the Arabs into contact with the Classical and Late Antique traditions of emparking and the keeping of beasts; traditions that they were later to transmit to the rest of the western World (Masseti 2009a;

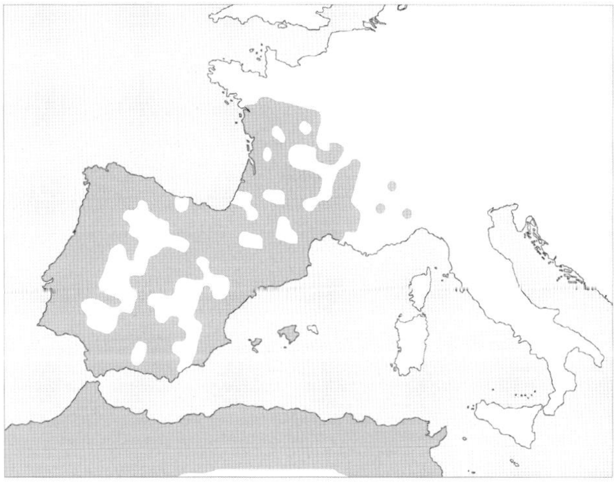

Figure 14.6. Present European distribution of the common genet, Genetta genetta *(L., 1758).*

2016). Literature attests in fact these hunting practices since, at least, the 1st century AD, when Pliny the Elder (*Naturalis historia* VIII, 211) observes that 'Fulvius Luppinus was the first person of Roman nationality who invented preserves for wild boars and the other kinds of game: he introduced keeping wild animals in the district of Tarquinia; and he did not long lack imitators, Lucius Lucullus and Quintus Hortensius' (see Rackham, 1940). Also the Greek scholar Claudius Aelian (*c.* 175–*c.* 235 AD) refers in his *On the Nature of Animals* of the existence of islands used as enclosures for wild goats, gazelles and hares (see Scholfield 1959; McNamee 2011).

Towards the end of the early Middle Ages, links between the Christian oecumene and the Islamic world became increasingly close, resulting in considerable political and cultural exchange (Loud 2003; *cf.* Cassady 1986). This provided ample opportunity for the exchange of ideas and goods, such as the spread of Arab architectural influence and of other aspects of Muslim culture. It is amply established that, after the 8th century, the Iberian Peninsula became a focal point for the transmission of Islamic contributions to the rest of Europe, a model and an example which was universally admired (Aubé 2006), playing a key role in the transmission of Islamic knowledge (Lewis 1993). This knowledge comprised also structures of the *viridarium* and hunting park type. The faunistic and ecological rebalancing continued in the periods immediately following those of Islamic influence on the western World. Since Early medieval times, Muslim cultural connections with Christian Europe, possibly borrowed from traditions deriving from Late Antiquity, may indeed have been one of the original sources of the by now global story of nature conservation, playing a fundamental role of cultural filter between the Late Antique and the Modern world (Masseti 2009a; 2016). It cannot be excluded that the conceptual foundations of the theory of 'nature conservation', which over recent centuries have been particularly developed by the Anglo-Saxon culture and disseminated to the rest of the world, were embedded in previous foreign traditions, inherited from far afield, possibly from the early Arabian culture that would pass it on to the Normans of Sicily and their cousins in Great Britain.

After dominating a large part of the Iberian Peninsula for almost eight centuries, Islam was eventually defeated by the *Reconquista* of the Catholic kings Isabella of Castile and Ferdinand of Aragon in 1492, a year that marks also the date of the discovery of America, paving the way for knowledge of the geography and fauna of the New World and the importation in Europe of fabulous zoological species hitherto unknown.

Acknowledgements

In the course of this study I have been fortunate to have the help of many friends and colleagues. I am particularly grateful to Umberto Albarella, Department of Archaeology of the University of Sheffield (UK); Chiara Assunta Corbino, Laboratorio di Scienze Applicate of the Parco Archeologico di Pompei, Napoli (Italy); Simon J.M. Davis, Instituto Portugueŝ de Arqueologia, Lisbon (Portugal); Arturo Morales-Muñiz, Departamento de Biologia, Facultad de Ciencias. UAM, Madrid (Spain); and Joao Mayol Serra, Conselleria de Medi Ambient, Govern de les Illes Balears, Palma de Mallorca (Spain). Special thanks are due to Joao Tereso and Cleia Detry for the critical reading of the manuscript and their suggestions.

References

Amari, G. (1937) Avifauna siciliana scomparsa. Il francolino. *Diana* 24, 1082–1084.

Amari, M. (1935) *Storia dei musulmani di Sicilia. Vol. 2*. Catania, Romeo Prampolini Editore.

Amari, M. and Schapparelli, C. (1883) L'Italia nel 'Libro di Ruggero'. Compilato da Edrisi. Memoria letta nella seduta del 17 dicembre 1876. Roma, ColTipografia Salviucci.

Amigues, S. (1999) Les belettes de Tartessos. *Anthropozoologica* 29, 55–64.

Andre, J. (1981) *L'alimentation et la cuisine à Rome*. Paris, Colléctions Etudes Anciennes, Les Belles Lettres.

Andre-Salvini, B., Demange, F., Juvin, C. and Cotty, M. (2010). *Routes d'Arabie. Archaeologie et histoire du Royame d'Arabie Saoudite*. Paris, Somogy editions d'art and Louvre editions.

Arnold, N. and Ovenden, D. (2004) A Field Guide to the Reptiles and Amphibians of Britain and Europe. London, Collins.

Arnold, E.N., Burton, J.A. and Ovenden, D. (1978) [C] *Field Guide to the Reptiles and Amphibians of Britain and Europe*. London, Collins.

Arrigoni Degli Oddi, E. (1929) *Ornitologia italiana*. Milano, Ulrico Hoepli.

Aubé, P. (2006) *Les empires normands d'Orient*. Paris, Editions Perrin.

Baccetti, N. (1992) Francolino. In P.A. Brichetti, N. Baccetti and and P. De Franceschi (eds) *Fauna d'Italia. Aves I. Gavidae-Phasianidae*, 792–799. Bologna, Edizioni Calderini.

Bader, T., Riegler, C. and Grillitsch, H. (2009) The herpetofauna of the Island of Rhodes (Dodecanese, Greece). *Herpetozoa* 21(3/4), 147–169.

Baldacci, U. (1964) *Il francolino, sua reintroduzione in Europa*. Pisa, Nistri-Lischi.

Bannerman, D.A. and Bannerman, W.M. (1983) *The Birds of the Balearics*. London and Camberra, Croom Helm.

Basso, R. and Calasso, C. (1991) I rettili della Penisola Salentina. Contributi alla conoscenza dell'ambiente e della fauna salentina. *Quaderni del Museo Civico di Storia Naturale del Salento* (Edizioni del Grifo, Lecce) 1, 1–63.

Blasco, M., Cano J., Crespillo, E., Escudero, J.C., Romero, J. and Sanchez, J.M. (1985) *El camaleon comun* (*Chamaleo chamaleon*) *en la peninsula iberica*. Monografias 43. Madrid, Ministerio de Agricultura, Pesca y Alimentacion. Instituto Nacional para la Conservacion de la Naturaleza.

Bobis, L. (2000) *Une histoire du chat. De l'Antiquité à nos jours*. Paris, Librairie Arthème Fayard.

Boffa, C.J. (1966) *The Islets of Comino and Filfla*. Malta, Lux Press.

Bon, M. and Masseti, M. (1995) Wild mammal remains from a Roman pit at Oderzo (Treviso, North-Eastern Italy). *Bollettino del Museo Civico di Storia Naturale di Venezia* 44, 153–162.

Borghini, A., Giannarelli, E., Marcone, A. and Ranucci, G. (1983) *Gaio Plinio Secondo. Storia naturale. II. Antropologia e zoologia. Libri 7–11*. Torno, Giulio Einaudi editore.

Bossard-Beck C. (1984) Le mobilier ostéologique et botanique. In J.M. Pesez (ed.) *Brucato. Histoire et archéologie d'un habitat médiéval en Sicile. Vol. 2.*, 615–671. Collection de l'École Française de Rome 78. Rome, École française de Rome.

Bresc, H. (2003) Venuti dal nord. La Sicilia normanna. In F. Benigno and G. Giarrizzo (eds) *Storia della Sicilia. 1. Dalle origini al Seicento*, 83–93. Roma-Bari, Editori Laterza.

Brown, P. (1971) *The World of Late Antiquity AD 150–750*. London, Thames and Hudson.

Bru de Ramon, J.B. (1784) Colección de láminas que representan los animales y monstruos del Real Gabinete de Historia Natural de Madrid. Madrid, En la Imprenta de Andres de Soto.

Cabrera, A. (1914) *Fauna Ibérica*. Madrid, Museo Nacional de Ciencias Naturales.

Calonghi, F. (1950) *Dizionario Latino-Italiano. Vol. I*. Tirino, Rosenberg & Sellier.

Campanella, L. and Wilkens, B. (2004) Una mangusta egiziana (*Herpestes icneumon*) dall'abitato fenicio di Sant'Antioco. *Rivista di Studi Fenici* 32(1), 25–48.

Capitani, U. and Garofano, I. (1986). Gaio Plinio Secondo. Storia naturale. IV. Medicina e farmacologia. Libri 28–32. Torino, Einaudi.

Cardoso, J.L. and Fernandes, I.C.F. (2012). A economia alimentar dos Muçulmanos e dos Cristãos do Castelo de Palmela; un contributo. *Arqueologia Medieval* 1340, 211–233.

Cassady, R.F. (1986) *The Norman Achievement, 1050–1100*. London, Guild Publishing.

Clutton-Brock, J. (1981) *Domesticated Animals from Early Times*. London, Heinemann/British Museum (Natural History).

Clutton-Brock, J. (2001) *Storia naturale della domesticazione dei mammiferi*. Torino, Bollati Boringhieri editore.

Corbet, G.B. and Hill, J.E. (1991) *A World List of Mammalian Species*. Third edition. London, Natural History Museum Publications/Oxford University Press.

Corbino, M.C. (2009) I reperti faunistici e l'ambiente della rocca. In G. Vannini (ed.) *Rocca Ricciarda, dai Guidi ai Ricasoli. Storia e archeologia di un castrum medievale nel Pratomagno aretino*, 317–328. Firenze, S.E.F.

Corti, C. and Lo Cascio, P. (2002) *The Lizards of Italy and Adjacent Areas*. Frankfurt am Mein, Edition Chimaira.

Crouch, D. (2002) *The Normans. The History of a Dynasty*. London, Hambledon Continuum.

Davis, S.J.M. (2003) Faunal remains from Alcáçova de Santarém, Portugal. *Trabalhos do CIPA* 43, 1–72.

Davis, S.J.M. (2005) Agricultural improvements in gharb al-Andaluz – zoo-archaeological evidence from the sheep bones. In *Proceedings of the Fourth Conference on Iberian Archaeology*, 219–235. Faro, DECAP.

Davis, S.J.M. (2008) Zooarchaeological evidence for Moslem and Christian improvements of sheep and cattle in Portugal. *Journal of Archaeological Science* 35, 991–1010.

Davis, S.M.J. (2017) Zooarchaeological evidence for Muslim improvement of sheep (*Ovis aries*) in Portugal. In U. Albarella (ed.) *The Oxford Handbook of Zooarchaeology*, 178–194. Oxford, Oxford University Press.

Delibes, M. (1977) Sobre las Ginetas de la isla de Ibiza (*Genetta genetta isabelae* n. ssp.). *Doñana, Acta Vertebrata* 4, 139–160.

Delibes, M. and Carrasco-Rus, J. (2015) Radiocarbon dating of introduced mammals: a genet (*Genetta genetta*) from a prehistoric site dated in Islamic times. *Galemys* 27, 59–62.

Detry, C., Bicho, N., Fernandes, H. and Fernandes, C. (2011) The emirate of Córdoba (756–929 AD) and the introduction of the Egyptian mongoose (*Herpestes ichneumon*) in Iberia: the remains from Muge, Portugal. *Journal of Archaeologial Science* 38(12), 3518–3523.

Detry, C., Cardoso, J.L., Heras Mora, J., Bustamante-Álvarez, M., Silva, A.M., Pimenta, J., Fernandes, I. and Fernandes, C. (2018) Did the Romans introduce the Egyptian mongoose (*Herpestes ichneumon*) into the Iberian Peninsula? *The Science of Nature* 105, 63.

Diaz, M., Asensio, B. and Telleria, J.L. (1996) *Aves Ibérucas. 1. No Paseriformes*. Madrid, J.M. Reyero Editor.

Diffloth, P. (1924) Zootechnie colonial. Guide de l éleveur en Algérie, Tunisie, Afrique occidentale, Madagascar, Indo-Chine, II. Chevaux, chameuax, moutons, porcs, éléphants, autruches. Paris, Librairie J.-B. Baillière et fils.

Epstein, H. (1971) *The Origin of the Domestic Animals of Africa. Vol. I*. New York/Munich/London, Africana Publishing Corporation.

Ferrer, X., Martinez-Vilalta, A. and Muntaner, J. (1986) *Historia natural dels Paisos Catalans. Vol. 12. Ocells*. Barcelona, Enciclopèdia Catalana S.A.

Gabrieli, F. (1966) Ibn Hawqal e gli Arabi di Sicilia. In *L'Islam nella storia*, 57–67. Bari, Dedalo.

Garcia y Bellido, A. (967) *Veinticinco estampas de la Espana Antigua*. Madrid, Espasa-Calpe.

Gaubert, P., Machordon, A., Morales, A., López-Bao, J.V., Veron, G., Amin, M., Barros, T., Basuony, M., Djagoun, C.A.M.S., Do Linh San, E., Fonseca, C., Geffen, E., Ozkurt, S.O., Cruaud, C., Couloux, A. and Palomares, F. (2011) Comparative phylogeography of two African carnivorans presumably introduced into Europe: disentangling natural versus human-mediated dispersal across the Strait of Gibraltar. *Journal of Biogeography* 38, 341–358.

Harrison, D.L. (1981) *The Mammals of the Arabian Gulf*. London, George Allen & Unwin.

Harrison, D.L. and Bates, P.J.J. (1991) *The Mammals of Arabia*. Sevenoaks (England), Harrison Zoological Museum.

Hemmer, H. (1990) *Domestication. The Decline of Environmental Appreciation*. Cambridge, Cambridge University Press.

Hemmer, H. (1999) *Felis silvestris* Schreber, 1775. In A.J. Mitchell-Jones, W. Bogdanowicz, B. Krystufek, P.J.H. Reijnders, F. Spitzenberger, C. Stubbe, J.B.M. Thissen, V. Vohralík and J. Zima (eds) *The Atlas of European Mammals*. 358–359. London, Academic Press.

Heptner, V.G. and Sludskii, A.A. (1992) *Mammals of the Soviet Union. Vol. II. Part 2. Carnivora (Hyaenas and Cats)*. Leiden, E.J. Brill.

Hilzheimer, M. (1909) *Die Haustiere in Abstammung und Entwicklung, Eine natur-und kulturgeschichtliche Darstellung*. Stuttgard.

Iapichino, C. and Massa, B. (1989) *The Birds of Sicily. AnAannotated Checklist*. Peterborough, BOU Check-list 11. British Ornithologists' Union.

Ingrams, H. (1966) *Arabia and the Isles*. 3rd edition. London, John Murray Publishers.

Jiménez, J. (2013) El curioso caso del francolin: extinguido tras 600 años de presencia en España. *Quercus* 330, 16–23.

Johnsgard, P.A. (1988) *The Quails, Partridges, and Francolins of the World*. Oxford/New York/London, Oxford University Press.

Lentacker, A. and Ervynck, A. (1999) The archaeofauna of the late Medieval, Islamic harbour town of Saltés (Huelva, Spain). *Archaeofauna* 8, 141–157.

Lever, C. (1985) *Naturalized Mammals of the World*. London and New York, Longman.

Lewis, B. (1993) *Les Arabes dans l'histoire*. Paris, Flammarion.

Linnaeus, C. (1758) *Systema Naturae. Regum Animale*. Anastatic edition of the 1st volume of the 10th edition, edited by G. De Beer. London (1956), British Museum (Natural History).

Llorente-Rodríguez, L., Nores-Quesada, C., Gaubert, P. and Morales-Muñiz, A. (2015) Nativo o foráneo? El meloncillo de Tabacalera y su aportación a la historia de la especie. In C. Ochoa (ed.) *La fábrica de tabacos de Gijón: arqueología e historia de un espacio milenário*, 246–251. Gijòn, Ayuntamiento de Gijòn.

Llabrès Ramis, J. and Vallespir Soler, J. (1983) *Els nostres arts i oficis d'antany. IV*. Ciutat de Mallorca, Estudis monogràfics del Museu Arqueòlogic La Porciùncula.

Lombard, M. (1971) L'Islam dans sa première grandeur (VIIIe–XIe siècle). Paris, Flammarion.

Loud, G.A. (2003) The Kingdom of Sicily and the Kingdom of England 1066–1266. *History* 88, 540–567.

Lo Valvo, M., Massa, B. and Sarà, M. (1993) Uccelli e paesaggio in Sicilia alle soglie del terzo millennio. *Il Naturalista siciliano* 17, Suppl., 1–374.

Lupo, F. (2006–2007) *Analisi del complesso faunistico di Palazzo Steri*. Tesi di laurea in Scienze Naturali. Palermo, Università degli Studi di Palermo.

MacKinnon, M. (2004) Production and consumption of animals in Roman Italy: integrating the zooarchaeological and textual

evidence. *Journal of Roman Archaeology, Supplementary Series* 54, 1–264.
Malek, J. (1993) *The Cat in Ancient Egypt*. London, British Museum Press.
Maluquer, J.S. and Travé, E.A. (1961) Presencia y extincion del Francolin en la Peninsula Iberica e Islas Baleares. *Ardeola* 7, 129–156.
Mason, E. (1984) Mongoose. In I.I. Mason (ed.) *Evolution of Domesticated Animals*, 234–236, London, Longman.
Masseti, M. (2002) Uomini e (non solo) topi. Gli animali domestici e la fauna antropocora. Firenze, Firenze University Press.
Masseti, M. (2009a) In the gardens of Norman Palermo, Sicily (twelfth century AD). *Anthropozoologica* 44(2), 7–34.
Masseti, M. (2009b) The mongoose of the cave of Nerja, southern Spain, is not the oldest Egyptian mongoose of Europe. *Archaeofauna* 18, 65–68.
Masseti, M. (2010a) Zoologia storica e archeologica dei Felidi italiani. In E. Randi, B. Ragni, L. Bizzarrini, N. Agostini and G. Tedaldi (eds) *Biologia e conservazione dei Felidi in Italia. Atti del convegno – Santa Sofia (FC) 7–8 Novembre 2008*, 9–28. Santa Sofia (FC), Ente Parco Nazionale Foreste Casentinesi.
Masseti, M. (2010b) Holocene mammals of Libya: A biogeographical, historical and archaeozoological approach. *Journal of Arid Environments* 74, 794–805.
Masseti, M. (2010c) Mammals of the Macaronesian islands (the Azores, Madeira, the Canary and Cape Verde islands): redefinition of the ecological equilibrium. *Mammalia* 74, 3–34.
Masseti, M. (2012) I 'boschetti con animali' e le raffigurazioni zoomorfiche. In J. Hayez and D. Toccafondi D. (eds) *Palazzo Datini a Prato. Una casa fatta per durare mille anni. Tomo I*, 145–153. Firenze, Edizioni Polistampa.
Masseti, M. (2014) Herpetological enigmas from the Arabian seas, with particular reference to the Sarso island racer, *Platyceps insularis* Mertens, 1965 (Farasan archipelago, Saudi Arabia). In *Studies on Amphibians and Reptiles in Honour of Benedetto Lanza Scripta Herpetologica*, 99–116. Laatina, Edizioni Belvedere.
Masseti, M. (2016) *Zoologia della Sicilia araba e normanna (827–1194)*. Palermo, Edizioni Danaus.
Masseti, M. (2018) The long-term redefinition of the non-volant vertebrate horizons of the Eastern Mediterranean islands. In S. Sfenthourakis, P. Pafilis, A. Parmakelis, N. Poulakakis and K.A. Triantis (eds) *Biogeography and Biodiversity of the Aegean*, 191–222. Nicosia (Cyprus), Broken Hill Publishers Ltd.
Masseti, M. (2019) Migrazioni umane e variazioni faunistiche nell'Alto Medioevo mediterraneo. In *Le migrazioni nell'Alto Medioevo: pratiche, simboli, ideologie*. Sessantaseiesima Settimana di Studio della Fondazione Centro Italiano di Studi sull'Alto Medioevo, Spoleto (in press).
McNamee, G. (2011) *Aelian's On the Nature of Animals*. San Antonio, Trinity University Press.
Molina, F.M. (1983) *Almuñécar, Arqueología e Historia. Vol. III*. Granada, Ayuntamiento de Almuñécar.
Mongitore, D.A. (1743) *Della Sicilia ricercata nelle cose più mirabili*. Palermo, Francesco Valenza.
Morales, A. (1994) Earliest genets in Europe. *Nature* 370, 512–513.
Morales, A. and Rofes, J. (2007) Early evidence for the Algerian hedgehog in Europe. *Journal of Zoology* 274, 1–4.
Morales Muñiz, A., Ruti Moreno, N. and Cereijo Pecharromán, M.A. (1988) Clatrava la Vieja: primer infrome sobre la fauna de vertebrados recuperada en el yacimiento almohade, primera parte: mamíferos. *Boletín de Arqueología Medieval* 2, 7–48.
Morales Muñiz, A., Riquelme, J.A. and Llesau Von Lettow-Vorbeck, C. (1995) Dromedaries in antiquity: Iberia and beyond. *Antiquity* 69(263), 368–375.
Morales, J. (1996) El registro fosil de los carnivoros ibéricos. In R. Garcia Perea, R.A. Baquero, R. Fernandez-Salvador and J. Gisbert (eds) *Carnivoros. Evolucion, ecologia y conservacion*, 117–135.
Madrid, Museo Nacional de Ciencias Naturale, C.S.I.C.
Muntaner, J., Ferrer, X. and Martínez-Vilalta, A. (1983) *Atlas dels ocells nidificants de Catalunya i Andorra*. Barcelona, Ketres.
O'Connor, T.P. (1982) *Animal Bones from Flaxengate, Lincoln, c. 870–1500*. Archaeology of Lincoln 18/1. London, Council for British Archaeology.
O'Connor, T.P. (1992) Pets and pests in Roman and medieval Britain. *Mammal Review* 22(2), 107–113.
Orlando, C. (1958) Il Francolino *Francolinus francolinus* (L.). *Venatoria Sicula* 12, 328.
Ottoni, C., Van Neer, W., De Cupere, B., Daligault, J., Guimaraes, S., Petres, J., Spassov, N., Prendergast, M.E., Boivin, N., Morales-Muñiz, A., Bălăuşescu, A., Becker, C., Bonecke, N., Boroneant, A., Buitenhuis, H., Chahoud, J., Crowther, A., Llorente, L., Manaseryan, N., Monchot, H., Onar, V., Osypińska, M., Putelat, O., Quintana Morales, E.M., Studer, J., Wierer, U., Decorte, R., Grange, T. and Geigl, E.-M. (2017) The palaeogenetics of cat dispersal in the ancient world. *Nature Ecology & Evolution* 1, 0139 (2017) | DOI: 10.1038/s41559-017-0139 | www.nature.com/natecolevol.
Paulo, O. (2014) On the origin of the Iberian Chamaleons. In C. Detry and R. Dias (eds) *Proceedings of the First Zooarchaeology Conference in Portugal*, 139–144. BAR International Series 2662. Oxford, Archaeopress.
Pecci, G.A. (1760*) Memorie Storiche delle Città, Terre e Castella dello Stato di Siena. Tomo III degli Abbozzi* (1^–248^). Codice manoscritto n. 110 (*carta* 106 *recto* e *verso*) of the Pecci Collection, Biblioteca Moreniana of Florence (Italy).
Perco, F. (1981) Dopo centocinquanta anni il francolino è ritornato a volare in Toscana: ma per la legge non esiste. *Airone* 5, 49.
Pereira, E., Davis, S.M.J., Pereira, L., Mcevoy, B., Bradley, D.G. and Amorim, A. (2006) Genetic signatures of a Mediterranean influence in Iberian Peninsula sheep husbandry. *Molecular Biology and Evolution* 23(7), 140–146.
Pleguenzuelos, J.M. and Feriche, M. (2003) *Granada. Guías de la naturaleza. Anfibios y reptiles*. Granada, Diputación de Granada.
Pocock, R.I. (1951) *Catalogue of the genus* Felis. London, Trustees of the British Museum (Natural History).
Pratesi, F. (1976) Tre storie di uccelli: il francolino, la quaglia tridattila e la gallina prataiola. In F. Pedrotti (ed.) *SOS Fauna. Animali in pericolo in Italia*, 357–372. Camerino, Edizioni W.W.F.
Rackham, H. (1940) *Pliny Natural History, with an English translation in ten volumes. Vol. III. Libri VIII–XI*. CambridgeMA, Harvard University Press/London, William Heinemann Ltd.
Ragni, B. and Possenti, M. (1991) Genetica e problemi di conservazione in popolazioni italiane di gatto selvatico. *Supplemento alle Ricerche di Biologia della Selvaggina* 18, 165–173.

Riquelme, J.A. (1992) La fauna de epoca califal procedente de la catedral de Granada. *Boletín de Arqueología Medieval* 6, 193–207.

Riquelme Cantal, J.A. (1993) Estudio faunistico del yacimeinto medieval de Plaza España, Motril (Granada). *Arqueologia Medieval* 2, 243–260.

Riquelme-Cantal, J.A., Simon-Vallejo, M.D., Palmqvist, P. and Cortes-Sanchez, M. (2008) The oldest mongoose of Europe. *Journal of Archaeological Science* 35, 2471–2473.

Rodriguez de la Fuente, F. (1970) *Fauna iberica. Lo animales cazadores*. Madrid, Editorial Prensa Española.

Rodríguez-Picavea Matilla, E. (2006) *La Corona de Aragón en la Edad Media*. Madrd, Ediciones Akal.

Rowley, T (1999) *The Normans*. Stroud, Tempus.

Sarà, M. (1997) Resti ossei di animali dallo scavo del Museo 'A. Salinas'. *Quaderni del Museo Archeologico Regionale 'Antonino Salinas'* 3, 77–79.

Sarà, M. (2005) Resti faunistici dal castro normanno di Calathamet (XIII sec. d. C., Sicilia nord-occidentale). In I. Fiore, G. Malerba and S. Chilardi (eds) *Atti del 3e Convegno Nazionale di Archeozoologia (Siracusa, 3–5 November 2000)*, 493–499. Roma, Istituto Poligrafico e Zecca dello Stato.

Scholfield, A.F. (1959) *Aelian. On Animals II. Books VI–XI*. Cambridge (MA), Harvard University Press/London, William Heinemann.

Talavera, R.R. and Sanchiz, B. (1985) Restos holocénicos de Camaleón común, *Chamaeleo chamaeleon* (L.) de Málaga. *Boletin de la Real Sociedad Espanola de Historia Natural (Geologia)* 81(1–2), 81–84.

Telles Antunes, M. (1996) Alimentação de origem animal em regime islâmico. Alcaria Longa e Casa II da Alcáçova de Mértola. *Arqeuologia Medieval* 4, 267–276.

Toynbee, J.M.C. (1973) *Animals in Roman Life and Art*. Ithaca and New York, Cornell University Press.

Vigne, J.-D., Guilaine, J., Debue, K., Haye, L. and Gèrard, P. (2004) Early taming of the cat in Cyprus. *Science* 304, 259.

Villari, P. (1995) *Le faune della tarda preistoria della Sicilia orientale*. Siracusa, Ente Fauna Siciliana.

Volpe, G., Buglione, A. and De Venuto, G. (2012) Lane, pecore e pastori in Puglia fra Tardoantico e Medioevo: novità dai dati archeozoologici. In M.S. Busana (ed.) *La lana nella Cisalpina romana, Economia e società*. Studi in onore di Stefania Pesavento Mattioli, 243–268. Padova, Università degli Studi di Padova.

Wozencraft, W.C. (2005) Carnivora. In D.E. Wilson and D.-A. Reeder (eds) *Mammal Species of the World. A Taxonomic and Geographic Reference. Vol. I*. Third edition, 532–628. Baltimore, The John Hopkins University Press.

Zeuner, F.E. (1950) The cat. *Oryx* 1(2), 65–71.

Zeuner, F.E. (1963) *A History of Domesticated Animals*. London, Hutchinson & Co. Ltd.

15

Hovering over hawking in Early medieval Iberia

Laura Llorente Rodríguez, Arturo Morales Muñiz, Leif Jonsson and Evelyne Browaeys

*The first archaeozoological evidence on the use of falconry in Western Europe occurs by the late 4th century AD. Among the short-winged hawks, goshawk (*Accipiter gentilis *L. 1758) and sparrowhawk (*Accipiter nisus *L. 1758) females were apparently more appreciated than males. Females are bigger and thus able to cope with a larger range of prey. However, the female sparrowhawk and the male goshawk overlap in size; therefore, the osteological identification can be difficult.*

We here present the identification of hawk remains from one Visigoth site (Buzanca) and one Byzantine site (Begastri) in Iberia through an osteometrical analysis in which modern Iberian and European reference specimens with known sex of both goshawk and sparrowhawk were used. These remains not only represent the earliest archaeozoological evidence of falconry in the Peninsula (6th c. *AD) but result in a case in point as both seem to represent small size goshawks. Within this context, we discuss the evidence of low flight hawking in these two sites and the practice of that type of hunting in Iberia, especially during Medieval times.*

Keywords: Falconry, goshawk, sparrowhawk, osteometry, Iberian Peninsula.

Introduction: hawking and falconry in Iberia

Hunting with birds of prey has fascinated humans since at least the beginning of the first millennium BC, as evidenced by a hunting scene with a bird of prey on an Assyrian bas-relief (Epstein 1943). Not only is the activity a means to procure game; it also denotes a very close human-bird relationship. Probably for this reason, hawkers and falconers consider themselves a most selective group of hunters. Bird care and hawking are such demanding tasks that it is said that the everyday life of the falconer changes until one of them – the human or the bird – die (Ceballos and Justribó 2011). Such human-bird liaison in hawking is referred by many authors as an art within hunting, to the extent of this activity being recognized by the UNESCO as an Intangible Cultural Humanity Heritage since 2010.

In Western Europe, Pliny the Elder (AD 23–79; *Nat. Hist*. 10.10) mentions the use of birds of prey in places such as Thracia (Turkey) but it is not until the 4th century AD when various mosaics from Piazza Armerina (Sicily), Tunisia and Argos (Greece) depict scenes of generally two hunters with a bird of prey which apparently acted as a decoy to facilitate catching thrushes (Wallis 2017). The earliest zooarchaeological hawking evidence has been suggested to appear in Britain during the 3rd century AD at Great Holts Farm (Albarella 2003). However, the practice of hunting with birds of prey is generally accepted to not be really established and widespread earlier than the 4th century AD when the Germanic tribes started to spread throughout the subcontinent and the later fall of the Roman Empire (Epstein 1943; Prummel 1997; 2013; Cherryson 2002; Dobney and Jaques 2002; Wallis 2017). The *Lex salica* (5th century AD), for example, makes indirect references to falconry through fines and punishments imposed for the theft or killing of hawks. This same assumption on the origin of falconry is generally accepted for Iberia; however, there exists a hypothesis of an earlier hunting practice with tamed birds of prey by the Iberian Culture elite during the second half of the Iron Age, based on ceramic and numismatic figurative scenes (3rd–1st century BC; Marín Ceballos 1994). At any rate, in the subsequent Roman *Hispania*, no records – whether pictorial, documentary or archaeozoological – testify to the practice of falconry.

The Visigoth king Euricus (*c.* AD 420–484) is the first Iberian monarch whose interest in falconry has been documented (Juarros 1997). Such attribution is provided in the *Kitab Dawârî at-tayr,* a later falconry treatise by al Gitrif ibn Qudama al-Gassani (8th century AD). In it, it is stated that the Visigoth king and the members of his court made peregrine falcons fly – not hunt – while riding. On one occasion a falcon pounced on a bird and ever since then Euricus decided to train his falcons for high flight falconry. Another piece of evidence on the Visigoth practice of hawking is provided by a 6th-century AD mosaic from *Mirtyla Iulia* (Mértola, Portugal) (Fig. 15.1). The depicted scene includes different gaming birds and a horseman with a raptor in his left hand that Fradejas-Rueda suggests it represents a female goshawk (Fradejas-Rueda 2006–2017).

These references to falconry in Visigoth Iberia contrast with the dearth of evidence from Byzantine Iberia, not being until the Umayyad conquest of *Hispania* that documentary and figurative evidences of hawking become common. The first of these references appears in the AD 961 *El Calendario de Córdoba* (Juarros 1997). This astronomic and agricultural treatise documents the practice of hawking for hunting aquatic birds by Arab elites on the Guadalquivir River, and also includes sections on the commerce of falcons, their migration and breeding seasons, the adequate season for hawking, and when and how to keep birds during moulting (Fradejas-Rueda 2006–2017). In the ensuing centuries hunting treatises as well as figurines, tapestry scenes and poetry from Al-Andalus all inform about the common practice of hawking (Juarros 1997).

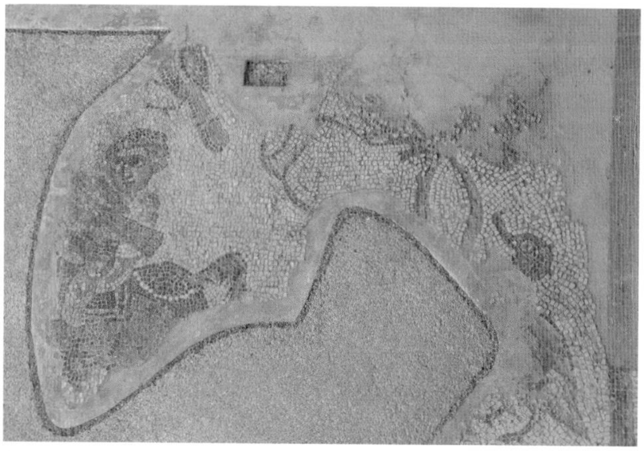

Figure 15.1. Mosaic floor from Mirtyla Iulia depicting a hunting scene with a bird of prey (6th century AD, Mértola, Lusitania, Portugal). Photo by Carole Raddato (2014) under a Creative Commons licence.

According to various authors, medieval Christian falconry is rooted in a Visigoth tradition although it exhibits influences from the Islamic territories where the activity was widespread (Juarros 1997). Within the Christian kingdoms, the earliest unequivocal documentary references to the use of birds of prey, mainly goshawks, date from 11th century AD documents archived in the Cathedral of Leon (Juarros 1997; Fradejas-Rueda 2006–2017). These early references basically focused on *ofertiones*, donations or means to seal commercial transactions where valued items – such as goshawks – were used.

Despite historical sources highlighting the esteem falconry commanded in medieval Iberia, little is known in terms of archaeozoological evidence. Even though the number of bird bones per thousand of mammals increased in medieval times compared with previous periods as noted by Simon Davis (2006), few sites in the Iberian Peninsula report raptors (Morales Muñiz 2002). Of these, the only two suggested as evidence of falconry derive from 15th–16th-century AD deposits at Beja (Portugal) and the earlier but more ambiguous goshawk third phalanx from a 9th-century AD level on the castle of Mértola (Moreno-García and Pimenta 2010; 2012). In this paper we present two hawk (*Accipiter* sp.) records we believe constitute the earliest archaeozoological evidence of falconry in Iberia. These derive from two different cultural 6th-century AD contexts, a rural Visigoth settlement (Buzanca 2) and an urban Byzantine city previously occupied by the Visigoths (Begastri). Given the different roots (*i.e.* Germanic *vs.* Roman) and geographical ranges (*i.e.* Central Europe *vs.* Eastern Mediterranean) of these two synchronic cultures, we additionally investigate whether any of these finds might constitute specimens brought from territories outside the Iberian Peninsula or else represent local tamed birds.

Buzanca 2 and Begastri specimens

The site of Buzanca 2 is located at 581 m.a.s.l. in the fertile fields of the Jarama meadows nearby the municipality of Ciempozuelos (Madrid) (Fig. 15.2). This is one of many rural medieval settlements of this region some 40 km away from Toledo which was the capital of the Kingdom in Visigoth times (Vigil-Escalera 2005; Penedo-Cobo 2006). Different radiocarbon and thermoluminiscence datings set the chronology between the 6th and 7th centuries. The palaeo-environmental data from Buzanca 2 document an open landscape where anthropic action is evidenced by a substantial structural simplification of the vegetation with a preponderance of meadows and pastures (Penedo Cobo and Sanguino Vázquez 2009). The site was divided into four areas but two of them, areas C (*c.* 12,000 m^2) and E (*c.* 12,500 m^2) accumulated most habitat structures and the associated faunal assemblages (Penedo-Cobo 2006; Morales *et al.* 2007; Penedo-Cobo and Sanguino-Vázquez 2014). Although the

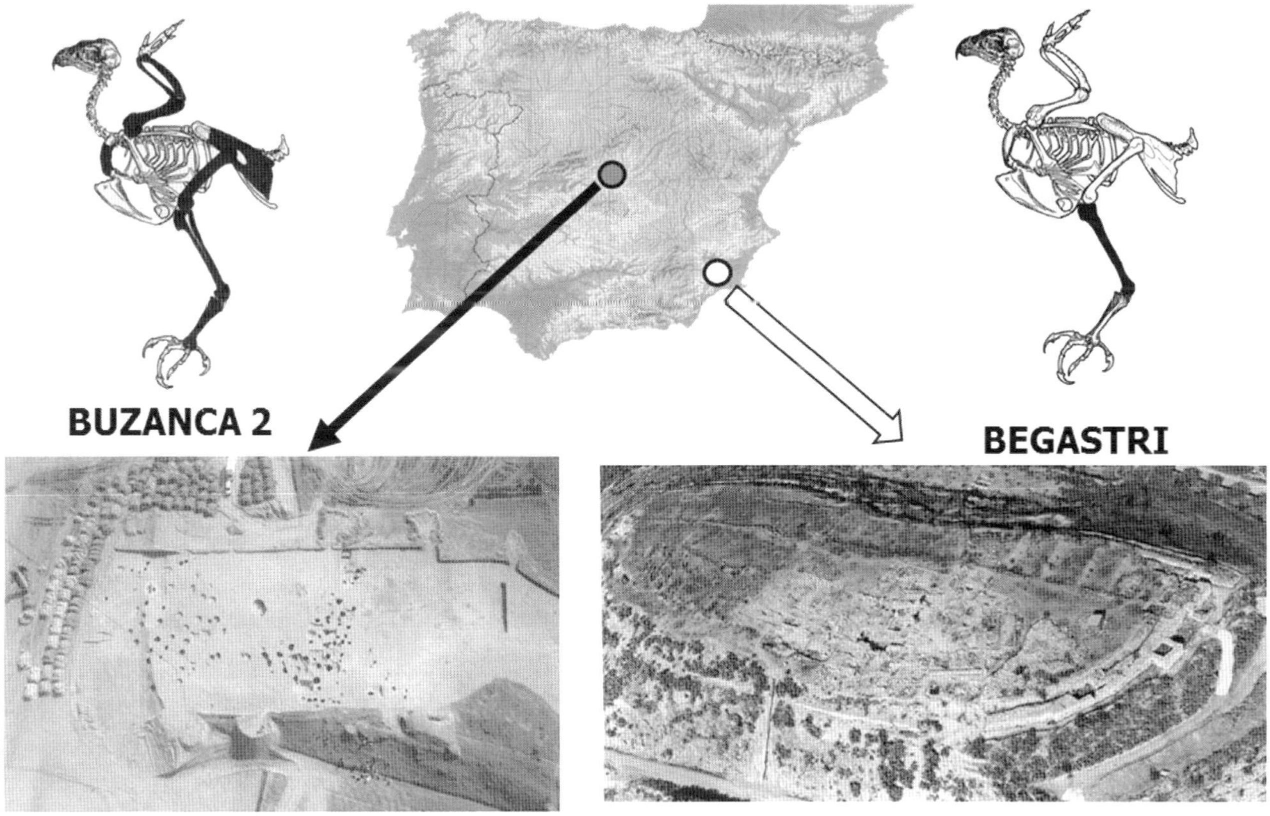

Figure 15.2. Location of Buzanca 2 and Begastri with reference to the hawk elements retrieved in each site.

excavation was hand-collected, a large animal assemblage of *c.* 5,000 specimens representing *c.* 1,300 individuals, were retrieved (Morales *et al.* 2007). Noteworthy at Buzanca 2 are the substantial dog and equine collections that do not fit domestic contexts (*i.e.* farming, transhumant herding), whose morphometric diversities raise the possibility of non-local dog and horse breeds brought in by the Visigoths. Within such context 13 bones from a presumably articulated hawk skeleton appeared in UE 3671, representing a silo (*i.e.* unit filled with consumption refuse) on the structures of Area E (Fig. 15.2; Morales *et al.* 2007).

The site of Begastri (Cehegín, Murcia) is located on a 550 m.a.s.l. hill bordering the Quípar River (López Bermúdez 1984; Fig. 15.2). Begastri is located on the semi-desert landscape that characterises the south-easternmost corner of the Iberian Peninsula, devoid of trees for centuries. A Roman municipality since 212 AD, Begastri was occupied by the Visigoths in the 5th century AD and later by the Byzantine Empire (AD 554 to AD 633) before returning to Visigoth hands in AD 633. The Muslims occupied the city from AD 711 until its abandonment in the 12th century AD (González Blanco 1998). It is in one of the Byzantine (6th–7th centuries AD) structures from the West trench outside the city wall (Sector 1-N) where a distal tibiotarsus of a hawk was found (Fig. 15.2) (Martínez Cavero 1984; González Blanco *et al.* 1994). The associated faunal assemblage is mostly constituted by domesticates but four additional unidentified bird bones complete the ornithological record (González *et al.* 1993; Hernández Carrasquilla 1993; Morales and Roselló 2002). Again, in this excavation the materials were collected by hand, and therefore we may expect a biased faunal record with a scarcity of bird remains that may explain the single hawk remain. Small mammals were also rare despite the area constituting an optimal ground for species such as the rabbit.

Methodology

Remains were identified with AMM reference collection housed at the Laboratorio de Arqueozoología of the Universidad Autónoma de Madrid (LAZ-UAM). It is usually a rapid size comparison, not morphological features, that can sometimes determine the taxonomic identification between goshawk and sparrowhawk. Due to the substantial sexual dimorphism both species exhibit, despite goshawk being larger than sparrowhawk, female sparrowhawks may turn out to be not so easy to set apart from male goshawks. At the time, we had a restricted sample of comparative specimens in terms of sex and regional provenience, thus could not ascertain size differences down to sexual or geographic provenience (*i.e.* whether the finds represented local or translocated birds). For

Table 15.1. Number of modern goshawk and sparrowhawk specimens used in the osteometrical analysis (see Appendix 1 for details).

Region	*Iberia*		*North Europe*	
Species	Female	Male	Female	Male
Goshawk	8	8	7	4
Sparrowhawk	7	1	18	17

that reason, all bones were reported as belonging to adults of the Genus *Accipiter* (Hernández Carrasquilla 1993; Morales and Roselló 2002; Morales *et al.* 2007).

To determine species ascription and geographical provenance, measurements were taken on 27 modern adult goshawk (*Accipiter gentilis* L.) and 43 sparrowhawk (*Accipiter nisus* L.) of known sex and age from the Iberian Peninsula and North-Western Europe (Table 15.1). In addition to AMM's reference collection, those of L. Jonsson (LJ Osteology), Göteborg Natural History Museum, the Cultural Heritage Agency of the Netherlands (RCE, Amersfoort) and the Groningen Institute of Archaeology (GIA, Groningen) were consulted (Table 15.1). All measurements were taken with a calliper with ± 0.5 mm estimated error and follow Von den Driesch (1976) and Schmidt-Burger (1982). Because of the restricted skeletal representation at both sites, only selected measurements from the tarsometatarsus (*i.e.* greatest length (GL) and proximal breadth (Bp)) and tibiotarsus (*i.e.* distal breadth (Bd) and distal depth (Td)) were considered. In order to enlarge the dataset, we incorporated measurements published by Schmidt-Burger (1982) and Moreno-García and Pimenta (2010) who provided mean values of North European goshawk and Iberian sparrowhawk samples respectively (see Appendix 15.1).

In addition to taxonomic identification, other archaeological evidence needs to be explored to postulate the use of birds as tamed or captive. According to Prummel (1997) these include: (1) presence of falconer's equipment, (2) bird of prey burial – preferentially with a human, (3) presence of raptor remains among the faunal assemblage, (4) dominance of females within the raptor assemblage, and (5) presence of hawked game. Only the last three types of evidences could be investigated in this paper as no equipment was found in the sites and contexts do not represent burials. However, we have incorporated two additional lines of evidence. We thus (1) examined traces to determine the taphonomic history of the specimens (bones examined under a Wild MZ binocular), and (2) hawk finds were framed from a palaeoenvironmental standpoint to evaluate the suitability of the local landscapes as conductive for the existence of these birds (Mulkeen and O'Connor 1997).

Results

Morphological features proved to be insufficient to determine hawk species. On the contrary, the osteometrical analysis undisputedly discriminates both *Accipiter* species and to some extent inter- and intra-populations (*i.e.* inter-regional and inter-sexual; Fig. 15.3). In the case of goshawk, male and female measurements of the two elements never overlap within a particular population, yet an overlap exists between female Iberian goshawks and North European male goshawks. In sparrowhawk, males and females can be discriminated with the selected osteometrical parameters selected, though not on an inter-regional (*i.e.* geographic) basis. In other words, these measurements set apart sex but not geographical provenance.

With such provisos in mind, we evaluated our archaeological specimens. In the case of that from Buzanca 2, our analysis evidences the presence of an Iberian male goshawk, Gl being the most discriminant measurement to set them apart from their European counterparts (Fig. 15.3). In the case of Begastri, surprisingly the tibiotarsus fell within an empty gap where no goshawk or sparrowhawk from neither Iberian nor North European populations are represented (Fig. 15.3). One possibility would be that the find could represent an even smaller goshawk specimen, the only population smaller than the Iberian (*A.g. gentilis*) today being that from the Mediterranean isles of Sardinia and Corsica (*A.g. arrigonii*, Kleinschmidt, 1903).

Four of the wild taxa from Buzanca 2 are regularly hunted with goshawk. These include rabbit (*Oryctolagus cuniculus* L.), Iberian hare (*Lepus granatensis* Rosenhauer, 1856), geese *sensu lato* (*i.e. Anser* sp.) and great bustard (*Otis tarda* L.) (Morales *et al.* 2007). Goshawk catches great bustard in low flight falconry because of its large size (López de Ayala 1385–86). In the case of Begastri, only the rabbit qualifies as an item hunted with hawks. The gypsum film covering most Buzanca 2 bones precluded the preservation of taphonomic marks on the bone surfaces. No traces of beak and talon damage or digestion marks attributable to a bird of prey were found either. Such facts notwithstanding, a peculiar pathology was detected on the left goshawk tarsometatarsus from Buzanca 2 (Fig. 15.4). This bone exhibits a slight bowing which is associated with environmentally influenced pathologies as described by Brothwell (1993). Specifically, the aetiology conforms best with an early life growth disorder documented in caged birds associated with nutritional deficiencies (Wise 1975; Brothwell 1993; Samour 1999). This suggests that the animal could have been taken as a chick and defectively fed.

The Buzanca 2 specimen hints at an articulated carcass that was partially recovered due to defective retrieval. Not qualifying as an intrusive or a beast of burden, the only category likely to appear as articulated carcasses on archaeological deposits is that of pets (Gautier 1987).

Finally, according to the palaeoenvironmental data available from both sites, it seems evident that these environments were non-conductive or minimally suitable habitats for such forest species as both goshawk and sparrowhawk are.

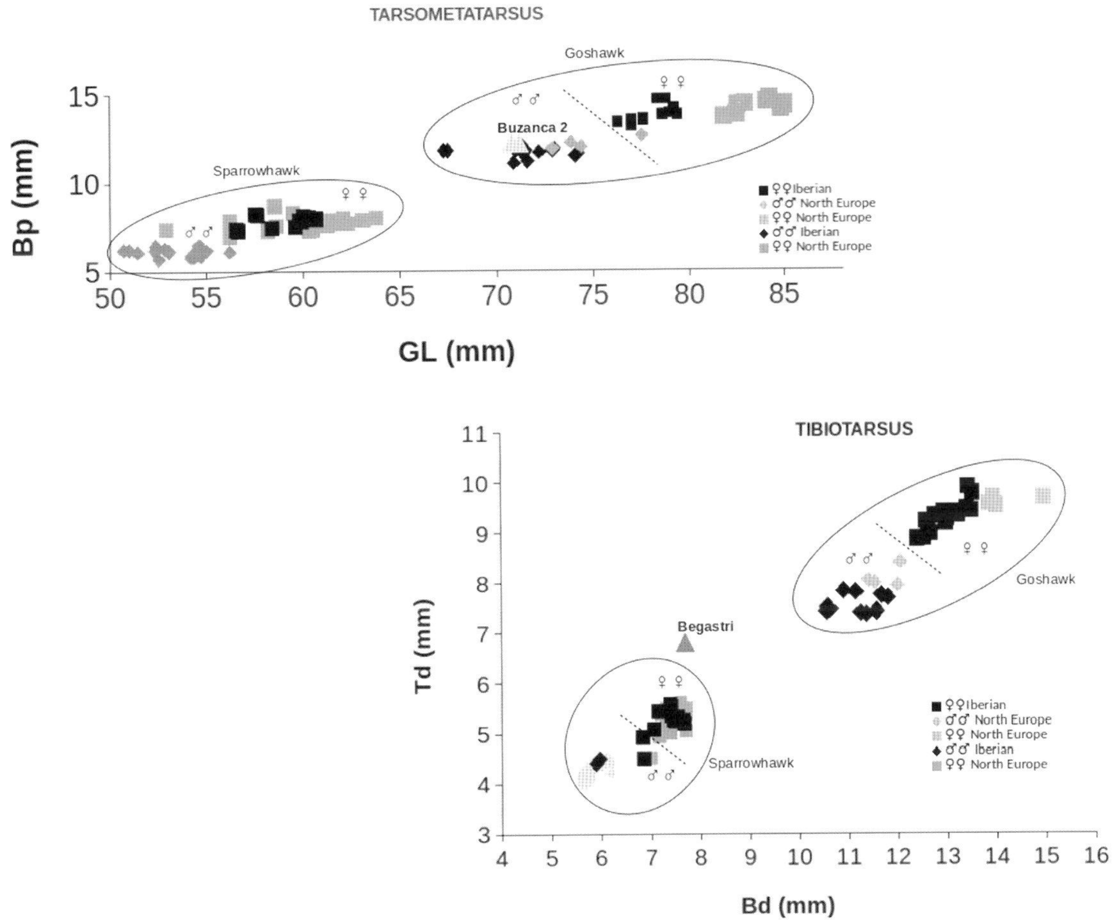

Figure 15.3. Scatterplots of tarsometatarsus' Bp and Gl (above) and tibiotarsus' Td and Bd values (below) from present day and archaeological specimens (Male Iberian sparrowhawk values fall outside these ranges being smaller than sparrowhawk females; see Appendix 15.1).

Discussion

The contextual information available for these finds is circumstantial. In this way, given the forested environment on which goshawk thrives, its presence at Buzanca 2 and Begastri hints at a human mediated translocation given its eminently forest niche (Mulkeen and O'Connor 1997; Zuberogoitia and Martínez 2015). Begastri is located in the region that features the lowest densities for the species within the Iberian Peninsula today (0–20% per 50 km^2; Carrascal *et al.* 2006). Likewise, although at times goshawk can visit rural and anthropic landscapes in certain seasons to hunt or take advantage of the micro-environmental conditions, it is not considered an urban scavenger and always breeds outside anthropized landscape (Boev 1993; O'Connor 1993). Given the low arboreal cover at Buzanca 2 and Begastri, and the fact that goshawk prefers to hunt living prey rather than scavenge, it seems implausible that the bird represents an intrusive item in the assemblage (Gautier 1987). The presence of a presumably articulated skeleton along with a tarsometatarsal pathology associated with inadequate feeding also suggest that the Buzanca 2 goshawk was a captive bird. The incidence rates of osteopathologies are usually low in wild bird populations (0.076%; Mlíkovský and Lukáš 1991; Table 15.2), probably due to the restricted 'error range' imposed to the avian skeleton by flight. In fact, most pathologies found in wild raptors are injuries produced when capturing prey (collision into fixed objects and immobilisation of prey, for example), developmental disorders being a minor fraction within pathologies (Goodman and Glynn 1988). The latter incidence rates skyrocket in domestic and captive birds, the care provided by humans allowing the survival of animals that would be quickly eliminated under natural conditions (Table 15.2). Indeed, other instances of goshawk with pathologies do exist in the archaeozoological literature such as those reported at Faccombe Netherton (Sadler 1990 *apud* Cherryson 2002). Here, one of the pathologies from what is thought to be a captive bird appeared on a tarsometatarsus but this apparently was an exostosis from a trauma perhaps produced by jesses.

Acknowledgements

We want to express our gratitude to the organisers of the conference 'From the Mediterranean to the Atlantic: Themes in Old World Zooarchaeology' for their invitation to this fruitful and friendly conference. It was an honour to have been part of such event to celebrate Simon Davis' archaeozoological legacy and friendship all around the world. Thanks Simon for inspiring generations of archaeozoologists with your knowledge and kindness.

This paper benefited from the funding support of a Marie-Curie-Sklodowska-If fellowship for Career development (L. Llorente; project FISHARC-IF 658022). Claire Zubiaurre and Begoña López Arias participated in the mammal archaeofaunal analyses discussed in this paper. Finally, we would like to thank Eduardo Penedo Cobo and Antonio González Blanco the opportunity to study the material from Buzanca 2 and Begastri respectively.

References

Albarella, U. (2003) Animal bone. In M. Germany (ed.) *Excavations at Great Holts Farm, Boreham, Essex, 1992–94*. Essex County Council, East Anglian Archaeology Report No. 105.

Bartosiewicz, L. (2012) Show me your hawk, I'll tell you who you are. In D.C.M. Raemaekers, K. Esser, R.C.G. Lauwerier and J.T. Zeiler (eds) *A Bouquet of Archaeozoological Studies. Essays in Honour of Wietske Prummel*, 180–189. Barkhuis (Groningen), Groningen Archaeological Studies, vol. 21.

Bochenski, Z.M., Tomek, T., Wertz, K., and Wojenka, M. (2016) Indirect evidence of falconry in Medieval Poland as inferred from published zooarchaeological studies. *International Journal of Osteoarchaeology* 26(4), 661–669.

Boev, Z. (1993) Archeo-ornithology and the synanthropisation of birds: a case study for Bulgaria. *Archaeofauna* 2, 145–153.

Boev, Z. and Beech, M.J. (2007) The bird bones. In A.G. Poulter (ed.) *Nicopolis ad Istrum. A Late Roman and Early Byzantine City*, 242–253. London, Oxbow Books.

Brandwood, A., Jayes, A.S. and Alexander, R.M. (1986) Incidence of healed fracture in the skeletons of birds, molluscs and primates. *Journal of Zoology* 208, 55–62.

Brothwell, D. (1993) Avian osteopathology and its evaluation. *Archaeofauna* 2, 33–43.

Carrascal, L.M., Palomino, D. and Lobo, J.M. (2006) *Mapas de distribución de la avifauna terrestre española (Peninsula Iberica)*. Sociedad de Amigos del MNCN – Museo Nacional de Ciencias Naturales – CSIC, Sociedad Española de Ornitología: Autora de la aplicación web map service: Stefanie Weykam https://lmcarrascal.eu/atlas/mapasalfab.html.

Ceballos, J. and Justribó, J.H. (2011) *Manual Básico y Ético de Cetrería*. Madrid, Avium www.avium.es.

Cherryson, A.K. (2002) The identification of archaeological evidence for hawking in medieval England. *Acta Zoologica Cracoviensia* 45, 307–314.

Davis, S.M.J. (2006). Faunal remains from Alcáçova de Santarém, Portugal. *Trabalhos de Arqueologia* 43, 6–140.

Dobney, K. and Jacques, D. (2002) Avian signatures for identity and status in Anglo-Saxon England. *Acta Zoologica Cracoviensia* 45, 7–21.

Driesch, A. von den (1976) *A Guide to the Measurement of Animal Bones from Archaeological Sites*. Cambridge MA, Peabody Museum Press.

Epstein, H.J. (1943) The origin and earliest history of falconry. *Isis* 34(6), 497–509.

Fradejas Rueda, J.M. (2006–2017) *Archivo Iberoamericano de Cetrería*. Universidad de Valladolid, ISSN 1887–5343. http://www.aic.uva.es/ Last accessed: 18-07-2018.

Gautier, A. (1987) Taphonomy groups: How and why? *Archaeozoologia* 2, 47–52.

González Blanco, A. (1998) *Historia de Murcia en las épocas Tardorromana, Bizantina y Visigoda*. Murcia, Diputación de Murcia.

González Blanco, A., Amante Sánchez M. and Martínez Villa, M.A. (1994) Memoria de los trabajos arqueológicos realizados en el yacimiento romano del «Cabezo Roenas» (Cehegín, Murcia) durante el año 1994. *Memorias de Arqueología* 6, 252–256.

González Blanco, A., Morales Muñiz, A. and de Miguel, J. (1993) El foso con cenizas, huesos y materiales aparecidos en Begastri en la campaña de 1989. *Memorias de Arqueología* 4, 211–215.

Goodman, S.M. and Glynn, C. (1988) Comparative rates of natural osteological disorders in a collection of Paraguayan birds. *Journal of Zoology* 214(1), 167–177.

Gotfredsen, A.B. (2014) Birds in subsistence and culture at Viking Age sites in Denmark. *International Journal of Osteoarchaeology* 24(3), 365–377.

Hernández Carrasquilla, F. (1993) Catálogo provisional de los yacimientos con aves del Cuaternario de la Península Ibérica. *Archaeofauna* 2, 231–275.

Juarros, F.J. (1997) La cetrería en la iconografía andalusí. *Anales de Historia del Arte* 7, 67–86.

Kroll, H. (2012) Animals in the Byzantine Empire: an overview of the archaeozoological evidence. *Archaeologia Medievale* 39, 93–121.

López Bermúdez, F. (1984) Begastri. *Antigüedad y Cristianismo* 1, 27–29.

López De Ayala, P. (1385–1386) *El libro de las aves de caça con las Glosas del Conde de Alburquerque*. Madrid, published in 1869 by Sociedad de Bibliófilos.

Marín-Ceballos, M.C. (1994) Cetrería en el mundo ibérico. In S.M. Ordóñez Agulla and P. Sáez Fernández (eds) *Homenaje al profesor Presedo*, 267–282. Sevilla, Universidad de Sevilla.

Martínez Cavero, P. (1984) Estratigrafía y cronología arqueológica de Begastri. *Antigüedad y Cristianismo* 1, 89–92.

Mlíkovský, J. (2005) Evidence for hawking in early medieval Czechia. *Buteo* 14, 53–56.

Mlíkovský, J. and Lukáš, J. (1988) Osteological disorders in late Pleistocene birds from the Schusterlucke, Lower Austria. *Annalen des Naturhistorischen Museums in Wien* 92, 101–103.

Morales Muñiz, D.C. (2002) Las aves cinegéticas en la Castilla medieval según las fuentes documentales. In *La caza en la Edad Media*, 129–150. Tordesillas, Seminario de Filología medieval.

Morales, A. and de Miguel, J. (1993) El foso con cenizas, huesos y materiales aparecidos en Begastri en la campaña de 1989. *Memorias de Arqueología* 4, 211–215.

Morales, A. and Roselló, E. (2002) Los restos animales del yacimiento de Begastri (Cehegín, Murcia). Análisis de la fauna de las siete primeras campañas (1980–1986) *Alquipir* 12, 111–154.

Morales, A., Llorente, L. and Zubiaurre, C. (2007) *Informe faunístico del yacimiento de Buzanca 2 (Ciempozuelos, Madrid)*. LAZ/UAM 2007/1 interim report.

Moreno García, M. and Pimenta, C. (2010) Beyond chicken: avian biodiversity in Portuguese late medieval urban site. In W. Prummel, J.T. Zeiler and D.C. Brinkhuizen (eds) *Birds in Archaeology: Proceedings of the 6th Meeting of the ICAZ Bird Working Group in Groningen (23.8–27.8. 2008),* 261–275. Barkhuis, Gronigen University Library.

Moreno García, M. and Pimenta, C.M. (2012) Ossos no lixo: o contributo arqueozoológico para oestudo da alimentação na Mértola islâmica. In S. Gómez Martínez (ed.) *Memórias dos sabores mediterrânicos,* 159–181. Mértola, Campo Arqueológico de Mértola.

Mulkeen, S. and O'connor, T. (1997) Raptors in towns: towards an ecological model. *International Journal of Osteoarchaeology* 7(4), 440–449.

O'Connor, T. (1993) Birds and the scavenger niche. *Archaeofauna* 2, 155–162.

Parmalee, P.W. (1977) Avian pathologies from Arikara sites in South Dakota. *Wilson Bullettin* 89, 628–632.

Penedo Cobo, E. (2006) El yacimiento visigodo de Buzanca 2 (Ciempozuelos, Madrid). *Zona Arqueológica* 8, 595–604.

Penedo Cobo, E. and Sanguino Vázquez, J. (2009) Documentación de aldeas altomedievales en el sur de Madrid. In J.A. Quirós Castillo (ed.) *The Archaeology of Early Medieval villages in Europe,* 341–354. Bilbao, Servicio de Publicaciones-Universidad del País Vasco.

Pimenta, C., Moreno García, M. and Lourenço, A. (2015) O registo ornito-arqueológico em Portugal: inventários, comentários e mapas *Revista Portuguesa de Arqueologia* 18, 289–312.

Prummel, W. (1997) Evidence for hawking (falconry) from bird and mammal bones. *International Journal of Osteoarchaeology* 7, 333–338.

Raddato, C. (2014) from FRANKFURT, Germany [CC BY-SA 2.0 (https://creativecommons.org/licenses/by-sa/2.0)], via Wikimedia Commons.

Samour, J. (1999) *Avian Medicine*. London, Harcourt-Brace.

Schmidt-Burger, P. (1982) *Vergleichend morphologische Untersuchungen an Einzelknochen in Zentraleuropa vorkommender mittelgrosser Accipitridae* II: Becken und Hinterextremität. Unpublished PhD thesis, Universität München.

Tyrberg, T. (2002) The archaeological record of domesticated and tamed birds in Sweden. *Acta Zoologica Cracoviensia* 45, 215–231.

Vigil-Escalera, A. (2005) Nuevas perspectivas sobre la arqueología madrileña de época visigoda. In *Actas de las Primeras Jornadas de Patrimonio Arqueológico en la Comunidad de Madrid,* 169–181. Madrid, Comunidad de Madrid, Consejería de Cultura y Deportes. Dirección General de Patrimonio Histórico.

Wallis, R.J. (2017) 'As the Falcon Her Bells' at Sutton Hoo? Falconry in Early Anglo-Saxon England. *Archaeological Journal* 174(2), 409–436.

Wise, D.R. (1975) Skeletal abnormalities in table poultry – a review. *Avian Pathology* 4, 1–10.

Zeiler, J. (2010) Hunting the hunters: owls and birds of prey as part of the falconers' game bag. In W. Prummel, J.T. Zeiler and D.C. Brinkhuizen *Birds in Archaeology: Proceedings of the 6th Meeting of the ICAZ Bird Working Group in Groningen (23.8–27.8. 2008),* 163–168. Barkhuis, Gronigen University Library.

Zuberogoitia, I. and Martínez, J.E. (2015) Azor común – *Accipiter gentilis*. In A. Salvador and M.B. Morales (eds) *Enciclopedia Virtual de los Vertebrados Españoles*. Madrid, Museo Nacional de Ciencias Naturales. http://www.vertebradosibericos.org/.

Appendix 15.1. Tarsometatarsus measurements of reference and archaeological goshawk and sparrowhawks following codes from Driesch (1976) and Schmidt-Burger (1982).

Species	*ID code/Reference*	*SEX*	*Gl*	*Bp*	*Tp*	*Sc*	*Bd*	*Td*	*Location and observations*
Accipiter gentilis	AMM–LAZ/UAM	F	85	14.3	11.8	6.00	15.4	9.5	Scandinavia
Accipiter gentilis	AMM–LAZ/UAM	F	85	14.4	12.00	–	–	–	Scandinavia
Accipiter gentilis	AMM–LAZ/UAM	F	82.5	13.8	11.5	5.9	14.7	9.4	Scandinavia
Accipiter gentilis	AMM–LAZ/UAM	F	81.8	13.7	11.4	–	–	–	Scandinavia
Accipiter gentilis	AMM–LAZ/UAM	F	84.2	14.7	11.7	6.00	14.9	9.3	Scandinavia
Accipiter gentilis	AMM–LAZ/UAM	F	84.8	14.1	–	6.00	15.00	9.2	Scandinavia
Accipiter gentilis	Schmidt-Burger, 1982	F	87.6	16.00	12.8	6.2	16.2	10.3	Scandinavia (n=1)
Accipiter gentilis	Schmidt-Burger, 1982	F	82.9	14.4	11.92	5.92	15.4	9.46	Germany (n=10)
Accipiter gentilis	GIA	F	84.06	14.54	–	–	–	–	Netherlands
Accipiter gentilis	GIA	F	82.54	14.34	–	–	–	–	Netherlands
Accipiter gentilis	AMM–LAZ/UAM	F	0.00	13.5	11.3	–	–	–	Iberia
Accipiter gentilis	AMM–LAZ/UAM	F	0.00	13.5	11.2	–	–	–	Iberia
Accipiter gentilis	AMM–LAZ/UAM	F	79.4	13.8	11.5	6.00	14.6	9.6	Iberia
Accipiter gentilis	AMM–LAZ/UAM	F	79.2	14.2	11.4	6.9	14.9	9.4	Iberia
Accipiter gentilis	AMM–LAZ/UAM	F	77.6	13.5	11.00	5.8	14.2	8.6	Iberia
Accipiter gentilis	AMM–LAZ/UAM	F	77.00	13.2	11.2	5.7	14.1	8.5	Iberia
Accipiter gentilis	AMM–LAZ/UAM	F	77.6	13.6	10.9	5.5	15.00	9.1	Iberia
Accipiter gentilis	AMM–LAZ/UAM	F	76.3	13.4	10.6	5.5	15.00	9.00	Iberia
Accipiter gentilis	AMM–LAZ/UAM	F	78.3	14.7	11.8	6.4	15.2	9.2	Iberia
Accipiter gentilis	AMM–LAZ/UAM	F	78.8	14.7	11.7	6.3	15.00	9.3	Iberia
Accipiter gentilis	AMM–LAZ/UAM	F	0.00	0.00	–	5.6	14.5	9.1	Iberia
Accipiter gentilis	AMM–LAZ/UAM	F	77.00	13.5	11.7	5.5	14.7	8.9	Iberia
Accipiter gentilis	AMM–LAZ/UAM	F	79.00	13.9	11.6	5.9	15.1	9.00	Iberia
Accipiter gentilis	AMM–LAZ/UAM	F	78.6	13.8	11.6	5.7	15.00	9.1	Iberia
Accipiter gentilis	Schmidt-Burger, 1982	M	77.55	12.65	10.4	5.15	13.1	7.9	Scandinavia (n=2)
Accipiter gentilis	Schmidt-Burger, 1982	M	74.41	12.00	9.87	4.75	12.39	7.57	Germany (n=14)
Accipiter gentilis	GIA	M	73.87	12.24	–	–	–	–	Netherlands
Accipiter gentilis	GIA	M	72.9	11.9	–	–	–	–	Netherlands
Accipiter gentilis	AMM–LAZ/UAM	M	71.4	11.6	–	5.00	12.5	7.8	Iberia
Accipiter gentilis	AMM–LAZ/UAM	M	71.5	11.8	9.8	4.9	12.3	7.8	Iberia

(Continued)

Appendix 15.1. (Continued)

Species	*ID code/Reference*	*SEX*	*Gl*	*Bp*	*Tp*	*Sc*	*Bd*	*Td*	*Location and observations*
Accipiter gentilis	AMM–LAZ/UAM	M	72.9	11.8	9.8	4.6	12.5	7.6	Iberia
Accipiter gentilis	AMM–LAZ/UAM	M	73.00	11.9	9.7	4.7	12.7	7.6	Iberia
Accipiter gentilis	AMM–LAZ/UAM	M	67.5	11.8	9.6	5.1	12.5	7.9	Iberia
Accipiter gentilis	AMM–LAZ/UAM	M	67.3	11.8	9.6	5.1	12.3	7.8	Iberia
Accipiter gentilis	AMM–LAZ/UAM	M	71.3	12.1	9.9	5.3	13.00	7.7	Iberia
Accipiter gentilis	AMM–LAZ/UAM	M	71.00	11.8	9.7	5.3	13.00	7.7	Iberia
Accipiter gentilis	AMM–LAZ/UAM	M	72.2	11.7	9.4	4.9	12.00	–	Iberia
Accipiter gentilis	AMM–LAZ/UAM	M	0.00	11.4	9.3	5.4	–	–	Iberia
Accipiter gentilis	AMM–LAZ/UAM	M	71.6	11.2	9.4	4.5	12.5	7.4	Iberia
Accipiter gentilis	AMM–LAZ/UAM	M	70.9	11.1	9.5	4.8	12.4	7.5	Iberia
Accipiter gentilis	AMM–LAZ/UAM	M	74.1	11.5	9.7	4.9	12.5	7.3	Iberia
Accipiter gentilis	AMM–LAZ/UAM	M	74.2	11.6	9.8	5.00	12.7	7.3	Iberia
Accipiter nisus	Moreno, Pimenta 2010	F	59.6	7.5	–	3.00	–	7.8	Iberia (LARC–DGPC; n=6)
Accipiter nisus	AMM–LAZ/UAM	F	60.05	8.12	6.37	2.9	8.01	4.41	Iberia
Accipiter nisus	AMM–LAZ/UAM	F	56.56	7.37	5.84	2.67	7.24	4.31	Iberia
Accipiter nisus	AMM–LAZ/UAM	F	56.67	7.31	6.25	2.73	7.35	4.21	Iberia
Accipiter nisus	AMM–LAZ/UAM	F	57.55	8.27	6.54	3.14	7.92	4.55	Iberia
Accipiter nisus	AMM–LAZ/UAM	F	57.64	8.2	6.71	3.09	7.94	4.6	Iberia
Accipiter nisus	AMM–LAZ/UAM	F	60.69	7.97	6.32	3.1	7.79	4.59	Iberia
Accipiter nisus	AMM–LAZ/UAM	F	60.27	8.09	6.24	2.86	7.29	4.47	Iberia
Accipiter nisus	AMM–LAZ/UAM	F	60.62	7.95	6.19	2.98	7.33	4.19	Iberia
Accipiter nisus	AMM–LAZ/UAM	F	58.4	7.47	6.14	2.91	7.21	4.03	Iberia
Accipiter nisus	AMM–LAZ/UAM	F	59.79	7.85	6.39	2.81	7.87	4.49	Iberia
Accipiter nisus	AMM–LAZ/UAM	F	59.93	7.83	6.35	2.93	7.66	4.48	Iberia
Accipiter nisus	AMM–LAZ/UAM	M	–	6.19	5.23	2.2	–	–	Iberia
Accipiter nisus	AMM–LAZ/UAM	M	–	6.05	5.27	2.45	–	–	Iberia
Accipiter nisus	LJ –Gibraltar	M	52.4	6.3	–	2.1	6.1	–	Iberia
Accipiter nisus	Moreno & Pimenta 2011	M	52.8	6.00	–	2.1	–	6.1	Iberia (n=4)

(Continued)

Appendix 15.1. Tarsometatarsus measurements of reference and archaeological goshawk and sparrowhawks following codes from Driesch (1976) and Schmidt-Burger (1982). (Continued)

Species	*ID code/Reference*	*SEX*	*Gl*	*Bp*	*Tp*	*Sc*	*Bd*	*Td*	*Location and observations*
Accipiter nisus	GIA	F	59.94	7.74	–	–	–	–	Netherlands
Accipiter nisus	GIA	F	58.53	8.71	–	–	–	–	Netherlands
Accipiter nisus	Amersfoort	F	62.3	7.7	6.5	2.8	8.03	4.75	Netherlands
Accipiter nisus	Amersfoort	F	62.1	7.97	6.53	2.8	8.02	4.6	Netherlands
Accipiter nisus	Amersfoort	F	60.3	7.3	6.17	2.8	7.7	4.25	Netherlands
Accipiter nisus	Amersfoort	F	60.5	7.35	6.16	2.77	7.56	4.25	Netherlands
Accipiter nisus	LJ 1032 f	F	63.1	7.9	–	2.8	8.1	–	Norway
Accipiter nisus	GNM 1247 f	F	63.8	8	–	3.0	7.8	–	Norway
Accipiter nisus	GNM 9132 f	F	59.5	8.3	–	2.9	7.7	–	Norway
Accipiter nisus	GNM 9136 f	F	61.7	7.7	–	2.8	7.7	–	Norway
Accipiter nisus	GNM 9149 f	F	56.2	7	–	2.8	7.2	–	Norway
Accipiter nisus	GNM 10657 f	F	58.2	7.3	–	2.6	7.3	–	Norway
Accipiter nisus	GNM 10677 f	F	61.3	7.6	–	2.9	7.6	–	Norway
Accipiter nisus	GNM 9260 f	F	61.1	7.6	–	2.9	7.9	–	Norway
Accipiter nisus	GNM 10711 f	F	56.2	7.9	–	2.6	7.6	–	Norway
Accipiter nisus	GNM 15605 f	F	60.4	7.7	–	2.7	7.6	–	Norway
Accipiter nisus	GNM 16220 f	F	52.9	7.4	–	2.7	7.1	–	Norway
Accipiter nisus	GNM 16551 f	F	61.4	7.9	–	2.9	7.8	–	Norway
Accipiter nisus	GNM 16870 f	F	58.6	7.6	–	2.6	7.9	–	Norway
Accipiter nisus	GNM 18609 f	F	61.5	7.9	–	2.8	7.8	–	Norway
Accipiter nisus	GNM 18262 f	F	60.7	7.9	–	2.9	7.6	–	Norway
Accipiter nisus	GNM 8633 m	M	56.2	6.1	–	2.0	6.4	–	Norway
Accipiter nisus	GNM 8635 m	M	54.2	5.9	–	2.0	5.8	–	Norway
Accipiter nisus	GNM 11413 m	M	54.7	5.9	–	2.1	6.1	–	Norway
Accipiter nisus	GNM 12925 m	M	54.2	5.8	–	2.0	5.9	–	Norway
Accipiter nisus	GNM 13257 m	M	51.4	6.1	–	2.2	6.1	–	Norway
Accipiter nisus	GNM 14118 m	M	52.3	6.2	–	2.2	5.7	–	Norway
Accipiter nisus	GNM 15519 m	M	52.5	5.7	–	2.0	5.9	–	Norway
Accipiter nisus	GNM 15620 m	M	52.8	6.3	–	2.1	6.1	–	Norway
Accipiter nisus	GNM 16609 m	M	52.3	6.2	–	2.1	6.4	–	Norway
Accipiter nisus	GNM 17068 m	M	54.6	6.5	–	2.0	6.9	–	Norway

(Continued)

Appendix 15.1. (Continued)

Species	*ID code/Reference*	*SEX*	*Gl*	*Bp*	*Tp*	*Sc*	*Bd*	*Td*	*Location and observations*
Accipiter nisus	GNM 18797 m	M	55	6.2	–	2.1	6.1	–	Norway
Accipiter nisus	1958–9747 m	M	54.7	6.4	–	2.1	6.1	–	Norway
Accipiter nisus	1958–9447 m	M	54.3	6	–	2.2	6.1	–	Norway
Accipiter nisus	GIA	M	54.34	5.8	–	–	–	–	Netherlands
Accipiter nisus	GIA	M	52.35	6.46	–	–	–	–	Netherlands
Accipiter nisus	Amersfoort	M	51.17	–	–	2.06	–	–	Netherlands
Accipiter nisus	Amersfoort	M	51.6	–	–	2.07	5.77	3.5	Netherlands
Accipiter nisus	Amersfoort	M	52.97	6.19	5.2	2.0	6.09	3.51	Netherlands
Accipiter nisus	Amersfoort	M	53.05	6.15	5.19	2.04	6.06	3.52	Netherlands
Accipiter nisus	Amersfoort	M	50.98	6.22	5.26	2.13	6.23	3.62	Netherlands
Accipiter nisus	Amersfoort	M	50.68	6.23	5.14	2.2	6.12	3.7	Netherlands
Buzanca D	Archaeological, Visigoth	**X**	71.2	12.2	9.9	4.7	12.2	7	San Martín de la Vega, Madrid, Spain
Buzanca S	Archaeological, Visigoth	**X**	70.8	12.2	–	4.5	12.2	7	San Martín de la Vega, Madrid, Spain
Accipiter nisus	Great holts farm	**X**	–	–	–	–	–	6.2	Albarella, 1997 (Roman)

AMM-LAZ/UAM: Arturo Morales Muñiz-Laboratorio de Arqueozoología/Universidad Autónoma de Madrid; LARC-DGPC: Laboratório de Arqueôciencias-Direção-Geral do Património Cultural; GIA: Groningen Institute of Archaeology; GNM: Göteborgs Naturhistoriska Museum; M: male; F: female.

Appendix 15.2. Tibiotarsus measurements of reference and archaeological goshawk and sparrowhawks following codes from Driesch (1976) and Schmidt-Burger (1982).

Species	*Collection/Reference*	*SEX*	*Gl*	*La*	*Dip=Dp*	*Bd*	*Dd=Td*	*Location and observations*
Accipiter gentilis	Schmidt-Burger. 1982	M	101.66	99.62	14.86	11.53	7.99	Germany (n=11)
Accipiter gentilis	GIA	M	–	–	–	12	7.95	Netherlands
Accipiter gentilis	GIA	M	–	–	–	11.4	8.05	Netherlands
Accipiter gentilis	Schmidt-Burger. 1982	F	114.75	112.08	18.27	13.92	9.69	Germany (n=20)
Accipiter gentilis	GIA	F	–	–	–	13.84	9.56	Netherlands
Accipiter gentilis	GIA	F	–	–	–	14	9.53	Netherlands
Accipiter gentilis	Schmidt-Burger. 1982	M	106.95	104.3	15.55	12.05	8.4	Scandinavia (n=2)
Accipiter gentilis	Schmidt-Burger. 1982	F	120.42	117.66	19.26	14.94	10.32	Scandinavia (n=5)
Accipiter gentilis	AMM-LAZ/UAM	F	108.59	106.78	17.45	12.74	9.33	Iberia
Accipiter gentilis	AMM-LAZ/UAM	F	109.09	107.19	16.92	12.84	9.32	Iberia
Accipiter gentilis	AMM-LAZ/UAM	F	109.02	106.69	17.51	13.49	9.43	Iberia
Accipiter gentilis	AMM-LAZ/UAM	F	(109.20)	(107.06)	17.62	13.39	9.48	Iberia
Accipiter gentilis	AMM-LAZ/UAM	F	109.78	107.57	16.95	12.4	8.87	Iberia
Accipiter gentilis	AMM-LAZ/UAM	F	109.64	107.75	16.54	12.52	8.89	Iberia
Accipiter gentilis	AMM-LAZ/UAM	F	109.57	107.3	17.16	12.66	8.98	Iberia
Accipiter gentilis	AMM-LAZ/UAM	F	109.08	106.9	16.96	12.61	9.08	Iberia
Accipiter gentilis	AMM-LAZ/UAM	M	97.88	95.59	14.26	10.58	7.51	Iberia
Accipiter gentilis	AMM-LAZ/UAM	F	117.5	115.31	17.93	13.41	9.9	Iberia
Accipiter gentilis	AMM-LAZ/UAM	F	117.72	115.65	17.76	13.5	9.78	Iberia
Accipiter gentilis	AMM-LAZ/UAM	F	115.63	113.26	17.16	12.79	9.31	Iberia
Accipiter gentilis	AMM-LAZ/UAM	F	115.95	113.02	17.12	12.92	9.4	Iberia
Accipiter gentilis	AMM-LAZ/UAM	F	116.46	114.13	17.49	13.22	9.33	Iberia
Accipiter gentilis	AMM-LAZ/UAM	F	116.98	114.59	17.66	12.97	9.18	Iberia
Accipiter gentilis	AMM-LAZ/UAM	F	110.74	108.15	17.49	12.99	9.27	Iberia
Accipiter gentilis	AMM-LAZ/UAM	F	110.02	107.94	17.63	13.12	9.38	Iberia
Accipiter gentilis	AMM-LAZ/UAM	M	92.94	90.91	13.81	11.57	7.42	Iberia
Accipiter gentilis	AMM-LAZ/UAM	M	92.55	90.8	13.7	11.58	7.42	Iberia
Accipiter gentilis	AMM-LAZ/UAM	M	100.3	98.18	14.04	11.8	7.7	Iberia
Accipiter gentilis	AMM-LAZ/UAM	M	100.22	97.7	13.88	11.67	7.75	Iberia
Accipiter gentilis	AMM-LAZ/UAM	M	95.56	93.29	13.49	11.24	7.39	Iberia
Accipiter gentilis	AMM-LAZ/UAM	M	95.41	93.31	13.87	11.35	7.36	Iberia

(Continued)

Appendix 15.2. (Continued)

Species	Collection/Reference	SEX	Gl	La	Dip=Dp	Bd	Dd=Td	Location and observations
Accipiter gentilis	AMM-LAZ/UAM	F	108.37	106.11	17.57	13.11	9.41	Iberia
Accipiter gentilis	AMM-LAZ/UAM	F	108.3	105.41	17.29	12.57	9.23	Iberia
Accipiter gentilis	AMM-LAZ/UAM	M	97.01	95.54	13.86	10.63	7.46	Iberia
Accipiter gentilis	AMM-LAZ/UAM	M	97.27	95.77	14.06	10.56	7.41	Iberia
Accipiter gentilis	AMM-LAZ/UAM	M	99.83	98.12	14.65	10.9	7.83	Iberia
Accipiter gentilis	AMM-LAZ/UAM	M	100.37	98.36	14.73	11.14	7.81	Iberia
Accipiter nisus	Moreno. Pimenta 2010	F	70.3	69.2	9.1	7.4	5.1	Iberia (n=6)
Accipiter nisus	Moreno. Pimenta 2010	M	60.1	59.00	7.2	5.9	4.4	Iberia (n=4)
Accipiter nisus	AMM-LAZ/UAM	F	72.11	71.00	9.44	7.5	5.24	Iberia
Accipiter nisus	AMM-LAZ/UAM	F	67.04	66.06	8.83	6.82	4.91	Iberia
Accipiter nisus	AMM-LAZ/UAM	F	66.98	66.09	8.84	6.85	4.48	Iberia
Accipiter nisus	AMM-LAZ/UAM	F	69.32	68.22	8.89	7.4	5.54	Iberia
Accipiter nisus	AMM-LAZ/UAM	F	69.24	67.9	9.41	7.39	5.59	Iberia
Accipiter nisus	AMM-LAZ/UAM	F	71.79	70.45	9.41	7.41	5.36	Iberia
Accipiter nisus	AMM-LAZ/UAM	F	71.67	70.1	9.36	7.14	5.45	Iberia
Accipiter nisus	AMM-LAZ/UAM	M	61.03	60.38	7.43	5.98	4.48	Iberia
Accipiter nisus	AMM-LAZ/UAM	M	60.62	60.16	7.54	5.97	4.48	Iberia
Accipiter nisus	AMM-LAZ/UAM	F	72.07	70.79	9.27	7.64	5.26	Iberia
Accipiter nisus	AMM-LAZ/UAM	F	72.04	70.7	9.02	7.68	5.2	Iberia
Accipiter nisus	AMM-LAZ/UAM	F	68.52	67.25	8.7	7.06	5.09	Iberia
Accipiter nisus	AMM-LAZ/UAM	F	–	–	8.71	–	–	Iberia
Accipiter nisus	AMM-LAZ/UAM	F	73.14	72.17	9.34	7.53	5.32	Iberia
Accipiter nisus	AMM-LAZ/UAM	F	73.38	72.17	9.35	7.46	5.26	Iberia
Accipiter nisus	LJ	M	60.4	59.2	7.2	6.3	4.4	Iberia
Accipiter nisus	GIA	F	–	–	–	7.14	5.02	Netherlands
Accipiter nisus	GIA	F	–	–	–	7.71	5.04	Netherlands
Accipiter nisus	Amersfoort	F	73.81	75.2	8.6	7.7	5.4	Netherlands
Accipiter nisus	Amersfoort	F	73.74	75.2	8.6	7.7	5.43	Netherlands
Accipiter nisus	Amersfoort	F	70.8	72.2	8.3	7.17	5	Netherlands
Accipiter nisus	Amersfoort	F	70.77	71.92	8.32	7.16	4.94	Netherlands
Accipiter nisus	Amersfoort	M	57.7	58.0	5.9	5.6	4	Netherlands
Accipiter nisus	Amersfoort	M	57.83	58.6	6.1	5.7	4	Netherlands
Accipiter nisus	Amersfoort	M	58.49	59.51	5.81	5.92	4.33	Netherlands
Accipiter nisus	Amersfoort	M	58.59	59.6	5.66	6.0	4.34	Netherlands

(Continued)

Appendix 15.2. Tibiotarsus measurements of reference and archaeological goshawk and sparrowhawks following codes from Driesch (1976) and Schmidt-Burger (1982).(Continued)

Species	*Collection/Reference*	*SEX*	*Gl*	*La*	*Dip=Dp*	*Bd*	*Dd=Td*	*Location and observations*
Accipiter nisus	Amersfoort	M	58.18	59.3	5.26	6.2	4.4	Netherlands
Accipiter nisus	Amersfoort	M	57.95	59.1	5.26	6.12	4.52	Netherlands
Accipiter nisus	GIA	M	–	–	–	5.72	4.3	Netherlands
Accipiter nisus	GIA	M	–	–	–	6.09	4.41	Netherlands
Accipiter nisus	LJ 1032 f	F	74.8	73.6	9.6	7.6	5.7	Norway
Accipiter nisus	GNM 1247 f	F	75.6	74.1	9.7	7.7	5.5	Norway
Accipiter nisus	GNM 1422 f	F	70.7					Norway
Accipiter nisus	GNM 9132 f	F	70.7	69.3	9.4	7.2	5.4	Norway
Accipiter nisus	GNM 9136 f	F	74.6	73.3	9.6	7.6	5.6	Norway
Accipiter nisus	GNM 9149 f	F	67.2	66	8.9	7	4.5	Norway
Accipiter nisus	GNM 10657 f	F	69.5	68.6	8.9	7.2	5	Norway
Accipiter nisus	GNM 10677 f	F	71.5	70	9.6	7.4	5	Norway
Accipiter nisus	GNM 9260 f	F	72.5	71.6	9.5	7.3	5.5	Norway
Accipiter nisus	GNM 10711 f	F	68.7	67.7	9.5	7.4	5.2	Norway
Accipiter nisus	GNM 15605 f	F	72.3	71.1	9.2	7.2	5.4	Norway
Accipiter nisus	GNM 16220 f	F	69.9	68.6	9.2	7.5	5.3	Norway
Accipiter nisus	GNM 16551 f	F	71.1	69.8	9.4	7.4	5.3	Norway
Accipiter nisus	GNM 16870 f	F	69.5	68.2	9.3	7.4	5.4	Norway
Accipiter nisus	GNM 18609 f	F	73	71.4	9.5	7.5	5.4	Norway
Accipiter nisus	GNM 18262 f	F	72	70.6	9.6	7.3	5.1	Norway
Accipiter nisus	GNM 8633 m	M	61.8	61	7.4	6	4.4	Norway
Accipiter nisus	GNM 8635 m	M	60.1	59.3	7.1	5.8	4.1	Norway
Accipiter nisus	GNM 11413 m	M	63.2	61.3	7.4	5.8	4.3	Norway
Accipiter nisus	GNM 12925 m	M	60.4	59.7	7	5.7	4.2	Norway
Accipiter nisus	GNM 13257 m	M	57.9	57	7.2	5.6	4.1	Norway
Accipiter nisus	GNM 14118 m	M	59.8	58.8	6.8	5.8	4.1	Norway
Accipiter nisus	GNM 15519 m	M	59.6	58.5	7	5.6	4.2	Norway
Accipiter nisus	GNM 15620 m	M	59.8	59.1	7.5	6.2	4.2	Norway
Accipiter nisus	GNM 16609 m	M	59.6	58.6	7.5	5.8	4.4	Norway
Accipiter nisus	GNM 17068 m	M	59.8	58.6	7.3	5.9	4.4	Norway
Accipiter nisus	GNM 18797 m	M	62.9	61.7	7.3	6.1	4.4	Norway
Accipiter nisus	1958-9747 m	M	61.7	60.8	7.4	6	4.4	Norway
Accipiter nisus	1958-9447 m	M	60.5	59.3	7.1	5.8	4.2	Norway
Begastri	Archaeological. Byzantine	X	–	–	–	7.7	6.8	Cehegín. Murcia

AMM-LAZ/UAM:Arturo Morales Muñiz-Laboratorio de Arqueozoología/ Universidad Autónoma de Madrid; LARC-DGPC: Laboratório de Arqueôciencias-Direção-Geral do Património Cultural; GIA: Groningen Institute of Archaeology; GNM: Göteborgs Naturhistoriska Museum; M: male; F: female.

16

Launceston twenty-five years after: the zooarchaeology of Cornwall

Polydora Baker

In 1996 Simon Davis and Umberto Albarella published the analysis of animal bones from Launceston Castle (13th–20th c.*), North Cornwall, South-West England. The assemblage provided a wealth of information on elite diet, medieval hunting and gifting, the role of animal remains as proxy for changing function and economic fortunes, and drove a major synthesis of comparative sites. Further, the biometric data provided novel insight into the timing of agricultural improvements in England. The assemblage remains exceptional for Cornwall with over 9000 identified bones spanning ten centuries. Despite having one of the richest archaeological records in Britain, the county of Cornwall has comparatively few such sizeable faunal assemblages, due in large part to the acidic soils and poor potential for bone preservation. Nonetheless, a number of Cornish Bronze Age to Early medieval assemblages (*e.g. *Gunwalloe, Trevelgue, Duckpool, Atlantic Road, Tintagel) have been reported on more recently which, along with much earlier faunal analyses for the county, allow us to expand our understanding of livestock and resource use in the Cornish peninsula through time, with Launceston remaining a key comparison for the region.*

Keywords: Launceston, Cornwall, synthesis, subsistence, husbandry.

Introduction

The animal bone assemblage from Launceston Castle (Albarella and Davis 1996) remains unique in the quantity and quality of zooarchaeological data available for a Cornish site, informing on the economic and social history of the castle and wider agrarian history and trends in social behaviour. The assemblage numbering over 9000 identified bones revealed changing husbandry strategies, a sequence of livestock 'improvement', the role of domestic and wild animals as status markers, and access to non-local resources. This paper provides a review of the zooarchaeology of the wider Cornish peninsula from later prehistory through to the post-medieval period. It considers whether and how the available data for mammals and birds can contribute to our understanding of animal use beyond Launceston, and how the evidence compares to wider patterns of animal exploitation across Southern England, with the Launceston results and contribution highlighted throughout. The time period covers the Bronze Age to post-medieval period. Though integral to subsistence at many sites, fish and molluscs are excluded from the present paper.

Launceston's legacy

The assemblage from Launceston Castle remains unprecedented in the amount of information obtained on medieval elite diet, animal husbandry in Cornwall, and in evidencing diachronic change from the medieval to post-medieval periods. This was and remains significant for the county but is invaluable within a wider research context also, with the Launceston analysis identifying in particular not only the rise and fall in status through diet but also developments and changes in livestock industries and husbandry, and clarifying the timing of these changes. These research avenues were driven and supported by detailed comparison to written sources, and are questions relevant well beyond Cornwall, and actively pursued since the Launceston analysis for different regions and periods (see Allen 2017a; 2017b; Holmes 2011; Poole 2011; Sykes; 2007).

Initially founded in the 11th century, Launceston Castle reached the height of its royal status and occupancy in the 13th century under Richard, Earl of Cornwall. Food remains revealed conspicuous consumption of meat in the high pig proportions and access to a range of luxury foods, including kid meat, venison, numerous wildfowl species and marine mammals. Launceston reduced in importance from the late 13th century when Royal administration was moved elsewhere. The status of the castle constables declined and the castle was used for the Justice's circuit from the 14th–16th century with later sole occupation by the low status gaoler and prisoners until the mid-19th century. Waste accumulated in the later periods (from mid-17th century), including possibly from the wider town area. This process is tracked through the decline in pigs and wild taxa, and variable shifts in cattle and sheep frequencies.

Wider changes in agricultural regimes were also evidenced in the assemblage, including a possible emerging dairy and veal industry after the 15th century, revealed through the increased cull of young cattle. The emergence of specialisation has been increasingly documented across England, notably at Norwich Castle (Albarella *et al.* 1997) and through new and varied analyses (Fisher and Thomas 2012). The particular contribution of Launceston to the study of livestock improvements is the pinpointing of the different 'effects and timing' of the Agricultural revolution in animal husbandry, notably in cattle and sheep (Albarella and Davis 1996, 1). In summary, amongst historical agricultural writers there were different viewpoints regarding the onset of livestock 'improvements', whether occurring at the end of the Late medieval period in the 16th century, or more recently in the post-medieval period around the 18th century. The Launceston research revealed a two-stage process with cattle increasing rapidly in size between the 15th–16th/17th century with evidence for new breeds possibly obtained through cross-breeding or imports, while size increase in sheep and possibly pigs occurred more gradually and mainly later from the 16th–18th century (Albarella and Davis 1996, 57). Not only was this shown for Launceston but these patterns found parallels elsewhere across England.

Since the Launceston analysis, research has continued on this question with cattle 'improvements' evident even earlier in the 14th century, for example at Dudley Castle (Thomas 2005; Fisher and Thomas 2012) and London (Thomas *et al.* 2013), and on the effect of enclosure on rural livestock husbandry (Hamilton and Thomas 2012; Fraser 2020). Recent and current research is further targeting livestock management in the Late medieval and post-medieval periods with new work combining biometric, geometric morphometric, biochemical and genetic analyses (Fisher and Thomas 2012; Salvagno pers. comm. 2020).

Cornish zooarchaeology in context

At the time of the Launceston analysis the number of zooarchaeological reports available for Cornwall was limited, impeding the possibility of chronological or synthetic reviews for the region or county itself (*e.g.* see Clutton-Brock 1976; Coy 1987; Levitan 1987). In comparison to the mainland, analyses of zooarchaeological data from the Scillies were slightly more abundant though the data equally limited, except in few cases. Since these early works and the research on Launceston, a number of new analyses and assessments, and publication of old excavations and assemblages have expanded the datasets available (Fig. 16.1). New syntheses produced as part of assemblage analyses or regional research agendas include summaries of bone frequencies and biometry, notably Hammon (2007a; 2007b; 2011), Ratcliffe and Straker (1996), Straker (2007), Webster (2011) and Johns (2012). Zooarchaeological data are referred to in wider archaeological and landscape surveys in Cornwall, including Rose and Preston-Jones (1995), Pearce (2004) and Turner (2006), though focussing on individual sites as key examples. The reviews of environmental data by Straker (2011) for the Cornish mainland and Ratcliffe and Straker (1996) for the Scilly Isles are more detailed, as these are restricted to Cornwall alone, the Scillies review remaining the most exhaustive and current today (with the addition of Ingrem 2006 and Locker 1996a; 1996b; 1996c; 1996d; 1997a; 1997b). Where they meet the regional review parameters (*e.g.* sample size), Cornish sites are also included in the Historic England regional reviews for Southern England (Hambleton 2008; 2009; Allen *et al.* 2012; Holmes 2017a; 2017b; Serjeantson 2011a; 2011b).

Nature of the evidence – opportunities and constraints

The acidic soils of Cornwall, including the head/rab (granite-derived soils) of the Scillies, mean that animal bone is either not preserved or only the more robust elements survive. However, at some sites conditions more suitable to bone preservation are present ensuring the survival of moderate to large assemblages. Amongst these are the calcareous and shelly sands, and shell middens which occur along the coast, for example at Atlantic Rd, Newquay (Reynolds 2001), and especially on the Scillies (Ratcliffe and Straker 1996; Straker 2006), and middens where the concentration of organic-rich material and depth of deposits may have enhanced preservation, notably at Launceston and Tintagel (Straker 2011; Campbell *et al.* 2018). For this reason, most of the available assemblages derive from coastal sites, with no data available from the uplands or inland locations apart from Launceston, limiting geographical and topographical considerations.

Information about recovery methods is not always stated, in particular for early excavations. However, many of the assemblages excavated in recent years have sieved samples making it possible to assess biases to a lesser or greater degree in the hand-collected portions. Recording methods also vary between specialists and research era which create additional challenges for comparative analysis. Some early reports include selective quantification data,

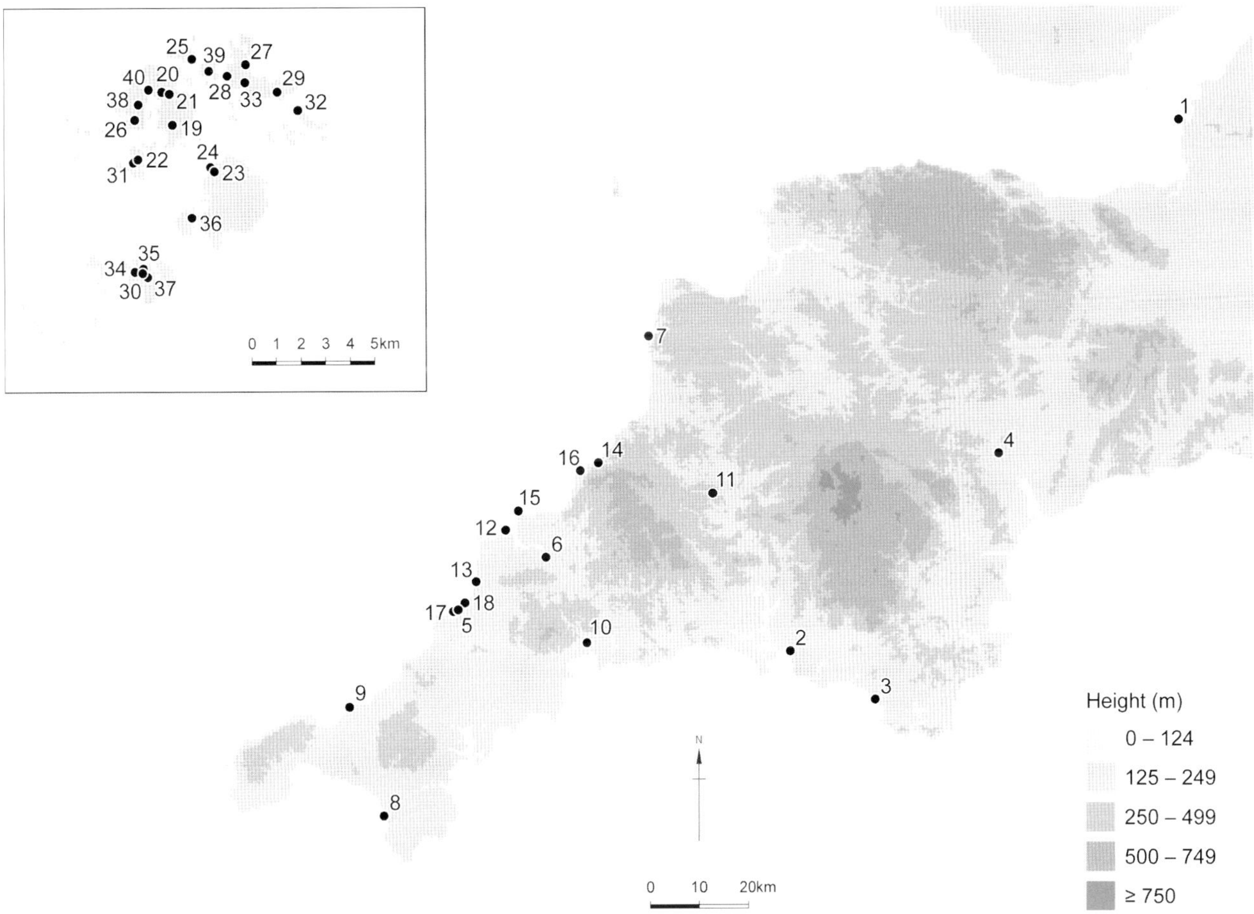

Figure 16.1. Location of Bronze Age to post-medieval sites in Cornwall with zooarchaeological assemblages, and of comparative sites in Somerset and Devon mentioned in the text. The Isles of Scilly, shown in the inset, are located c. *45 km south-west of the Cornish mainland. The site names are listed along with map numbers in Tables 16.1 and 16.2. © Sharon Soutar, Historic England.*

species presence/absence or derived data only. Furthermore, few protocols for quantification, recording age at death data, measurements or species distinction (*e.g.* sheep and goat; horse and donkey; cetacea) were available or cited in early research. We can surmise that, within the work of a single specialist, biometric data can be compared between assemblages but not necessarily with data recorded by other specialists following published standards. For example, measurements recorded prior to publication of Driesch's biometric guide (1976) must be considered with caution. Despite these limitations, the early zooarchaeological data can provide valuable insight into resource use and can now be integrated within the context of more recent research.

Methods and materials

For this synthesis all assemblages with identified taxa (from assessment data or analyses) were included, regardless of assemblage size or availability of quantified data. This was deemed essential given the nature of preservation in this region, the lack of previous coverage, and the opportunity to gain a more holistic view of the dataset at this time. The sites and phases span the Bronze Age (excluding Bronze Age monuments, *i.e.* barrows, cairns, cists, see Webster 2007; Nowakowski and Johns 2015) to post-medieval period. Earlier prehistory, the Mesolithic and Neolithic, is excluded given the dearth of assemblages from the region, despite evidence for settlement, subsistence (from artefacts) and mortuary practice (Webster 2007; Randall 2010); a single calf mandible, from the Isles of Scillies, is known for the Neolithic period (Ratcliffe and Straker 1996; Serjeantson 2011a). Assemblage chronology is as published (except for prehistoric Gunwalloe where Bronze and Iron Age deposits are distinguished), and the assemblages are grouped into broader periods following FISH (nd). Assemblage references are provided in Tables 16.1–16.2. Scientific names of taxa mentioned in the text and tables are included in Appendix 16.1.

The dataset includes 86 phased assemblages from 37 sites, including 14 mainland sites and 23 sites on the Scillies (Fig. 16.1; Tables 16.1–16.2). Multiple entries, including separate hand-collected and sieved assemblages may derive from a single site. Here the focus is mainly on mainland sites but with key comparisons to the Scillies in particular regarding wild species. While Launceston Castle still remains by far the largest dataset, there are a number of assemblages from several periods with over 100–1000 identified specimens, though many are also under NISP 100. Each period is dominated by one or two larger assemblages, Trevelgue for the Iron Age, Atlantic Rd for the Roman period, Gwithian and to a lesser extent Gunwalloe for the Bronze and Early medieval periods.

Presence and taxonomic abundance, mainly Number of Identified Specimens (NISP) but sometimes Minimum Number of Individuals (MNI) were included for mammals and birds, and presence only for microfauna (small mammals and herpetofauna) and fish. The presence of foetal/neonate bones, juvenile and young animals were noted, but there were too few data to warrant conversion and comparison of raw tooth/mandible wear stages. Raw measurements were compared to the dataset from Launceston Castle (Albarella and Davis 1996), to Exeter (Maltby 1979) and to summaries in Allen *et al.* (2017a; 2017b), Maltby (2010) and Poole (2011). Metric data from 11th–12th-century Launceston Castle, and Roman and medieval Exeter were obtained from the publications as well as unpublished datasets and reworked where required for comparative purposes. Withers heights are calculated following Driesch and Boessneck 1974 (Teichert for sheep, and Matolsci 1970 for cattle using the female multiplication factor), and for equids using corrected factors after Johnstone (2004, table 3.5).

Data overview

In most of the quantified assemblages (*i.e.* with NISP or MNI) from the mainland, cattle, sheep/goat and pig generally constitute over 95% of mammal NISP. Sheep make up between 30% to over 80% of the main domestic taxa and cattle *c.* 20–60%, while pigs rarely exceed 5–15%. These taxa are considered in more detail below in domestic species representation and use over time. Goat remains are present in small numbers in only one to two sites per period. At Launceston the presence of juvenile goats (kid meat) was in keeping with the high status of the castle at this time. Historic documents (16th–19th c.) further suggested that goats were once more common than at present, and said to have been best suited to hilly and barren country as found in Cornwall and Devon. Recent research confirms the general scarcity of goat (other than horncores) in the zooarchaeological record. Horse/equid remains are reported from Bronze Age to post-medieval sites across the mainland and Scilllies, with donkey tentatively identified at Roman Kilhallon. In mainland sites, equids become increasingly common from the Bronze Age (when they are present in one of four sites) to the Roman and medieval periods when they are present across most sites, but generally in small numbers. Dogs are present in one prehistoric site (Bronze Age Gwithian) and half of Roman and medieval sites, while cat (wild/domestic) is slightly less common.

Wild mammal taxa and domestic and wild birds are much less abundant than livestock. Deer are ubiquitous, including red and roe deer, but in Cornwall fallow deer are known only from Launceston Castle. Other wild mammals including species not commonly eaten, *e.g.* mustelids and small mammals, occur occasionally and in small numbers. A range of cetaceans, including large whale species and smaller dolphins and porpoise have been identified at mainland and in particular Scillonian sites, where they are present in most periods, while seal is known only from the Scillies. The defining characteristic of Cornish bird bone assemblages, which are generally small, is the marked presence of waterfowl and seabirds in the Isles of Scilly.

In general, most sites have evidence for juvenile, immature, subadult and adult animals, suggesting mixed use of cattle and sheep, and kill-off of subadult or younger pigs. Few sites have evidence of specialised husbandry. Many sites yielded remains of neonatal/perinatal animals, as well as calves, lambs and piglets and/or specimens described as very young, providing important husbandry and seasonality information (Table 16.3). The presence of perinatal and very young animals in over half the coastal mainland sites suggests that calving, lambing (and farrowing) were timed, at least in part, to occur at the settlements perhaps prior to moving flocks/herds to upland pastures and rough ground.

Most of the larger assemblages include skeletal elements from heads, main limbs and distal extremities, suggesting presence of whole carcasses and movement of animals on the hoof. Some assemblages reveal a distinction in food processing and waste disposal practice, as at Trevelgue, Atlantic Rd, Newquay and Tintagel, while fracturing for marrow is evident at Gunwalloe and on the Scillies at at Halangy Down and Little Bay. Specialist processing is suggested from a deposit of goat horncores identified at Mawgan Porth, and craft-working from bone and antler objects and offcuts at numerous sites.

Cattle, sheep and pigs: use over time

There is little doubt that cattle would have provided a large part of the meat supply along with secondary products and labour in most periods in the Cornish peninsula. However, the role of sheep appears to have been substantial at various times, and their use of 'long-standing' tradition, *i.e.* central to farming systems from the Bronze Age (*e.g.* Randall 2010; Table 16.1; Fig. 16.2). Some broad trends may be apparent, with the fluctuating role of sheep and representation of pig remains of particular interest with regards to wider trends in Southern England.

Table 16.1. List of assemblages with identified mammal bones by site, location (Mainland/Isles of Scilly) and chronology, based on NISP, MNI.

	Site and date	Recovery	Total cattle sheep/goat, pig	Bird total	Fish total	Amphibia	Marine shell	%Cattle	%Sheep/goat	%Pig	%Deer	%Bird	Cattle	Sheep/goat	Sheep	Goat	Pig	Large mammal	Medium mammal	Equid	Dog	Cat
	Cornwall mainland																					
	Trethellan Farm MBA	h	76	-	-	-	p	*90.8*	*7.9*	*1.3*	*1.3*	-	69	6	-	-	1	25	10	-	-	-
BA	Gwithian assess E-M/L BA	h	1328	6	3	-	p	*50.2*	*42.8*	*6.9*	*p*	*0.4*	667	569	p	-	92	-	-	p	p	-
	Gwithian 2005 assess M/LBA	hs	20	1	1	-	p	*p*	*p*	*p*	-	*p*	7	11	-	-	2	-	-	-	-	-
	Gunwalloe LBA	h	37	-	-	-	p	*21.6*	*16.2*	*62.2*	*7.5*	-	8	6	-	-	23	20	15	-	-	-
	Trevelgue (6) 6th-5th c BC	h	5	-	-	-	p	*p*	*p*	-	-	-	3	2	-	-	-	-	1	-	-	-
	Trevelgue (7) 5th c BC-1st c AD	h	327	1	1	-	p	*57.2*	*28.4*	*14.4*	*3.5*	*0.3*	187	78	11	4	47	49	40	16	-	-
	Trevelgue (7/8) 5th c BC-4th c AD	h	9	-	-	-	-	*p*	*p*	-	-	-	6	3	-	p	-	2	1	-	-	-
IA	Gunwalloe EIA-RB	h	39	-	-	-	p	*46.2*	*43.6*	*10.3*	*2.5*	-	18	17	-	-	4	33	23	2	-	-
	Gunwalloe EIA-RB	s	1	-	-	-	p	*p*	-	-	-	-	1	-	-	-	-	3	2	-	-	-
	Atlantic Rd c 4th c BC-1st c AD	hs	30	-	-	-	p	*36.7*	*53.3*	*10.0*	-	-	11	14	2	-	3	5	6	-	-	-
	The Rumps* LIA 100 BC-AD100	i	-	3	-	-	p	*14.0*	*68.0*	*18.0*	*p?*	*p*	4	19	-	-	5	-	-	-	-	-
	Kilhallon 1st-3rd c	i	-	p	-	-	p	*p*	*p*	*p*	-	*p*	p	p**	p	-	p	-	-	p	p	-
	Trevelgue 1st-4th c	h	7	-	-	-	p	*p*	*p*	*p*	-	-	3	3	-	-	1	4	-	-	-	-
	Lellizick 1st-6th c	hs	202	1	-	-	p	*23.8*	*71.3*	*5.0*	-	*0.5*	48	144	p	-	10	p	p	-	-	-
	Atlantic Rd (1) 2nd-4th c	h	392	5	1	-	p	*20.4*	*72.7*	*6.9*	*0.3*	*1.3*	80	271	13	1	27	62	64	5	3	-
	Atlantic Rd (1) 2nd-4th c	s	226	3	81	-	p	*10.2*	*88.1*	*1.8*	*0.4*	*1.3*	23	185	14	-	4	17	82	-	-	-
RB	Atlantic Rd (2) 2nd-4th c	s	77	-	46	8	p	*16.9*	*72.7*	*10.4*	-	-	13	52	4	-	8	4	26	-	-	-
	Atlantic Rd (2) 2nd-4th c	h	92	1	2	1	p	*32.6*	*59.8*	*7.6*	-	*1.1*	30	53	2	-	7	24	15	1	-	-
	Atlantic Rd (3) 4th-5th c	h	624	4	-	-	p	*17.6*	*77.1*	*5.3*	*0.3*	*0.6*	110	470	11	-	33	117	125	5	2	4
	Atlantic Rd (3) 4th-5th c	s	21	-	-	-	p	*p*	*p*	*p*		-	2	18	-	-	1	1	4	-	1	-
	Duckpool (Heard) 2nd-4th c	h	78	3	-	-	p	*51.3*	*30.8*	*17.9*	*6.0*	*3.7*	40	22	2	-	14	169	78	2	-	-
	Duckpool (1992) 2nd-4th c	h	86	3	-	-	p	*50.0*	*23.3*	*26.7*	*3.4*	*3.4*	43	18	2	-	23	311	148	2	-	1
	Duckpool (1992) 2nd-4th c	s	22	1	-	1	p	*p*	*p*	*p*	-	*p*	4	7	1	-	10	10	47	-	1	-
	Tintagel (1) 5th c	hs	277	2	6?	-	p	*39.4*	*14.8*	*45.8*	*0.4*	*0.7*	109	38	2	1	127	42	40	1	-	-
	Tintagel (2) 8th-9th c	hs	77	2	-	-	p	*37.7*	*26.0*	*36.4*	-	*2.5*	29	19	1	-	28	3	11	3	-	-
EM	Tintagel 5th-7th c	h	9	-	-	-	p	*p*	*p*	*p*	-	-	6	1	-	-	2	-	-	-	-	-
	Tintagel 5th-7th c - 12th-13th c	hs	10	-	p	-	p	*p*	*p*	*p*	-	-	4	4	-	-	2	-	-	-	-	-
	Gwithian assess, 5th-8th c	h	1367	37	362	1	p	*59.5*	*32.6*	*8.0*	*0.1*	*2.6*	813	445	-	-	109	-	-	p	p	-
	Tintagel 8th-12th c	hs	13	-	-	-	p	*p*	*p*	*p*	-	-	7	4	-	-	2	2	3	-	-	-
	Mawgan Porth 9th-11th c	i	1120	p	-	-	p	*45.2*	*51.5*	*3.3*	-	*p*	506	577	-	p	37	-	-	18	p	p
	Crane Godrevy assess 9th-12th c	h	744	26	43	-	p	*33.7*	*52.2*	*14.1*	-	*3.4*	251	388	p	-	105	-	-	27?	p	1?
EM-MED	Duckpool (1992) 10th-12th c	hs	5	-	-	-	p	*p*	*p*	*p*	-	-	1	3	-	-	1	10	4	-	-	-
	Gwithian, Hockin's pit 10th-12th c	i	-	p	1	-	-	*p*	*p*	*p*	*p*	*p*	p	p	p	-	p	-	-	p	-	p
	Gwithian, Sandy Lane 11th-13th c	i	-	-	p	-	-	*p*	*p*	*p*	-	-	p	p	p	-	p	-	-	-	-	-
	Gunwalloe 10th-13th c	h	374	5	176	8	p	*42.2*	*52.7*	*5.1*	*0.5*	*1.3*	158	183	14	-	19	219	369	8	4	1
	Gunwalloe 10th-13th c	s	69	7	38	1	p	*11.6*	*84.1*	*4.3*	-	*9.2*	8	51	7	-	3	14	98	-	-	-
L-PM	Burmere, 14th-16th c	i	-	-	-	-	-	*p*	*p*	*p*	-	-	p	p	-	-	p	-	-	p	-	-
	Boscastle 17th/18th c	i	-	-	-	-	-	*p*	*p*	-	-	-	p	p	-	-	-	-	-	-	-	-

(Continued)

Table 16.1. List of assemblages with identified mammal bones by site, location (Mainland/Isles of Scilly) and chronology, based on NISP, MNI. (Continued)

	Site and date	Recovery	Red deer	Roe deer	Deer indet	Fox	Brown hare/ hare	Rabbit	Hare/rabbit	Beaver	Badger	Otter	Pine marten	Mustelid	Seal	Grey seal	Cetacea	Large cetacea	Porpoise	Common dolphin	Risso's? dolphin	Bottlenose dolphin	Small mammals	Reference	Map number
	Cornwall mainland																								
BA	Trethellan Farm MBA	h	1	-	-	-	-	-	-	-	-	-	-	-	-	-	-	-	-	-	-	-	-	Payne 1989	17
BA	Gwithian assess E-M/L BA	h	p	-	-	-	-	p	-	-	-	-	-	-	-	-	p	-	-	-	-	-	-	Hammon 2007a	9
BA	Gwithian 2005 assess M/LBA	hs	-	-	-	-	-	-	-	-	-	-	-	-	-	-	-	-	-	-	-	-	7	Hammon 2007b	9
BA	Gunwalloe LBA	h	2	1	-	-	-	-	-	-	-	-	-	-	-	-	-	-	-	-	-	-	-	Howard 2015	8
IA	Trevelgue (6) 6th-5th c BC	h	-	-	-	-	-	-	2	-	-	-	-	-	-	-	-	-	-	-	-	-	-	Hammon 2011	18
IA	Trevelgue (7) 5th c BC-1st c AD	h	11a	-	1a	1	-	8	2	-	-	-	-	-	-	-	-	-	-	-	-	-	-	Hammon 2011	18
IA	Trevelgue (7/8) 5th c BC-4th c AD	h	-	-	-	-	-	1	1	-	-	-	-	-	-	-	-	-	-	-	-	-	-	Hammon 2011	18
IA	Gunwalloe EIA-RB	h	-	1	-	-	-	-	-	-	-	-	-	-	-	-	-	-	-	-	-	-	-	Howard 2015	8
IA	Gunwalloe EIA-RB	s	-	-	-	-	-	-	-	-	-	-	-	-	-	-	-	-	-	-	-	-	-	Howard 2015	8
IA	Atlantic Rd c 4th c BC-1st c AD	hs	-	-	-	-	-	-	-	-	-	-	-	-	-	-	-	-	-	-	-	-	-	Ingrem 2000	5
IA	The Rumps* LIA 100 BC-AD100	i	-	p?	-	-	-	-	-	1	-	-	-	-	-	-	-	-	-	-	-	-	-	Chaplin & Coy 1964	15
RB	Kilhallon 1st-3rd c	i	-	-	-	-	-	-	-	-	-	-	-	-	-	-	-	-	-	-	-	-	-	Turk 1982	10
RB	Trevelgue 1st-4th c	h	-	-	-	-	-	-	-	-	-	-	-	-	-	-	-	-	-	-	-	-	-	Hammon 2011	18
RB	Lellizick 1st-6th c	hs	-	-	-	-	-	4	-	-	-	-	-	-	-	-	-	-	-	-	-	-	-	Grimm 2008	12
RB	Atlantic Rd (1) 2nd-4th c	h	1	-	-	-	-	-	-	-	1	-	1	-	-	-	-	-	-	-	-	-	1	Ingrem 2000	5
RB	Atlantic Rd (1) 2nd-4th c	s	-	1	-	-	-	-	-	-	-	-	-	-	-	-	-	-	-	-	-	-	29	Ingrem 2000	5
RB	Atlantic Rd (2) 2nd-4th c	s	-	-	-	-	-	-	-	-	-	-	-	-	-	-	-	-	-	-	-	-	1	Ingrem 2000	5
RB	Atlantic Rd (2) 2nd-4th c	h	-	-	-	-	-	-	-	-	-	-	-	-	-	-	-	-	-	-	-	-	9	Ingrem 2000	5
RB	Atlantic Rd (3) 4th-5th c	h	2	-	-	-	-	-	-	-	-	-	-	-	-	-	-	-	-	-	-	-	1	Ingrem 2000	5
RB	Atlantic Rd (3) 4th-5th c	s	-	-	-	-	-	-	-	-	-	-	-	-	-	-	-	-	-	-	-	-	1	Ingrem 2000	5
RB	Duckpool (Heard) 2nd-4th c	h	5	-	-	-	1	-	-	-	-	-	-	-	-	-	-	-	-	-	-	-	-	Powell & Serjeanston 1995	7
RB	Duckpool (1992) 2nd-4th c	h	3	-	-	-	-	-	-	-	-	-	-	-	-	-	-	-	-	-	-	-	-	Powell & Serjeanston 1995	7
RB	Duckpool (1992) 2nd-4th c	s	-	-	-	-	-	-	-	-	-	-	-	-	-	-	-	-	-	-	-	-	-	Powell & Serjeanston 1995	7
EM	Tintagel (1) 5th c	hs	-	1	-	1	-	-	-	-	-	-	-	-	-	-	-	-	-	-	-	-	2	Baker forthcoming	16
EM	Tintagel (2) 8th-9th c	hs	-	-	-	-	-	-	-	-	-	-	-	-	-	-	-	-	-	-	-	-	-	Baker forthcoming	16
EM	Tintagel 5th-7th c	h	-	-	-	-	-	-	-	-	-	-	-	-	-	-	-	-	-	-	-	-	-	Baker 2007	16
EM	Tintagel 5th-7th c - 12th-13th c	hs	-	-	-	-	-	2	-	-	-	-	-	-	-	-	-	-	-	-	-	-	p	Baker 2007	16
EM	Gwithian assess, 5th-8th c	h	-	1	-	-	-	-	-	-	-	-	-	2	-	-	-	-	-	-	-	-	2	Hammon 2007a	9
EM-MED	Tintagel 8th-12th c	hs	-	-	-	-	-	-	-	-	-	-	-	-	-	-	-	-	-	-	-	-	-	Baker forthcoming	16
EM-MED	Mawgan Porth 9th-11th c	i	-	-	-	-	-	-	-	-	-	-	-	-	-	-	-	-	-	-	-	-	-	Clutton-Brock 1976	13
EM-MED	Crane Godrevy assess 9th-12th c	h	-	-	-	1	1	10?	-	-	-	2	-	1	-	-	-	-	-	-	-	-	-	Hammon 2007a	9
EM-MED	Duckpool (1992) 10th-12th c	hs	-	-	-	-	-	-	-	-	-	-	-	-	-	-	-	-	-	-	-	-	-	Powell & Serjeanston 1995	7
EM-MED	Gwithian, Hockin's pit 10th-12th c	i	p	-	-	-	-	1	-	-	-	-	-	-	-	-	-	-	-	-	-	-	-	Turk 1969	9
EM-MED	Gwithian, Sandy Lane 11th-13th c	i	-	-	-	-	-	2	-	-	-	-	-	-	-	-	-	-	-	-	-	-	-	Turk 1969	9
EM-MED	Gunwalloe 10th-13th c	h	-	2	-	-	1	78	-	-	-	-	-	-	-	-	-	-	-	-	-	-	23	Howard 2015	8
EM-MED	Gunwalloe 10th-13th c	s	-	-	-	-	1	8	-	-	-	-	-	-	-	-	-	-	-	-	-	-	19	Howard 2015	8
L-PM	Burmere, 14th-16th c	i	-	-	-	-	-	-	-	-	-	-	-	-	-	-	-	-	-	-	-	-	-	Turk 1965	6
L-PM	Boscastle 17th/18th c	i	-	-	-	-	-	-	-	-	-	-	-	-	-	-	-	-	-	-	-	-	-	Turk 1965	14

(Continued)

Table 16.1. (Continued)

	Site and date	Recovery	Total cattle sheep/goat, pig	Bird total	Fish total	Amphibia	Marine shell	%Cattle	%Sheep/goat	%Pig	%Deer	%Bird	Cattle	Sheep/goat	Sheep	Goat	Pig	Large mammal	Medium mammal	Equid	Dog	Cat
MX	Tintagel (3-4) Med-Modern	h	10	-	-	-	-	*p*	*p*	-	-	-	6	4	-	-	-	1	-	1	-	-
	Tintagel Med-Modern	h	2	3	2	-	p	*p*	*p*	-	-	*p*	2	-	-	-	-	-	-	-	-	-
	Trethellan Farm Undated	h	8	2	-	-	-	*p*	*p*	*p*	-	*p*	1	3	2	-	2	2	-	-	-	-
	Cornwall Isles of Scilly																					
BA-IA	May's Hill, St Martin's 1700-1000 BC	i	-	6	p	-	p	*p*	*p*	*p*	*p*	*p*	2	p	-	-	p	-	-	-	-	-
	Porth Killier, St Agnes LBA	hs	25	32	140	29	p	*p*	*p*	*p*	-	*56.1*	6	18	-	-	1	30	8	-	1	-
	May's Hill, St Martin's 1700 BC-100 AD	i	-	-	p	-	p	*p*	*p*	-	-	-	p	p	-	-	-	-	-	-	-	-
	Nornour 2nd-late 1st millenium BC	i	-	p	-	-	p	*p*	*p*	*p*	*p*	*p*	p	p**	p	p	p	-	-	p	-	-
	Halangy Porth, St Mary's BA-IA	h	-	-	-	-	p	*p*	*p*	-	*p*	-	p	-	p	-	-	-	-	p	-	-
	Porth Killier, St Agnes BA-IA	i	3	p	-	-	p	*p*	*p*	-	*p*	*p*	p	-	3	-	-	-	-	1	-	-
	Little Bay, St Martin's c. 250 BC	s	108	11	422	-	p	*79.6*	*20.4*	-	-	*9.2*	86	22	-	-	-	31	59	-	-	-
	Hillside Farm, Bryher LIA-ERB	hs	-	14	548	-	p	*p*	*p*	*p*	-	-	p	50	-	-	p	-	-	-	2	-
RB	Halangy Down, St Mary's 2nd-4th c	h	744	70	301	-	p	*38.6*	*47.6*	*13.8*	*p*	*8.6*	287	354	p	-	103	233	137	4	-	7
	Halangy Down, St Mary's 3rd c	h	299	35	38	-	p	*34.1*	*47.5*	*18.4*	*p*	*10.5*	102	142	-	-	55	109	82	1	-	34
	Poynter's Garden, St Mary's 3rd c	i	-	p	p	-	p	-	-	-	-	*p*	-	-	-	-	-	-	-	-	-	-
	Parsonage Field, St Martin's LRB	i	-	-	p	-	p	-	-	-	-	-	-	-	-	-	-	-	-	-	-	-
	Dial Rocks, Tresco RB	i	82	12	p	-	p	*51.2*	*39.0*	*9.8*	-	*12.8*	42	25	7	-	8	-	-	2	-	-
RB-EM	May's Hill, St. Martin's 2nd-7th c	i	-	p	p	-	p	*p*	*p*	*p*	*p*	*p*	p	p	p	5	p	-	-	p	2	-
	Tean 2nd-8th c	i	-	-	-	-	p	*p*	*p*	*p*	-	*p*	p	-	p	p	p	-	-	-	p	-
EM-MED	East Porth, Samson 6th c	i	-	p	p	-	-	*p*	*p*	*p*	-	*p*	p	p	-	-	p	-	-	-	-	-
	East Porth, Samson 6th-9th c	i	-	p	-	-	-	*p*	*p*	-	-	*p*	1	p	-	-	-	-	-	-	-	-
	East Porth, Samson 9th c	i	-	1	p	-	-	*p*	*p*	*p*	-	*p*	p	p	p	-	p	-	-	p	-	-
	East Porth, Samson 12th c	i	-	p	p	-	-	*p*	*p*	*p*	-	*p*	p	p	-	-	p	-	-	-	-	-
	Hermitage, St Helen's 6th-12th c	i	-	-	-	-	p	-	-	-	-	-	-	-	-	-	-	-	-	-	-	-
	Periglis, St Agnes EM-LM	i	-	1	p	-	p	-	-	-	-	*p*	-	-	-	-	-	-	-	-	-	-
	Middle Town, St Agnes EM-MED	i	-	1	-	-	-	-	-	-	-	*p*	-	-	-	-	-	-	-	-	-	-
	Lower Town, St Martin's MED	s	2	121	128	-	p	-	*p*	-	-	*98.4*	-	2	-	-	-	-	-	-	-	-
PM	Dolphin Town, Tresco PM	i	16	18	-	-	p	*p*	*p*	*p*	-	*52.9*	5	6	1	2	2	-	-	1	-	-
	Neck of Samson, Samson 1833-55	hs	-	p	-	-	p	-	*p*	-	-	-	-	p	-	-	-	-	-	-	-	-
MX	Staddon, Bryher BA-Modern	i	6	2	-	-	p	*p*	*p*	-	-	*p*	2	2	1	1	-	-	-	-	-	-
	Abbey Gardens, Tresco EM-PM	i	1	-	-	-	p	-	*p*	-	-	-	-	-	1	-	-	-	-	-	-	-
	Tresco channel ?13th-14th/Unstrat	hs	56	10	142	-	p	*46.4*	*28.6*	*25.0*	*p*	*15.2*	26	16	-	-	14	p	p	-	-	1
	Site II, Perkilben, St Agnes Unstrat	i	-	-	-	-	-	*p*	*p*	-	-	-	p	p	-	-	-	-	-	-	-	-
	Site III, Top of Hill, St Agnes Unstrat	i	-	-	-	-	-	*p*	*p*	*p*	-	-	p	p	-	-	p	-	-	-	-	-

(Continued)

Table 16.1. List of assemblages with identified mammal bones by site, location (Mainland/Isles of Scilly) and chronology, based on NISP, MNI. (Continued)

	Site and date	Recovery	Red deer	Roe deer	Deer indet	Fox	Brown hare/ hare	Rabbit	Hare/rabbit	Beaver	Badger	Otter	Pine marten	Mustelid	Seal	Grey seal	Cetacea	Large cetacea	Porpoise	Common dolphin	Risso's? dolphin	Bottlenose dolphin	Small mammals	Reference	Map number
MX	Tintagel (3-4) Med-Modern	h	-	-	-	-	-	-	-	-	-	-	-	-	-	-	-	-	-	-	-	-	p	Baker forthcoming	16
	Tintagel Med-Modern	h	-	-	-	-	-	-	-	-	-	-	-	-	-	-	-	-	-	-	-	-	p	Baker 2007	16
	Trethellan Farm Undated	h	-	-	-	-	-	1	-	-	-	-	-	-	-	-	-	-	-	-	-	-	-	Payne 1989	17
	Cornwall Isles of Scilly																								
BA-IA	May's Hill, St Martin's 1700-1000 BC	i	3	-	-	-	-	-	-	-	-	-	-	-	-	-	-	-	-	-	-	-	-	Turk 1984b	29
	Porth Killier, St Agnes LBA	hs	-	-	-	-	-	2	-	-	-	-	-	-	5	3	-	-	-	-	-	-	10	Locker 1997b	35
	May's Hill, St Martin's 1700 BC-100 AD	i	-	-	-	-	-	-	-	-	-	-	-	-	-	-	-	-	-	-	-	-	-	Turk 1984b	29
	Nornour 2nd-late 1st millenium BC	i	p	p	-	-	-	p	-	-	-	-	-	-	-	p	-	-	p	p	p	-	-	Turk 1967, 1970, 1973b+	32
	Halangy Porth, St Mary's BA-IA	h	p	-	p	-	-	-	-	-	-	-	-	-	p	-	-	-	-	-	-	-	-	Gray 1972	24
	Porth Killier, St Agnes BA-IA	i	1	-	-	-	-	-	-	-	-	-	-	-	-	-	-	-	-	-	-	p	-	Turk 1991	35
	Little Bay, St Martin's c. 250 BC	s	-	-	-	-	-	-	-	-	-	-	-	-	-	-	-	-	-	-	-	-	1	Locker 1983	27
	Hillside Farm, Bryher LIA-ERB	hs	-	-	-	-	-	-	-	-	-	-	-	-	-	-	-	-	-	-	-	-	5	Ingrem 2006	26
RB	Halangy Down, St Mary's 2nd-4th c	h	1	-	-	-	-	57	-	-	-	-	-	-	-	-	p?	-	-	-	-	-	-	Locker 1996a	23
	Halangy Down, St Mary's 3rd c	h	1	-	-	-	-	155	-	-	-	-	-	-	-	-	-	-	-	-	-	-	-	Locker 1996a	23
	Poynter's Garden, St Mary's 3rd c	i	-	-	-	-	-	-	-	-	-	-	-	-	-	-	-	-	-	-	-	-	-	Dudley 1960-1	36
	Parsonage Field, St Martin's LRB	i	-	-	-	-	-	-	-	-	-	-	-	-	-	-	-	-	-	-	-	-	-	Pollard 1952	33
	Dial Rocks, Tresco RB	i	-	-	-	-	-	8	-	-	-	-	-	-	3	1	-	-	-	-	-	-	-	Turk 1991	20
RB-EM	May's Hill, St. Martin's 2nd-7th c	i	1	-	-	-	-	6	-	-	-	-	-	-	p	-	2	-	-	-	-	-	p	Turk 1984b	29
	Tean 2nd-8th c	i	-	-	-	-	-	-	-	-	-	-	-	-	-	-	-	-	-	-	-	-	-	Turk 1969	39
EM-MED	East Porth, Samson 6th c	i	-	-	-	-	-	-	-	-	-	-	-	-	-	-	-	-	-	-	-	-	-	Turk 1973b	22
	East Porth, Samson 6th-9th c	i	-	-	-	-	-	-	-	-	-	-	-	-	-	-	p	-	-	-	-	-	-	Turk 1973b	22
	East Porth, Samson 9th c	i	-	-	-	-	-	-	-	-	-	-	-	-	-	-	-	-	-	-	-	-	-	Turk 1973b	22
	East Porth, Samson 12th c	i	-	-	-	-	-	-	-	-	-	-	-	-	p	-	-	-	-	-	-	-	-	Turk 1973b	22
	Hermitage, St Helen's 6th-12th c	i	-	-	-	-	-	-	-	-	-	-	-	-	-	-	-	p	-	-	-	-	-	O'Neil 1964	25
	Periglis, St Agnes EM-LM	i	-	-	-	-	-	-	-	-	-	-	-	-	-	-	-	-	-	-	-	-	-	Turk 1991	34
	Middle Town, St Agnes EM-MED	i	-	-	-	-	-	-	-	-	-	-	-	-	-	-	-	-	-	-	-	-	-	Turk 1991	30
	Lower Town, St Martin's MED	s	-	-	-	-	-	-	-	-	-	-	-	-	-	4	-	-	-	-	-	-	-	Locker 1997a	28
PM	Dolphin Town, Tresco PM	i	-	-	-	-	-	-	-	-	-	-	-	-	3	1	-	-	-	-	-	-	-	Turk 1991	21
	Neck of Samson, Samson 1833-55	hs	-	-	-	-	-	-	-	-	-	-	-	-	-	p	-	-	-	-	-	-	-	Mason 1984	31
MX	Staddon, Bryher BA-Modern	i	-	-	-	-	-	-	-	-	-	-	-	-	1	1	-	-	-	-	-	-	-	Turk 1991	38
	Abbey Gardens, Tresco EM-PM	i	-	-	-	-	-	-	-	-	-	-	-	-	-	-	-	-	-	-	-	-	-	Turk 1991	19
	Tresco channel ?13th-14th/Unstrat	hs	1	-	-	-	-	36	-	-	-	-	-	-	-	-	-	-	-	-	-	-	-	Baker 2014	40
	Site II, Perkilben, St Agnes Unstrat	i	-	-	-	-	-	-	-	-	-	-	-	-	-	-	-	-	p	-	-	-	-	Turk 1968	37
	Site III, Top of Hill, St Agnes Unstrat	i	2	-	-	-	-	-	-	-	-	-	-	-	p	-	-	-	-	-	-	-	-	Turk 1968	37

(* The Rumps only) or presence (p). Bird total is total identified and/or unidentified bird bones; Fish total is total identified and/or unidentified fish bones. ** described as most numerous taxon; % Cattle, Sheep/goat, Pig is % of total cattle, sheep/goat, pig; %Deer is % of total cattle, sheep/goat, pig and deer. %Bird is % of total cattle, sheep/goat, pig and bird. Pig may include wild boar. Equid may include horse or donkey (reported at Kilhallon only). ? uncertain identification. + includes data from Turk (1978), Pernetta and Handford (1970). ^ Worked whale bone. BA Bronze Age, IA Iron Age, RB Romano-British, EM Early medieval, MED medieval, L–PM Late medieval–post-medieval, MX Mixed/Unstratified. For Launceston Castle see https://www.envarch.net/publications/circaea/12.1/12-1-whole.pdf.

Table 16.2. List of assemblages with bird bones by site, location (Mainland/Isles of Scilly) and chronology, based on NISP or presence (p).

	Site and date	Recovery	Birds total	N species	Seabirds present	Domestic fowl /?	Galliformes	Mallard/dom duck	Greylag/dom goose	Other Anser sp/?	Branta sp	Goose unspecified	Swan/?	Shelduck/?	Teal	Pochard?	Scaup	Long-tailed duck	Goldeneye?	Black grouse	Grey partridge	White stork	Bittern?	Grey heron/Heron/?	Corncrake/?	Moorhen/?	Coot/?
	Cornwall mainland																										
BA	Gwithian assess E-M/L BA	h	6	?	p	-	-	-	-	-	-	-	-	-	-	-	-	-	-	-	-	-	-	-	-	-	-
	Gwithian 2005 assess M/LBA	hs	1	1	np	-	-	-	-	-	-	-	-	-	-	-	-	-	-	-	-	-	-	-	-	-	-
IA	Trevelgue (7) M-LIA 5th c BC-1st c AD	h	1	-	np	-	-	-	-	-	-	-	-	-	-	-	-	-	-	-	-	-	-	-	-	-	-
	The Rumps LIA 100 BC-AD 100	i	3	-	np	-	-	-	-	-	-	-	-	-	-	-	-	-	-	-	-	-	-	-	-	-	-
RB	Kilhallon 1st-3rd c	i	p	2	np	-	-	p	p	-	-	-	-	-	-	-	-	-	-	-	-	-	-	-	-	-	-
	Lellizick 1st-6th c	hs	1	1	p	-	-	-	-	-	-	-	-	-	-	-	-	-	-	-	-	-	-	-	-	-	-
	Atlantic Rd (1) 2nd-4th c	h	5	1	np	-	-	-	1	-	-	4	-	-	-	-	-	-	-	-	-	-	-	-	-	-	-
	Atlantic Rd (1) 2nd-4th c	s	3	1	np	-	-	-	-	-	-	-	-	-	-	-	-	-	-	-	-	-	-	-	-	-	-
	Atlantic Rd (2) 2nd-4th c	h	1	1	p	-	-	-	-	-	-	-	-	-	-	-	-	-	-	-	-	-	-	-	-	-	-
	Atlantic Rd (3) 4th-5th c	h	4	2	np	1	-	-	-	-	-	1	-	-	-	-	-	-	-	-	-	-	-	-	-	-	-
	Duckpool (Heard) 2nd-4th c	h	3	-	np	-	-	-	-	-	-	-	-	-	-	-	-	-	-	-	-	-	-	-	-	-	-
	Duckpool (1992) 2nd-4th c	h	3	1	np	3	-	-	-	-	-	-	-	-	-	-	-	-	-	-	-	-	-	-	-	-	-
	Duckpool (1992) 2nd-4th c	s	1	-	np	-	-	-	-	-	-	-	-	-	-	-	-	-	-	-	-	-	-	-	-	-	-
EM	Tintagel (1) 5th c	hs	2	1	np	-	1	-	-	-	-	-	-	-	-	-	-	-	-	-	-	-	-	-	-	-	-
	Tintagel (2) 8th-9th c	hs	2	2	np	-	1	-	-	-	-	-	1	-	-	-	-	-	-	-	-	-	-	-	-	-	-
	Gwithian assess, 5th-8th c	h	37	?	p	p	-	-	-	p	-	-	-	-	-	-	-	-	-	-	-	-	-	-	-	-	-
EM-MED	Mawgan Porth 9th-11th c	i	p	1	np	p	-	-	-	-	-	-	-	-	-	-	-	-	-	-	-	-	-	-	-	-	-
	Crane Godrevy assess 9th-12th c	h	26	?	p	p	-	-	-	p	-	-	-	-	-	-	-	-	-	-	-	-	-	-	-	-	-
	Gwithian, Hockin's pit 10th-12th c	i	p	1	p	-	-	-	-	-	-	-	-	-	-	-	-	-	-	-	-	-	-	-	-	-	-
	Gunwalloe 10th-13th c	h	5	2	p	3	-	-	-	-	-	-	-	-	-	-	-	-	-	-	-	-	-	-	-	-	-
	Gunwalloe 10th-13th c	s	7	3	p	-	-	-	-	-	1	-	-	-	-	-	-	-	-	-	-	-	-	-	-	-	-
MX	Tintagel Med-Modern	h	3	2	np	1	-	-	-	-	-	-	-	-	-	-	-	-	-	-	-	-	-	-	-	-	-
	Cornwall Isles of Scilly																										
BA-IA	May's Hill, St Martin's 1700-1000 BC	i	6	3	p	-	-	-	-	-	-	-	1	-	-	-	-	-	-	-	-	-	-	-	-	-	-
	Porth Killier, St Agnes LBA	hs	32	6	p	-	-	-	-	-	-	-	-	-	-	1	-	-	-	-	-	-	-	-	-	-	-
	Nornour 2nd-late 1st millenium BC	i	p	29	p	p	-	p	-	p	-	-	-	-	-	-	-	-	p	p	p	p	-	p	-	-	-
	Porth Killier, St Agnes BA-IA	i	p	5	p	2	-	-	-	-	-	-	-	-	-	-	-	-	-	-	-	-	-	-	3	-	p
	Little Bay, St Martin's c. 250 BC	s	11	1	p	-	-	-	-	-	-	-	-	-	-	-	-	-	-	-	-	-	-	-	-	-	-
	Hillside Farm, Bryher LIA-ERB	hs	14	8	p	-	1	-	-	-	-	-	-	-	-	-	-	-	-	-	-	-	-	1	-	-	-
RB	Halangy Down, St Mary's 2nd-4th c	h	70	15	p	2?	-	-	-	-	2	-	-	-	-	-	-	-	-	-	2	-	-	-	-	-	2
	Halangy Down, St Mary's 3rd c	h	35	9	p	-	-	1	-	-	-	2	-	-	-	-	-	-	-	-	-	-	-	-	-	-	2
	Poynter's Garden, St Mary's 3rd c	i	p	?	np	-	-	-	-	-	-	-	-	-	-	-	-	-	-	-	-	-	-	-	-	-	-
	Dial Rocks, Tresco RB	i	12	4	p	-	1	-	-	-	-	-		1	-	-	-	-		-	-	-	-	-	-	-	-
RB-EM	May's Hill, St. Martin's 2nd-7th c	i	p	19	p	p	-	-	-	5	-	-	p	1	2	-	1	1		-	-	-	p	p	-	p	-
EM-MED	East Porth, Samson 6th c	i	p	1	p	-	-	-	-	-	-	-	-	-	-	-	-	-	-	-	-	-	-	-	-	-	-
	East Porth, Samson 6th-9th c	i	p	1	np	-	-	-	-	-	-	-	-	-	-	-	-	-	-	-	-	-	-	-	-	-	-
	East Porth, Samson 9th c	i	1	-	np	-	-	-	-	-	-	-	-	-	-	-	-	-	-	-	-	-	-	-	-	-	-
	East Porth, Samson 12th c	i	p	1	p	-	-	-	-	-	-	-	-	-	-	-	-	-	-	-	-	-	-	-	-	-	-
	Periglis, St Agnes EM-LM	i	1	1	p	-	-	-	-	-	-	-	-	-	-	-	-	-	-	-	-	-	-	-	-	-	-
	Middle Town, St Agnes EM-MED	i	1	-	np	-	-	-	-	-	-	-	-	-	-	-	-	-	-	-	-	-	-	-	-	-	-
	Lower Town, St Martin's MED	s	121	3	p	-	-	-	-	-	-	-	-	-	-	-	-	-	-	-	-	-	-	-	-	-	-
PM	Dolphin Town, Tresco PM	i	18	5	p	-	-	-	-	-	-	-	-	-	-	-	-	-	-	-	-	-	-	-	-	-	-
	Neck of Samson, Samson 1833-55	hs	p	1	p	-	-	-	-	-	-	-	-	-	-	-	-	-	-	-	-	-	-	-	-	-	-
MX	Staddon, Bryher BA-Mod	i	2	1	np	-	-	-	-	-	-	-	-	-	-	-	-	-	-	-	-	-	-	-	-	-	-
	Tresco channel ?13th-14th/Unstrat	hs	10	5	p	4	-	-	-	-	-	-	-	-	-	-	-	-	-	-	-	-	-	-	-	-	-

(Continued)

Table 16.2. List of assemblages with bird bones by site, location (Mainland/Isles of Scilly) and chronology, based on NISP or presence (p). (Continued)

	Site and date	Recovery	Stone curlew	Lapwing	Woodcock	Snipe	Black-tailed godwit	Curlew?	Redshank?	Knot	Ruff?	Great North. diver	Manx shearwater	Gannet	Cormorant	Cormorant/auk	Shag	Herring/L B B gull	Great B B gull	Other gull/?	Guillemot	Razorbill	Great auk	Puffin	Other seabird
	Cornwall mainland																								
BA	Gwithian assess E-M/L BA	h	-	-	-	-	-	-	-	-	-	-	-	-	-	-	-	-	-	p	-	-	-	-	p
BA	Gwithian 2005 assess M/LBA	hs	-	-	-	-	-	-	-	-	-	-	-	-	-	-	-	-	-	-	-	-	-	-	-
IA	Trevelgue (7) M-LIA 5th c BC-1st c AD	h	-	-	-	-	-	-	-	-	-	-	-	-	-	-	-	-	-	-	-	-	-	-	-
IA	The Rumps LIA 100 BC-AD 100	i	-	-	-	-	-	-	-	-	-	-	-	-	-	-	-	-	-	-	-	-	-	-	-
RB	Kilhallon 1st-3rd c	i	-	-	-	-	-	-	-	-	-	-	-	-	-	-	-	-	-	-	-	-	-	-	-
RB	Lellizick 1st-6th c	hs	-	-	-	-	-	-	-	-	-	-	-	1	-	-	-	-	-	-	-	-	-	-	-
RB	Atlantic Rd (1) 2nd-4th c	h	-	-	-	-	-	-	-	-	-	-	-	-	-	-	-	-	-	-	-	-	-	-	-
RB	Atlantic Rd (1) 2nd-4th c	s	-	-	-	-	-	-	-	-	-	-	-	-	-	-	-	-	-	-	-	-	-	-	-
RB	Atlantic Rd (2) 2nd-4th c	h	-	-	-	-	-	-	-	-	-	-	-	-	-	-	-	-	-	-	1	-	-	-	-
RB	Atlantic Rd (3) 4th-5th c	h	-	-	-	-	-	-	-	-	-	-	-	-	-	-	-	-	-	-	-	-	-	-	-
RB	Duckpool (Heard) 2nd-4th c	h	-	-	-	-	-	-	-	-	-	-	-	-	-	-	-	-	-	-	-	-	-	-	-
RB	Duckpool (1992) 2nd-4th c	h	-	-	-	-	-	-	-	-	-	-	-	-	-	-	-	-	-	-	-	-	-	-	-
RB	Duckpool (1992) 2nd-4th c	s	-	-	-	-	-	-	-	-	-	-	-	-	-	-	-	-	-	-	-	-	-	-	-
EM	Tintagel (1) 5th c	hs	-	-	-	-	-	-	-	-	-	-	-	-	-	-	-	-	-	-	-	-	-	-	-
EM	Tintagel (2) 8th-9th c	hs	-	-	-	-	-	-	-	-	-	-	-	-	-	-	-	-	-	-	-	-	-	-	-
EM	Gwithian assess, 5th-8th c	h	-	-	-	-	-	-	-	-	-	-	-	-	-	-	-	-	-	-	-	-	-	-	p
EM-MED	Mawgan Porth 9th-11th c	i	-	-	-	-	-	-	-	-	-	-	-	-	-	-	-	-	-	-	-	-	-	-	-
EM-MED	Crane Godrevy assess 9th-12th c	h	-	-	-	-	-	-	-	-	-	-	-	-	-	-	-	-	-	-	-	-	-	-	2
EM-MED	Gwithian, Hockin's pit 10th-12th c	i	-	-	-	-	-	-	-	-	-	-	-	-	-	-	-	-	-	p	-	-	-	-	-
EM-MED	Gunwalloe 10th-13th c	h	-	-	-	-	-	-	-	-	-	-	-	-	-	-	-	1	-	-	-	-	-	-	-
EM-MED	Gunwalloe 10th-13th c	s	-	-	-	2	-	-	-	-	-	-	-	-	-	-	-	1	-	-	-	-	-	-	-
MX	Tintagel Med-Modern	h	-	-	1	-	-	-	-	-	-	-	-	-	-	-	-	-	-	-	-	-	-	-	-
	Cornwall Isles of Scilly																								
BA-IA	May's Hill, St Martin's 1700-1000 BC	i	-	-	-	-	-	-	-	-	-	-	-	-	-	-	-	-	-	-	1	4	-	-	-
BA-IA	Porth Killier, St Agnes LBA	hs	-	-	-	-	-	-	-	-	-	-	-	-	1	-	-	-	-	-	10	10	-	-	-
BA-IA	Nornour 2nd-late 1st millenium BC	i	p	p	-	-	p	p	p	p	p	-	p	p	p	-	-	-	-	p	p	p	-	p	-
BA-IA	Porth Killier, St Agnes BA-IA	i	2	-	-	-	-	-	-	-	-	-	-	-	-	-	-	-	-	-	-	3	-	-	-
BA-IA	Little Bay, St Martin's c. 250 BC	s	-	-	-	-	-	-	-	-	-	-	-	-	-	-	-	-	-	-	1	-	-	-	-
BA-IA	Hillside Farm, Bryher LIA-ERB	hs	-	-	-	-	-	-	-	-	-	-	1	-	-	-	3	-	-	-	1	3	2	1	-
RB	Halangy Down, St Mary's 2nd-4th c	h	-	1	-	1	-	-	-	-	-	2	-	-	3	2	2	1	1	-	-	-	14	-	-
RB	Halangy Down, St Mary's 3rd c	h	-	-	1	1	-	-	-	-	-	-	-	-	1	-	3	-	-	-	-	-	1	-	-
RB	Poynter's Garden, St Mary's 3rd c	i	-	-	-	-	-	-	-	-	-	-	-	-	-	-	-	-	-	-	-	-	-	-	-
RB	Dial Rocks, Tresco RB	i	-	-	-	-	-	-	-	-	-	-	-	-	-	-	-	-	-	-	-	1	-	1	-
RB-EM	May's Hill, St. Martin's 2nd-7th c	i	-	-	-	1	-	-	-	-	-	-	-	-	-	-	-	-	-	-	p	p	-	p	-
EM-MED	East Porth, Samson 6th c	i	-	-	-	-	-	-	-	-	-	1	-	-	-	-	-	-	-	-	-	-	-	-	-
EM-MED	East Porth, Samson 6th-9th c	i	-	-	-	-	-	-	-	-	-	-	-	-	-	-	-	-	-	-	-	-	-	-	-
EM-MED	East Porth, Samson 9th c	i	-	-	-	-	-	-	-	-	-	-	-	-	-	-	-	-	-	-	-	-	-	-	-
EM-MED	East Porth, Samson 12th c	i	-	-	-	-	-	-	-	-	-	-	-	-	-	-	-	-	-	-	1	-	-	-	-
EM-MED	Periglis, St Agnes EM-LM	i	-	-	-	-	-	-	-	-	-	-	-	-	-	-	-	-	-	-	-	1	-	-	-
EM-MED	Middle Town, St Agnes EM-MED	i	-	-	-	-	-	-	-	-	-	-	-	-	-	-	-	-	-	-	-	-	-	-	-
EM-MED	Lower Town, St Martin's MED	s	-	-	-	-	-	-	-	-	-	-	-	-	-	-	31	-	-	-	-	15	-	19	-
PM	Dolphin Town, Tresco PM	i	-	-	-	-	-	-	-	-	-	-	-	-	-	-	-	-	-	3	6	2	-	6	-
PM	Neck of Samson, Samson 1833-55	hs	-	-	-	-	-	-	-	-	-	-	-	-	-	-	-	-	-	p	-	-	-	-	-
MX	Staddon, Bryher BA-Mod	i	-	-	-	-	-	-	-	-	-	-	-	-	-	-	-	-	-	-	-	-	-	-	-
MX	Tresco channel ?13th-14th/Unstrat	hs	-	-	-	-	-	-	-	-	-	-	2	-	2	-	-	-	-	1	-	-	-	1	-

(Continued)

Table 16.2. (Continued)

	Site and date	*Recovery*	*Pigeon*	*Barn owl*	*Buzzard/?*	*Chough*	*Carrion crow*	*Raven*	*Wren?*	*Blackbird*	*Song thrush/?*	*Thrush/Turdidae*	*Sparrow*	*Skylark*	*Corn bunting*	*Ssmall passerine*	*Bird unspecified*	*Bird indeterminate*	*Reference*	*Map number*
	Cornwall mainland																			
BA	Gwithian assess E-M/L BA	h	-	-	-	-	-	-	-	-	-	-	-	-	-	-	p	-	Hammon 2007b	9
BA	Gwithian 2005 assess M/LBA	hs	-	-	-	-	-	-	-	-	-	-	-	-	-	1	-	-	Hammon 2007a	9
IA	Trevelgue (7) M-LIA 5th c BC-1st c AD	h	-	-	-	-	-	-	-	-	-	-	-	-	-	-	-	1	Hammon 2011	18
IA	The Rumps LIA 100 BC-AD100	i	-	-	-	-	-	-	-	-	-	-	-	-	-	-	-	3	Chaplin & Coy 1964	15
RB	Kilhallon 1st-3rd c	i	-	-	-	-	-	-	-	-	-	-	-	-	-	-	-	-	Turk 1982	10
RB	Lellizick 1st-6th c	hs	-	-	-	-	-	-	-	-	-	-	-	-	-	-	-	-	Grimm 2008	12
RB	Atlantic Rd (1) 2nd-4th c	h	-	-	-	-	-	-	-	-	-	-	-	-	-	-	-	-	Ingrem 2000	5
RB	Atlantic Rd (1) 2nd-4th c	s	-	-	-	-	-	-	-	-	-	-	-	-	-	1	-	2	Ingrem 2000	5
RB	Atlantic Rd (2) 2nd-4th c	h	-	-	-	-	-	-	-	-	-	-	-	-	-	-	-	-	Ingrem 2000	5
RB	Atlantic Rd (3) 4th-5th c	h	-	-	-	-	-	-	-	-	-	-	-	-	-	-	-	2	Ingrem 2000	5
RB	Duckpool (Heard) 2nd-4th c	h	-	-	-	-	-	-	-	-	-	-	-	-	-	-	-	3	Powell & Serjeantson 1995	7
RB	Duckpool (1992) 2nd-4th c	h	-	-	-	-	-	-	-	-	-	-	-	-	-	-	-	-	Powell & Serjeantson 1995	7
RB	Duckpool (1992) 2nd-4th c	s	-	-	-	-	-	-	-	-	-	-	-	-	-	-	-	1	Powell & Serjeantson 1995	7
EM	Tintagel (1) 5th c	hs	-	-	-	-	-	-	-	-	-	-	-	-	-	-	1	-	Baker forthcoming	16
EM	Tintagel (2) 8th-9th c	hs	-	-	-	-	-	-	-	-	-	-	-	-	-	-	-	-	Baker forthcoming	16
EM	Gwithian assess, 5th-8th c	h	-	-	-	-	-	-	-	-	-	-	-	-	-	-	p	-	Hammon 2007b	9
EM-MED	Mawgan Porth 9th-11th c	i	-	-	-	-	-	-	-	-	-	-	-	-	-	-	-	-	Clutton-Brock 1976	13
EM-MED	Crane Godrevy assess 9th-12th c	h	1	-	-	-	-	-	-	-	-	-	-	-	-	-	2	-	Hammon 2007	9
EM-MED	Gwithian, Hockin's pit 10th-12th c	i	-	-	-	-	-	-	-	-	-	-	-	-	-	-	-	-	Turk 1969	9
EM-MED	Gunwalloe 10th-13th c	h	-	-	-	-	-	-	-	-	-	-	-	-	-	-	-	1	Howard 2015	8
EM-MED	Gunwalloe 10th-13th c	s	-	-	-	-	-	-	-	-	-	-	-	-	-	-	-	3	Howard 2015	8
MX	Tintagel Med-Modern	h	-	-	-	-	-	-	-	-	-	-	-	-	-	-	-	-	Baker 2007	16
	Cornwall Isles of Scilly																			
BA-IA	May's Hill, St Martin's 1700-1000 BC	i	-	-	-	-	-	-	-	-	-	-	-	-	-	-	-	1	Turk 1984b	29
BA-IA	Porth Killier, St Agnes LBA	hs	-	-	-	-	-	-	-	-	-	8	2	-	-	-	-	-	Locker 1997b	35
BA-IA	Nornour 2nd-late 1st millenium BC	i	p	-	p	-	-	p	p	p	p	-	-	p	-	-	-	-	Turk 1967, 1970, 1973b, 1978+	32
BA-IA	Porth Killier, St Agnes BA-IA	i	-	-	-	-	-	-	-	-	-	-	-	-	-	-	-	-	Turk 1991	35
BA-IA	Little Bay, St Martin's c. 250 BC	s	-	-	-	-	-	-	-	-	-	-	-	-	-	-	-	10	Locker 1983	27
BA-IA	Hillside Farm, Bryher LIA-ERB	hs	-	-	-	-	-	-	-	-	-	-	-	-	-	-	-	1	Ingrem 2006	26
RB	Halangy Down, St Mary's 2nd-4th c	h	-	12	-	-	-	-	-	-	1	1	1	-	-	-	-	20	Locker 1996a	23
RB	Halangy Down, St Mary's 3rd c	h	-	-	-	-	1	-	-	-	-	-	-	-	-	-	-	22	Locker 1996a	23
RB	Poynter's Garden, St Mary's 3rd c	i	-	-	-	-	-	-	-	-	-	-	-	-	-	-	p	-	Dudley 1960-1	36
RB	Dial Rocks, Tresco RB	i	-	-	-	-	-	-	-	-	-	-	-	-	-	-	-	p	Turk 1991	20
RB-EM	May's Hill, St. Martin's 2nd-7th c	i	1	-	-	1	-	4	-	-	3	-	-	-	1	-	-	p	Turk 1984b	29
EM-MED	East Porth, Samson 6th c	i	-	-	-	-	-	-	-	-	-	-	-	-	-	-	-	-	Turk 1973b	22
EM-MED	East Porth, Samson 6th-9th c	i	-	-	-	-	-	-	-	-	-	-	-	-	-	p	-	-	Turk 1973b	22
EM-MED	East Porth, Samson 9th c	i	-	-	-	-	-	-	-	-	-	-	-	-	-	-	-	p	Turk 1973b	22
EM-MED	East Porth, Samson 12th c	i	-	-	-	-	-	-	-	-	-	-	-	-	-	-	-	-	Turk 1973b	22
EM-MED	Periglis, St Agnes EM-LM	i	-	-	-	-	-	-	-	-	-	-	-	-	-	-	-	-	Turk 1991	34
EM-MED	Middle Town, St Agnes EM-MED	i	-	-	-	-	-	-	-	-	-	-	-	-	-	-	1	-	Turk 1991	30
EM-MED	Lower Town, St Martin's MED	s	-	-	-	-	-	-	-	-	-	-	-	-	-	-	-	56	Locker 1997a	28
PM	Dolphin Town, Tresco PM	i	-	-	-	-	-	-	-	-	1	-	-	-	-	-	-	-	Turk 1991	21
PM	Neck of Samson, Samson 1833-55	hs	-	-	-	-	-	-	-	-	-	-	-	-	-	-	-	-	Mason 1984	31
MX	Staddon, Bryher BA-Mod	i	-	-	1	-	-	-	-	-	-	-	-	-	-	-	-	1	Turk 1991	38
MX	Tresco channel ?13th-14th/Unstrat	hs	-	-	-	-	-	-	-	-	-	-	-	-	-	-	-	-	Baker 2014	40

Total bird is total identified and/or unidentified bird bones. ?uncertain identification. + includes data from Pernetta and Handford (1970). Herring/L B B gull is Herring/Lesser black-backed gull. Great B B gull is Great Black-backed gull. Chronological abbreviations in Table 16.1. For Launceston Castle see https://www.envarch.net/publications/circaea/12.1/12-1-whole.pdf

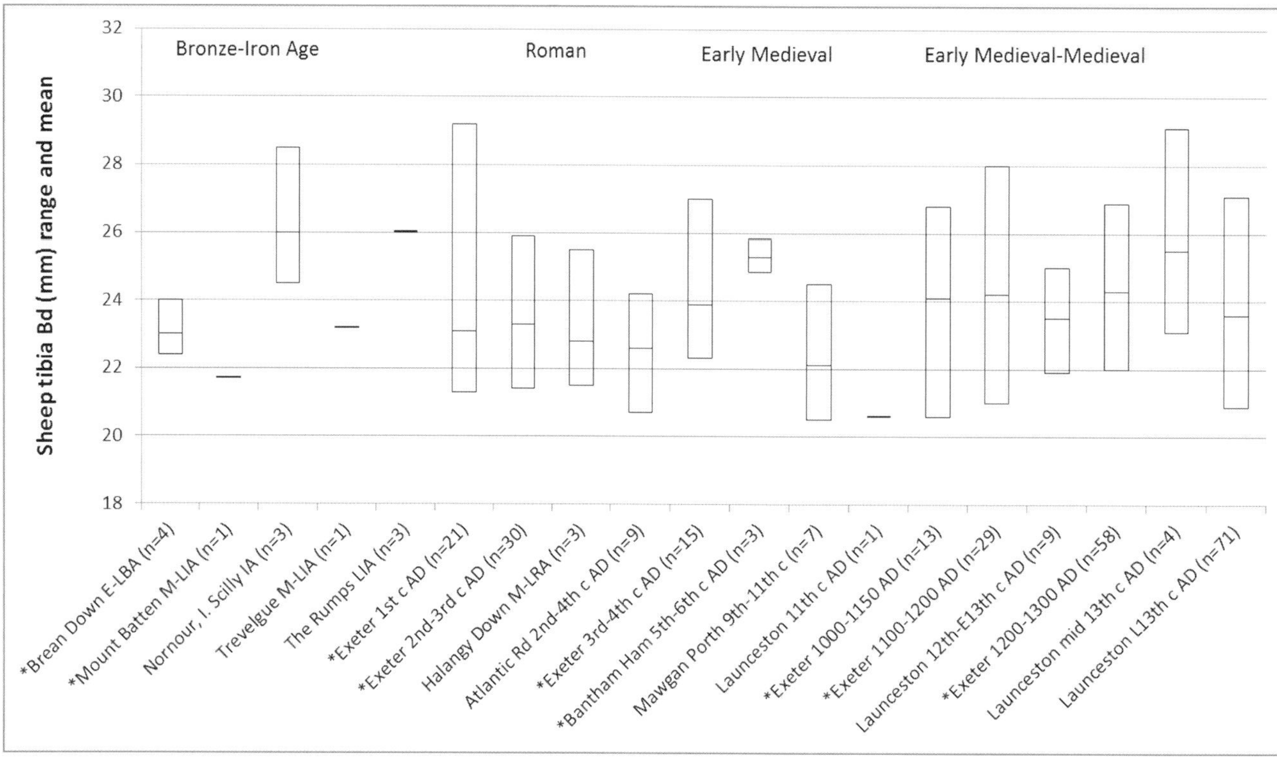

Figure 16.5. Sheep (and sheep/goat) astragalus length (GLl) in Cornish zooarchaeological assemblages and comparative sites in Devon. Measurements in mm.*

from Nornour (Scillies) (Table 16.4) and tibia breadths from Trevelgue, The Rumps and Nornour (Fig. 16.5) compare to Iron Age rural sites further east, including the larger measurements (Allen 2017a, 105, fig. 3.28), though the reliability of the early datasets (pre-1976, see Methods above) is uncertain.

Previous work has indicated that Romano-British sheep in the South-West, including Exeter were smaller on average than elsewhere in Southern Britain and in the Midlands, and showed little subsequent improvement (Maltby 1979; 2010). There is some variation however; withers heights estimates for Halangy Down and Tean (Scillies) are all around or less than 550 mm, with Halangy Down similar in slenderness to Soays (after Sykes 2007, fig. 51). In contrast three values from Atlantic Rd are all over 570 mm (Table 16.4), similar to the larger values from Exeter, as well as Late Roman rural and urban sites in the South and South-East (Allen 2017a, 105, fig 3.28; Maltby 2010, 294, table 2.234 and app. 1). The presence of these larger specimens as well as smaller animals compared to Trevelgue and Exeter sheep may, as noted by Hammon (2011), reflect the presence of different types/breeds of sheep as well as sexes, within and between sites (*e.g.* Figs 16.3–16.4).

Data are particularly scarce for the medieval periods. A few values from 5th-century occupation at Tintagel (not shown) exceed Roman Cornish measurements and Roman and Early medieval data from Exeter. Similarly, two metapodials from Bantham Ham reveal continuity albeit with some evidence for smaller animals as well (Higbee 2007; Figs 16.5–16.6; Table 16.4). The data from 8th–12th-century Tintagel, Gunwalloe, Gwithian (Turk 1969), Mawgan Porth and Launceston tend to compare to the lower Roman values with some exceptions (Figs 16.5–16.6 and archive). Maltby (1979) has suggested that little change occurred from the Roman–medieval period at Exeter but it is currently difficult to fully assess the evolution of sheep in the Cornish peninsula in the intervening period (as attempted in eastern regions, see Poole 2011 on Mid-Saxon breeds, Sykes 2007 on Late Saxon improvements). From the 12th to 13th centuries at Exeter sheep show a similar average and range in size and shape perhaps indicating consistency in the selection of animals for urban markets. The 12th–13th-century data from Launceston include some of the smallest withers heights and astragalus lengths and wider chronological variation in bone breadth, perhaps indicating different sourcing of animals compared to Exeter at this time. There are no biometric data for Cornish sheep in the Late medieval and post-medieval periods, other than at Launceston, which showed little change prior to the 16th century (Albarella and Davis 1996); Turk (1965) describes the sheep from Late medieval Burmere, Wadebridge as similar to 'dark-age', unimproved breeds.

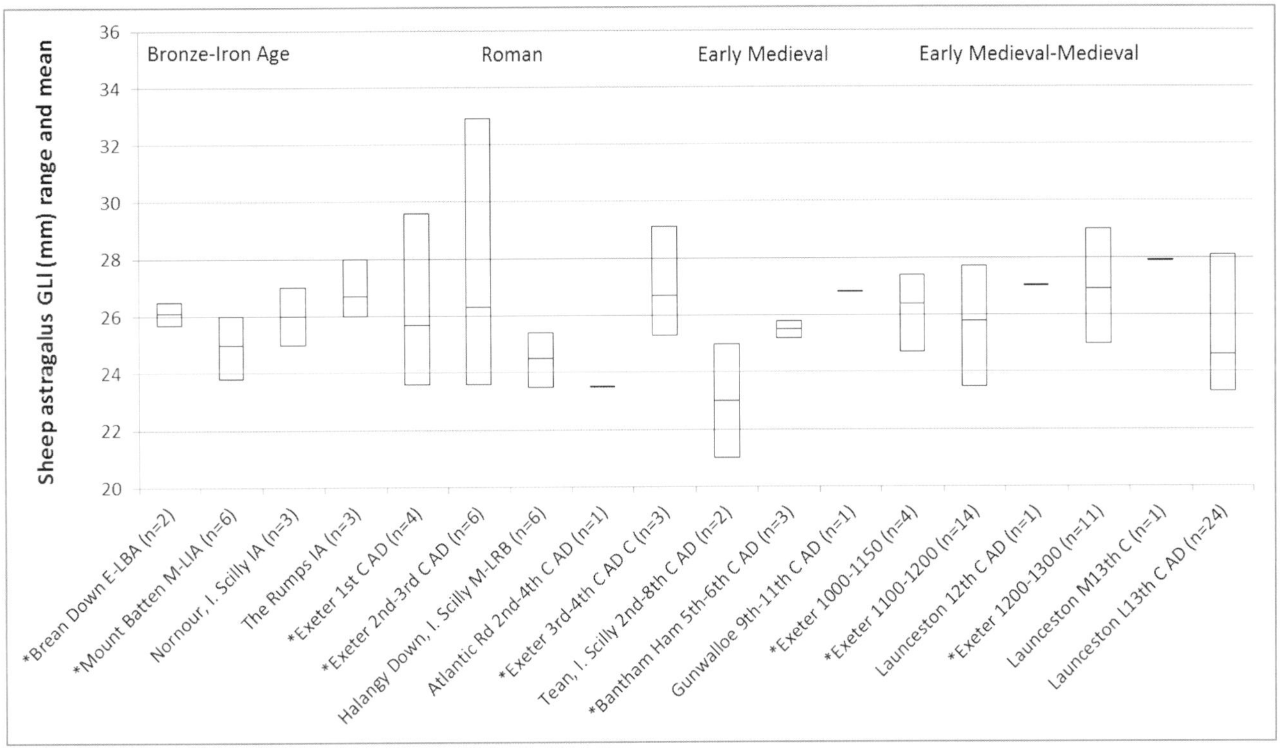

Figure 16.6. Sheep (and sheep/goat) tibia distal breadth (Bd) in Cornish zooarchaeological assemblages and comparative sites in Devon. Measurements in mm.*

Pigs

Syntheses of pig biometric data suggest there was little variation or change in pig size or conformation prior to the Late medieval period (Sykes 2007; Maltby 2010; Poole 2011). The Cornish data, albeit limited, seem to support this. The Iron Age values from Trevelgue and The Rumps, as well as the slightly larger Iron Age dataset from Mount Batten, Devon (Grant 1988) compare to Roman Cornish data (*e.g.* Atlantic Rd) and sites outside the region, suggesting little chronological or geographical diversity. That said pig size at Halangy Down, Scillies may have included smaller animals while Maltby (2010) has identified possible Roman size increase (Dorchester) and selection of larger types/individuals for urban meat supply. The limited data from 5th-century Tintagel are interesting in this respect in that the postcranial measurements (of which some from subadult animals) include some of the largest values recorded from Cornwall. Furthermore, the Tintagel dataset does not include unusually large tooth measurements, hinting perhaps at selection of larger carcass animals, but not necessarily 'improved' or new types. There are no comparative pig data from Cornish medieval sites for the intervening period up to the 12th century apart from Launceston where the few 12th-century data compare to 13th–15th-century values. Albarella and Davis (1996) observed an increase in bone size but decreased tooth size in the 16th and early 17th centuries suggesting a possible improvement and introduction of new breeds at this time, in line with changes in other livestock (Albarella and Davis 1996).

Horses

Horses in Bronze Age to medieval Cornwall appear to have been *c.* 12–14 hands, in other words ponies, similar to medieval and post-medieval horses at Launceston Castle. Kerridge (1968, 154) describes the 17th-century West Country equids as 'bred from the best of the small, sturdy, moorland ponies', but descriptions and data vary. An Iron Age–Roman period tibia from Gunwalloe provides a withers height of 1286 mm (12 hands 3 inches), similar to average values for Iron Age horses elsewhere (Johnstone 2004, 396). A radio-ulna from medieval Gunwalloe gives a withers height of 1324/1356 mm (*c.* 13 hands), within Roman and Saxon ranges (Johnstone 2004, 396; Maltby 2010; Poole 2011), and larger than a reference Exmoor pony (Howard 2015). First phalanges from Early medieval Tintagel (Baker in prep) and Gwithian (Turk 1969) are comparable in size and gracility to modern-day Exmoor ponies while a specimen from Kilhallon (Turk 1982) is smaller. Turk (1965, 54) provides additional descriptions noting that horses at Mays Hill (2nd–7th c. AD, Scillies) were smaller than average Dartmoor and Exmoor ponies, and that those from Late medieval Burmere, Wadebridge were considerably smaller than 'average' ponies.

Access to wild resources

One of the key markers of status in medieval assemblages is the diversity and abundance of resources that elites could call upon to display wealth and power. At Launceston this display is illustrated through consumption of large amounts of venison from fallow deer and the range of gamebirds and waterfowl, some deriving most likely from the Isles of Scilly. But what of the availability of these species in Cornwall through time? Wild mammals are present in sites of all periods and include mainly red and roe deer, and less commonly beaver, lagomorphs (hare and the often intrusive rabbit), occasional mustelids (*e.g.* otter, badger, pine marten), seal (including grey seal) and cetaceans (whale, common and bottle-nose dolphins, porpoise) (Table 16.1). These species attest to the range of environments exploited including forest and field-edge, marshland and coastal zones.

Deer

Red and roe deer are reported in all periods on mainland sites but disappear from the zooarchaeological record of the Scillies after the 7th/8th century or earlier; they are absent from the islands today (Table 16.1; see also Ratcliffe and Straker 1996). Fallow deer has not been recorded at Cornish sites other than Launceston, despite the large number of recorded deer parks (Herring 2007; https://historicengland.org.uk/listing/the-list/list-entry/1016509; see also Turk 1973a). As elsewhere, the pre-Norman expansion of fallow deer may have been limited while their scarcity in the Cornish faunal record is no doubt due to the lack or small size of assemblages, scarcity of elite sites and possibly difficulties of identification. At Launceston, red and/or roe deer were present from the late 11th and 12th centuries and throughout occupation, though more abundant (relative to cattle, sheep/goat, pig and deer) prior to the 13th century after which fallow deer became the dominant deer species. The ubiquity of red and roe deer postcranial bones and presence of butchered specimens, in addition to worked and unmodified antler, within the mainland dataset suggest deer were locally available and exploited by many communities for venison as well as antler (*e.g.* Hambleton 2008; Wood 2015, 73; Fig. 16.7). Given the small number of sites in each period it is not possible to plot the exploitation through time or across the social spectrum, though after the Romano-British period deer bones make up less than 0.5% of NISP on medieval sites, indicating most probably increasingly restricted access (*e.g.* Sykes 2007; Holmes 2017b; Table 16.1). The Romano-British site of Duckpool has a relatively high proportion of red deer (NISP 8 or 6% NISP cattle, sheep/goat, pig and deer), of which all but one

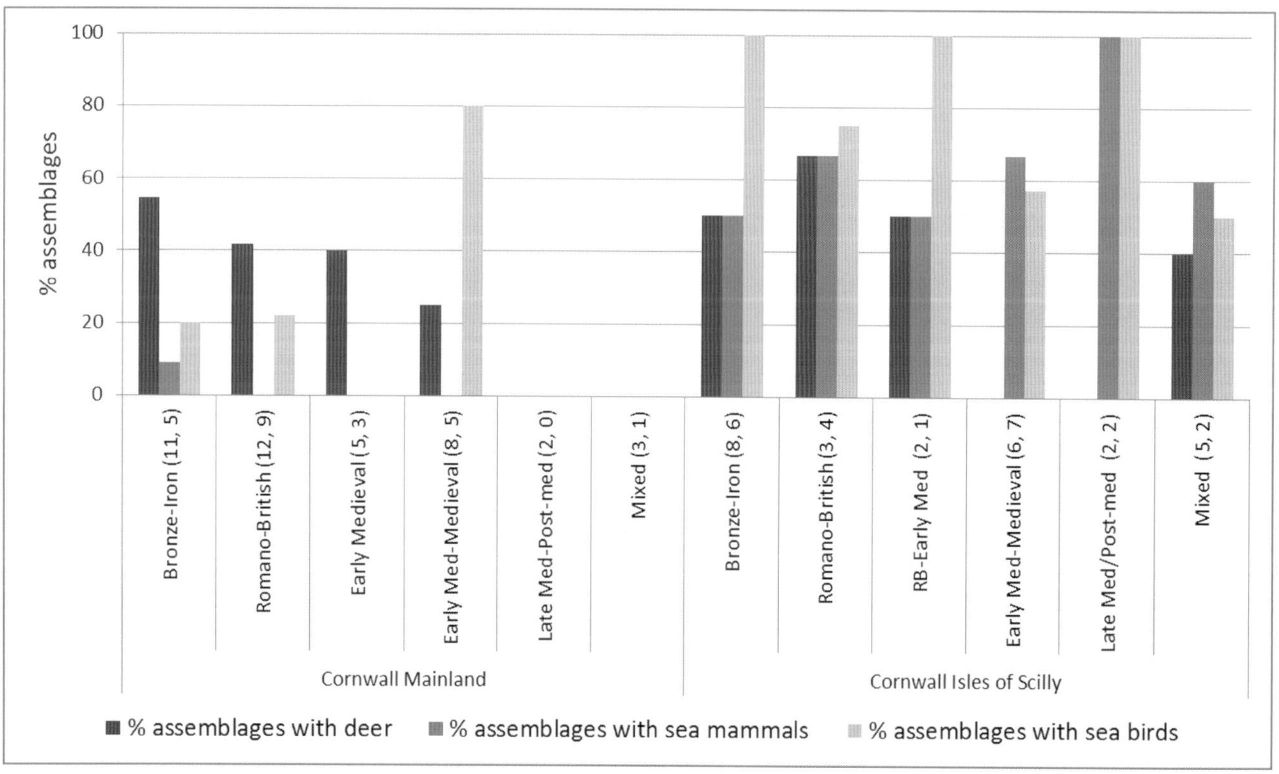

Figure 16.7. Proportion of assemblages by period with deer, sea mammals and sea birds. Numbers in ( ): n assemblages with mammals, n assemblages with birds. Romano-British: 1st–4th c.*; RB–Early medieval: 2nd–8th* c.*; Early medieval: 5th–9th* c.*; Early medieval–medieval 9th–13th* c.*; Late medieval–post-medieval: 14th–18th* c. *Launceston excluded.*

are postcranial bones (in particular distal limb bones). This industrial site may have traded with the local community, so perhaps venison or skins were traded commodities, if not hunted directly.

Marine mammals

The Launceston Castle assemblage included numerous remains of cetaceans including vertebrae of large whales (*Balaena/Balaeonoptera/Physeter* species size), one of which was probably used as a chopping block, and smaller cetaceans including common dolphin. Cetacean remains are present at only one other Cornish mainland site, Bronze Age Gwithian (NISP 2: cetacean vertebrae and vertebra/long bone end), while seals are altogether absent. Turk (1973a) reports grey seal remains from as early as the Bronze Age to medieval period on the Cornish mainland coast, but no records have been identified in the current or previous reviews (*e.g.* Webster 2007, 134; Worley 2016). In contrast, cetacean and seal are ubiquitous on the Isles of Scilly, species of one or the other group present in at least half of the sites in each period (Table 16.1; Fig. 16.7). Where identified to taxon, seal species include grey seal, and cetaceans include common, bottlenose and possible Risso's dolphin, and porpoise. Only Launceston has yielded clear evidence for larger whale species, though a worked whalebone (possible ploughshare) from medieval St Helens (Scillies; O'Neil 1964) is most probably also from a large species.

Cetacean remains are often interpreted as strandings, probably exploited for meat, bone, blubber and oil, but with access rights increasingly restricted to the elite (Gardiner 1997; Sykes 2007; Hambleton 2008; Holmes 2017b). In contrast, the presence at some sites of seal skeletal elements (some with butchery marks) from all bodyparts, including from immature animals suggests seals and their pups were actively hunted. Seals (and stranded cetaceans) may have provided oil for lighting (*e.g.* Davies 1956), and food where animal husbandry or agricultural production could not supply sufficient resources (Hambleton 2008). It is difficult to gauge their importance in past subsistence; they are rare in mainland sites but may represent up to *c.* 10% of mammal bones (more in the very small assemblages) on the Scillies (Table 16.1). Their occurrence in the archaeological record may be related to a species-preferred habitat, migration patterns, social and breeding behaviour (*e.g.* Summers 1974; Ratcliffe and Straker 1996; Couzens *et al.* 2017). For example, grey seal colonies, haul-outs and breeding grounds include remote caves, beaches and islands, making access difficult. These reasons alone do not fully explain the disparity between mainland and islands and it is probable that their presence and pursuit on the Scillies played an important role in island cultural life, in addition to subsistence, from the earliest times.

Birds

Currently it is difficult to assess the relative importance of birds in Cornish sites due to small assemblage sizes, in particular for mainland sites, and incomplete reporting. The effect of recovery bias is clear, with the larger assemblages recovered through sieving. A further issue to consider is the origin, whether anthropogenic or from natural death assemblages. For example, skeletal element distribution of seabird finds in an unstratified (possible 14th c.) wreck site may be intrusive (Baker 2014).

The most striking aspect of the bird bone assemblages is the diversity of species, in particular on the Scillies, with the earlier species distributions in Ratcliffe and Straker (1996) confirmed by more recent work (Locker 1996a; 1997a; 1997b; Ingrem 2006; Table 16.2). Domestic fowl/medium size Galliformes and domestic/wild duck and goose are present in most Roman and medieval mainland and island sites albeit in small numbers, while wild duck and goose species are present in the larger island assemblages. Other waterfowl in particular seabirds are commonly identified, especially across the Scillies, and in the larger bird bone assemblages (Table 16.2; Fig. 16.7). At Launceston Castle, seabirds such as Manx shearwater may have been obtained from the Scillies (Albarella and Davis 1996). Some have also suggested an earlier, Roman trade of seabirds to Exeter, while later medieval rents were sometimes defined in numbers of birds (Ashbee 1996, 137; Baldwin 2009). Mainland Bronze Age sites include occasional gull and cormorant size seabird, Roman sites include gannet and guillemot, while gulls and waders (snipe, woodcock) are present in the medieval period. Recovery may influence species distributions, for example mainly larger species are represented in hand-collected assemblages. However, availability and taste will have played a part, with seabirds more commonly identified than large wetland birds. Important colonies are documented historically in locations such as Godrevy, North Cornwall and the Isles of Scilly, which retain exceptional seabird populations today (Baldwin 2009; Evans *et al.* 2015). The Scillonian assemblages include a wide range of wetland birds including, marine and winter visiting ducks and divers and a host of seabirds. Like marine mammals, island locations offered suitable habitats for seabirds to forage and nest (*e.g.* Ingrem 2006; Straker 2006; Baldwin 2009), and their exploitation on the Scillies, for example of puffins, shags and cormorants, may reflect ready access and a long tradition of self-sufficiency (*e.g.* Hambleton 2008). It may also reflect a strong cultural element as suggested for seal exploitation. The wide species range resembles that cited for the Northern and Welsh islands, where seabird use for meat, eggs, oil and feathers has a long history and their procurement and consumption has held an important place in lifestyles and tradition (Serjeantson 1988; Locker 1996a Baldwin 2009; Yalden and Albarella 2009; Best and Mulville 2013).

A possible victim of over-exploitation is the Great Auk, absent after the Roman period while a new find of gannet from a Roman (1st–6th c.) site (Lellizick) adds to the record for this species in Southern England (Albarella and Davis 2012; Maltby and Hamilton-Dyer 2012). Both species are rare in the Cornish dataset, and their procurement, whether hunted, traded or perhaps storm-washed is uncertain (*e.g.* Locker 1996a; Albarella and Davis 2006). Manx shearwater is absent in zooarchaeological assemblages after the Roman period (at Hillside Farm, Bryher) but its presence at Launceston in the late 13th century suggests exploitation and an active trade in the medieval period.

Discussion

Since publication of Launceston Castle, new analyses have provided insight into animal exploitation and husbandry in Cornwall in earlier and contemporary periods. The three main themes identified at the beginning of the paper are revisited here. First, can we identify general patterns or recurrent strategies in animal husbandry in the Cornish peninsula? The data indicate that from the Bronze Age pastoralism was the mainstay of animal-based subsistence across coastal and lowland Cornwall. While providing the bulk of meat consumed, livestock held various roles and their proportions appear to have fluctuated through time. In the report on the 9th–11th-century settlement of Mawgan Porth, Clutton-Brock (1976, 380) stated 'Cornwall is not sheep country'. The variable but relatively abundant cattle remains across Bronze and Iron Age sites in the Cornish peninsula may provide support for Clutton-Brock's (1976) assertion, however it also reflects the vagaries of very small assemblage sizes in some cases, and recovery bias in others (*e.g.* Hammon 2011).

Sheep would have provided an important source of manure when cultivating or intensifying arable production, evidenced from the late 2nd millennium BC, as well as meat and wool suggested by the zooarchaeological data from Trevelgue, the Rumps and other Iron Age sites. The high proportions of sheep recorded in Cornish Romano-British sites suggest continuity from the Late Iron Age and an increase in sheep husbandry in line with that observed more widely in Southern Britain into the Early Roman period, though whether this reflects Romanisation is difficult to judge (*e.g.* Quinnell 2004, 236; Hambleton 2008; Allen 2017a, 87). The increased age at death of sheep at Atlantic Rd, Newquay and Duckpool, and presence of wool-related artefacts at Atlantic Rd (Reynolds 2001, 18), Kilhallon (Carlyon 1982), and other Cornish sites (*e.g.* Reawla, Penwith; Nowakowski 2011) may indicate the importance of wool production in/near the coastal area, albeit as part of a generalised economy (Maltby 1979; 2016a; Allen 2017a, 116).

Though high sheep counts over *c.* 50% are not encountered after the Roman period, this may reflect recovery bias in the available medieval assemblages, and possibly site type. More widely in Southern England sheep relative frequencies above 50% occur primarily in urban and ecclesiastical sites in the Early–Late medieval periods (Holmes 2017a; 2017b), site types absent from the quantified Cornish dataset. The consistent predominance of sheep may indicate a continuing engagement with wool production, which peaked more widely in the mid-14th century, however the relatively modest sheep frequencies compared to values reaching 50–80% elsewhere in Southern England at this time (mid-11th–mid 14th c.; Albarella and Davis 1996; Holmes 2017b, 71), may indicate that Cornwall engaged in a more mixed economy, with sheep remaining a valuable source of manure within a system of convertible farming (Herring *et al.* 2011a; 2011b). It is interesting to contemplate a possible rise of cattle farming from the medieval period, perhaps precursor to modern industries (Lobley 2006), though without further substantial assemblages for the medieval and post-medieval periods, like that from Launceston Castle, this remains speculation.

The role of pigs as identified previously through the Launceston work is highly variable, their abundance in high status sites clearly suggesting conspicuous consumption of meat. The presence of pigs in high or above average frequencies in Iron Age promontory sites, and Romano-British and Early medieval industrial and/or trading sites in Cornwall (and other South-West and Welsh sites) may support interpretation of its use in trade and possibly tribute. Romano-British industrial parallels may be found at Leicester and Ower, Poole, which may be explained by cultural preference (King 1999) though the link to industry is intriguing with these sites perhaps requiring a more diverse meat supply to serve non-food producing artisans. The current work at Tintagel further supports a post-Romano-British link between high status and conspicuous meat consumption (through pig abundance), well-illustrated in Saxon period sites to the east.

The second theme, related to the first is that of animal size and 'improvements'. Much of Cornish data conforms to previous conclusions regarding the smaller size of livestock compared to sites further east in Southern Britain and the Midlands. From the Iron Age in Cornwall the size of both sheep and cattle tended to the lower end of wider metric ranges. Nonetheless, there are exceptions and variability in size and/or shape within the datasets, in large part due to small sample size, but some of which may relate to regional variation *e.g.* between Isles of Scillies and mainland stock, to diversification of types for functional purposes such as sheep for wool production (*e.g.* Trevelgue; Atlantic Rd), to sources of animals for provisioning Launceston (*vs.* urban supply at Exeter), and to environmental and population pressure possibly evident in the gradual diminution in cattle size through to the late 13th century. It is not clear to what extent Romanisation may have influenced South-West animal husbandry. There are slight increases in cattle size at Exeter in the Late Roman period and persistence of large sizes in the 5th and 6th century at Bantham Ham, Devon, and possibly Tintagel but the Cornish data are inconclusive. Similarly for sheep, comparison of Trevelgue and Atlantic Rd showed variation in sizes at both sites with some rivalling livestock to the east but most were smaller than their eastern and southern counterparts. As noted above, animal size does not appear to drop off noticeably in the post-Roman period. However, it is interesting to contemplate the possible effect of changes in wider farming systems on animal size and demographics, for example of convertible husbandry established from the 8th century and intensification of agriculture from the 12th or 13th century (Herring *et al.* 2011a, 267; Herring *et al.* 2011b, 288). Sykes (2007) has previously identified the most dramatic size decrease in cattle size in the mid-7th/9th to mid-9th/11th centuries. Such a shift is potentially visible in the assemblages from Mawgan Porth (Clutton-Brock 1976, fig. 9.2) and Gunwalloe (Howard 2015), persisting perhaps in the slightly later assemblage from 13th-century Launceston (and its smaller 11th–12th c. datasets). The causes however remain elusive.

The third theme considered was access to wild mammals and birds. Exploitation of shellfish, fish, seabirds, and/or sea and terrestrial wild mammals, for example deer for meat and antler is attested on the mainland in all periods. While generally few in number the ubiquity of deer remains suggests a close relationship with the 'wild' and furthermore that suitable woodland and forest-edge environments existed for viable populations to survive. That said social relations and status may have increasingly restricted access to the medieval and post-medieval elites as at Launceston Castle. Such resources were present in abundance at Launceston at the height of its elite status, but with mainly deriving mainly from emparked (fallow) deer. Prior to this, wild resources are evident in many sites though a heavy reliance is not evident, except on the Scillies, where seal hunting and exploitation of seabirds may have been integral to subsistence and cultural life. In particular, the abundance of seabirds recalls the hunting and gathering economies of the Scottish Isles throughout prehistory and historic periods (Serjeantson 1988; Mulville 2002; Best and Mulville 2013). The lack of marine resource exploitation on the mainland other than fish and molluscs may partly be explained by difficulties of access as noted above. Seals and seabirds tend to prefer more remote locations for breeding and raising their young. There may well have been less of a need for exploitation of these resources on the mainland, where ample agricultural land and grazing would have been available including through seasonal transhumance to uplands, and through farming systems such as convertible husbandry. In contrast, on the Scillies, land was more limited, requiring a more diverse subsistence strategy, including at some sites a heavy reliance on marine foods. These no doubt became central to cultural life, engaging communities and individuals in their pursuit.

Challenges and future directions

A number of tantalising and challenging strands have emerged from this review which merit further investigation within the context of multidisciplinary studies including identifying geographical diversity in husbandry and subsistence strategies. For example, evidence of Bronze Age upland transhumance huts, Roman period cattle byres inland (Quinnell 2004) and establishment of 'milk houses' on the edge of rough ground in the medieval period (Herring *et al.* 2011a; 2011b) suggest cattle held an important role within Cornish pastoral economies, but the lack of assemblages from inland impede a clear understanding of the use of cattle and other livestock in different areas through time. The presence of neonatal and very juvenile animals throws light on animal management, and potentially scheduling of flock and herd transhumance and birthing. Transhumance huts and settlements have been identified along moorland slopes and uplands in various periods. Combining structural, artefactual and bioarchaeological (including biochemical analysis) data may help to better understand decision-making in animal husbandry. The role of marine resources in particular sea mammals and waterfowl, namely sea birds, appears to have held importance in the lifeways of the Scilly Isles inhabitants. The zooarchaeological data also provide important biogeographic information for wetland and marine fauna. Where available old assemblages would merit re-examination to verify species identification and taphonomy, in order to expand our understanding of assemblage formation, island traditions, past biodiversity and inform modern conservation efforts. Similarly, biometric data collected prior to publication of standard protocols is difficult to use with confidence; the older and larger collections, such as Mawgan Porth and important new or unstudied assemblages from this and other sites (*e.g.* Gwithian) would help to provide insight into husbandry within the peninsula which can be compared to new studies of the wider South-West and beyond.

Acknowledgements

My sincere thanks are due to the conference organisers for inviting me to contribute to this volume in honour of Simon Davis, and for their patience in waiting for a completed paper. Thanks are also due to Gill Campbell, for editing and encouragement, to Julia Cussans for her comments on an early draft of this review and to Sharon Soutar for the map image. Jacky Nowakowski and Alison Locker kindly provided information about Cornish sites including unpublished works and Mark Maltby provided copies of his raw metric data from Exeter. Foremost, I am indebted to Simon and Sebastian Payne for welcoming me, as PhD student and freelance specialist, to the English

Heritage (Ancient Monuments) laboratory in London in the 1990s, where I benefitted immensely from discussion and exchange with Simon over the years and where I also had the opportunity to meet Umberto Albarella. Following English Heritage's (now Historic England) move to Portsmouth and Simon's to Lisbon, along with excellent colleagues we have continued to develop Simon's legacy of reference collection and research, and I continue to be inspired by Simon's scholarship, his wide range of interests and creativity, and his and Umberto's friendship. Any limitations of the use, summary, and interpretation of data in this review are the author's own.

References

Albarella, U. and Davis, S. (1996, for 1994) Mammal and birds from Launceston Castle, Cornwall: decline in status and rise of agriculture. *Circaea* 12(1), 1–156.

Albarella, U. and Serjeantson, D. (2002) A passion for pork: meat consumption at the British Late Neolithic Site of Durrington Walls. In P. Miracle and N. Milner (eds) *Consuming Passions and Patterns of Consumption*, 33–49. Oxford, Oxbow Books.

Albarella, U., Beech, M. and Mulville, J. (1997) *The Saxon, Medieval and Post-Medieval Mammal and Bird Bones Excavated 1989–91 from Castle Mall, Norwich, Norfolk*. Ancient Monuments Laboratory Report 72/97. London, English Heritage

Allen, M.G. (2011) *New Perspectives on the Iron Age/Romano-British Transition*. Unpublished PhD thesis, University of Nottingham.

Allen, M. (2017a) Pastoral farming. In M. Allen, L. Lodwick, T. Brindle, M. Fulford and A. Smith *The Rural Economy of Roman Britain* 2, 85–141. Britannia Monograph Series 30. London, Society for the Promotion of Roman Studies.

Allen, M. (2017b) Animal Products. In M. Allen, L. Lodwick, T. Brindle, M. Fulford and A. Smith *The Rural Economy of Roman Britain* 2, 216–221. Britannia Monograph Series 30. London, Society for the Promotion of Roman Studies.

Allen, M., Baker, P. and Worley F. (2012, updated 2013) *Review of the Animal Bone Evidence from the Roman Period in Southern England* [data- set]. Portsmouth, English Heritage.

Allen, M., Blick, N., Brindle, T., Evans, T., Fulford, M., Holbrook, N., Lodwick, L., Richards, J.D. and Smith, A. (2018) *The Rural Settlement of Roman Britain: an online resource* [data-set]. York, Archaeology Data Service [distributor]. doi.org/10.5284/1030449.

Ashbee, P. (1972) Excavations at Halangy Down, St Mary's Isles of Scilly, 1965 and 1966. *Cornish Archaeology* 11, 20–27.

Ashbee, P. (1996) Halangy Down, St Mary's, Isles of Scilly, Excavations 1964–1977. *Cornish Archaeology* 35, 9–201.

Baker, P. (2007) Various reports. In C. Barrowman, C.E. Batey and C.D. Morris (eds) *Excavations at Tintagel Castle, Cornwall, 1990–1999*. London, The Society of Antiquaries of London.

Baker, P. 2014 (updated 2019) *Assessment of the Animal Bones from the Tresco Channel Exploratory Investigation*. Unpublished document. Portsmouth, English Heritage.

Baker, P., Campbell, G., Hazell, Z., Nowakowski, J. and Gossip, J. (2020) Exploring settlement and agricultural economy from the post-Roman to early medieval period at Tintagel, Cornwall: new discoveries and insights from recent excavations. Presentation at the EAA virtual conference, Aug 26–30, 2020.

Baldwin, J.R. (2009) Harvesting seabirds and their eggs on the Irish Sea Islands (Part 1: The Welsh Islands, Lundy and Scilly). *Folk Life* 47:1, 76–96. Doi: 10.1179/175967009X422837.

Barrowman, R.C., Batey, C.E. and Morris, C.D. (2007) *Excavations at Tintagel Castle, Cornwall, 1990–1999*. London, The Society of Antiquaries of London.

Best, J. and Mulville, J. (2013) Between the sea and sky: the archaeology of avian resource exploitation in Scottish island environments. In M.-Y. Daire, C. Dupont, A. Baudry, C. Billard, J.-M. Large, L. Lespez, E. Normand and C. Scarre (eds) *Ancient Maritime Communities and the Relationship between People and Environment along the European Atlantic Coasts*417–426. British Archaeological Reports International Series 2570. Oxford, Archaeopress.

Brooks, R.T. (1964) The Rumps, St Minver: interim report on the 1963 excavations. *Cornish Archaeology* 3, 26–34.

Bruce-Mitford, R. (1997) *Mawgan Porth: A Settlement of the Late Saxon Period on the North Cornish Coast*. English Heritage Archaeological Report 13. London, English Heritage.

Butcher, S.A. (1978) Excavations at Nornour, Isles of Scilly, 1969–73: the Pre-Roman settlement. *Cornish Archaeology* 17, 29–112.

Campbell, G., Baker, P., Hazell, Z., Nowakowski, J. and Gossip, J. (2018) Facing the sea, relying on the land: investigating the Tintagel middens. Unpublished paper delivered at the 39th AEA conference, Arhus, Denmark, Nov. 29th–Dec. 1st, 2018.

Carlyon, P.M. (1982) A Romano-British Site at Kilhallon, Tywardreath: excavation in 1975. *Cornish Archaeology* 21, 155–170.

Chaplin, R.E. and Coy, J.P. (1964) Appendix 2: Report on the animal bones. In R.T. Brooks, The Rumps, St Minver: interim report on the 1963 excavations, 31–34. *Cornish Archaeology* 3, 26–34.

Clutton-Brock, J. (1976) The animal resources. In D.M. Wilson (ed.) *The Archaeology of Anglo-Saxon England*, 373–92. London, Methuen and Co. Ltd.

Couzens, D., Swash, A., Still, R. and Dunn, J. (2017) *Britain's Mammals. A Field guide to the Mammals of Britain and Ireland*. Princeton, Princeton University Press.

Coy, J. (1987) Non-domestic faunal resources in South West England. In N.D. Balaam, B. Levitan and V. Straker (eds) *Studies in Palaeoeconomy and Environment in South West England*. British Archaeological Reports British Series 181, 9–29. Oxford, Archaeopress.

Davies, J.L. (1956) The grey seal at the Isles of Scilly. *Proceedings of the Zoological Society of London* 127(2), 160–6.

Davis, S. (1983) Morphometric variation of house mice *Mus domesticus* in Britain and Faroe. *Journal of Zoology* 199, 521–534.

Davis, S. (1987) *The Archaeology of Animals*. London, B.T. Batsford Ltd.

Driesch, A. von den (1976) *A Guide to the Measurement of Animal Bones from Archaeological Sites*. Peabody Museum Bulletin 1. Cambridge, MA, Harvard University.

Driesch, A. von den, Boessneck, J. (1974) Kritische Anmerkungen zur Widerristhöhenberechnung aus LängenmaSSen vor-

und fruhgeschichtlicher Tierknochen. *Säugetierkundliche Mitteilungen* 22, 325–348.

Dudley, D. (1960–1) Some cist-graves in Poynter's Garden, St Mary's, Isles of Scilly. *Proceedings of the West Cornwall Field Club* 2(5), 221–231.

Evans, J.C., Dall, S.R.X., Bolton, M., Owen, E. and Votier, S.C. (2015) Social foraging European shags: GPS tracking reveals birds from neighbouring colonies have shared foraging grounds. *Journal of Ornithology* 157, 23–32 (2016). Doi 10.1007/s10336-015-1241-2.

Fairnell, E. (2011) *Fur-bearing Species: A Zooarchaeological Meta-analysis of Their Presence and Use Across Three Regions of Britain*. Unpublished PhD thesis, University of York.

FISH nd *Chronology*. Forum on Information Standards in Heritage (FISH), http://www.heritage-standards.org.uk/chronology/ [accessed September 2020].

Fisher, A. and Thomas, R. (2012) Isotopic and zooarchaeological investigation of later medieval and post-medieval cattle husbandry at Dudley Castle, West Midlands. *Environmental Archaeology* 12, 151–67.

Fraser, T. (2020) *Livestock and Landscape: Livestock Improvement and Landscape Enclosure in Late and Post-Medieval England*. Unpublished PhD thesis, University of Sheffield.

Gardiner, M. (1997) The exploitation of sea-mammals in medieval England: bones and their social context. *Archaeology Journal* 154, 173–5.

Gerrard, C. (1987) A regional approach to faunal data from Somerset. In N.D. Balaam, B. Levitan, and V. Straker (eds) *Studies in Palaeoeconomy and Environment in South West England*, British Archaeological Reports British Series 181, 81–88. Oxford, Archaeopress.

Gilchrist, R. (1988) A reappraisal of Dinas Powys: local exchange and specialized livestock production in 5th- to 7th-century Wales. *Medieval Archaeology* 32, 50–62.

Grant, A. (1988) Bone deposition and animal husbandry: the animal remains. In B. Cunliffe, *Mount Batten, Plymouth, a Prehistoric and Roman Fort*. OUAC Monograph 26, 28–35. Oxford, Oxford University Committee for Archaeology.

Grant, A (2002) Food, status and social hierarchy. In P. Miracle and N. Milner (eds) *Consuming Passions and Patterns of Consumption*, 17–23. Oxford, Oxbow Books.

Gray, A. (Edited, with a commentary, by Paul Ashbee) (1972) Prehistoric habitation sites on the Isles of Scilly. *Cornish Archaeology* 11, 19–49.

Gray, A. (1983) The animal remains. In P. Ashbee, *Halangy Porth St Mary's, Isles of Scilly, Excavations 1975–76*, 38–39. *Cornish Archaeology* 22, 3–46.

Grimm, J. (2008) Lellizzick: Animal Bone Assessment (65312). Unpublished document. Salisbury, Wessex Archaeology.

Grove, J and Croft, B (eds) (2012) *South West Archaeological Research Framework. Research Strategy 2012–2017*. Taunton, Somerset Heritage Service.

Hambleton, E. (2008) *Review of Middle Bronze Age – Late Iron Age Faunal Assemblages from Southern Britain*. Research Report Series 71-2008. Portsmouth, English Heritage.

Hambleton, E. (2009) *A Review of Animal Bone Evidence from Southern England*. York, Archaeology Data Service [distributor] https://doi.org/10.5284/1000102.

Hamilton, J. and Thomas, R. (2012) Pannage, pulses and pigs: isotopic and zooarchaeological evidence for changing pig management practices in later medieval England. *Medieval Archaeology* 56, 234–59.

Hammon, A. (2007a) 12 Gwithian, Cornwall. Assessment of the vertebrate assemblage 2005 fieldwork (site GMXVII). In J. Nowakoski (ed.) *Excavations of a Bronze Age Landscape and a Post-Roman Industrial Settlement 1953–1961, Gwithian, Cornwall. Assessments of Individual Key Datasets 2003–2006. Vol II*, 81–2. Truro, Cornwall County Council.

Hammon, A. (2007b) 22 Gwithian, Cornwall. Assessment of the vertebrate remains. In J. Nowakowski (ed.) *Excavations of a Bronze Age Landscape and a Post-Roman Industrial Settlement 1953–1961, Gwithian, Cornwall. Assessments of Individual Key Datasets 2003–2006. Vol II*, 145–62. Truro, Cornwall County Council.

Hammon, A. (2011) Iron Age mammal and fish remains. In J. Nowakowski and H. Quinnell (eds), *Trevelgue Head, Cornwall: The Importance of CK Croft Andrew's 1939 Excavations for Prehistoric and Roman Cornwall*. Truro, Cornwall County Council.

Herring, P. (2007) An historic and archaeological survey of Cornish deer and ornamental parks. *The History, Ecology and Archaeology of Medieval Parks and Parklands. Landscape Archaeology and Ecology* 6, 60–64.

Herring, P. (2011) The Cornish landscape. *Cornish Archaeology* 50, 161–74.

Herring, P., Preston-Jones, A., Thorpe, C. and Wood, I. (2011a) Early medieval Cornwall. *Cornish Archaeology* 50, 263–86.

Herring, P., Kirkham, G., Cahill, N., Thomas, N., Berry, E., Preston-Jones, A., Langdon, A., Sharpe, A. and Thorpe, C. (2011b) *Later medieval Cornwall. Cornish Archaeology* 50, 287–314.

Higbee, L. (2007) Appendix 3 The faunal remains. In S. Reed and P. Bidwell (eds) *Excavation at Bantham, South Devon and Post-Roman trade in south-west England* [data-set]. York, Archaeology Data Service [distributor] https://doi.org/10.5284/1000305.

Holmes, M. (2017a) *A Review of Animal Bone Evidence from the Saxon to Post Medieval Periods in Southern Britain (6145)* [data-set]. York, Archaeology Data Service [distributor] https://doi.org/10.5284/1047191.

Holmes, M. (2017b) *Southern England: A Review of Animal Remains from Saxon, Medieval and Post-Medieval Archaeological Sites*. Research Department Report 8-2017. Portsmouth, Historic England.

Hooke, D. (1998) *The Anglo-Saxon Landscape of England*. Leicester, Leicester University Press.

Howard, W. (2015) Mammal, fish and invertebrate assemblage. In I. Wood (ed.) *Gunwalloe Through The Ages: Middle Bronze Age to the 12th century AD. Lizard Peninsula, Cornwall. Evaluation Report for works carried out 2011–2012*, 52–56. Helston, National Trust Penrose Estate.

Ingrem, C. (2000) *The Animal Bone from Romano-British Deposits at Atlantic Road, Newquay, Cornwall*. Unpublished client report. Truro, Cornwall Archaeological Unit.

Ingrem, C. (2006) Animal bone. In C. Johns, An Iron Age sword and mirror cist burial from Bryher, Isles of Scilly, 53–8. *Cornish Archaeology* 41–42 (2002–3), 1–79.

Johns, C. (2012) *Isles of Scilly Historic Envirnment Research Framework. Resource Assessment ad Research Agenda*. Truro, Cornwall Council.

Johnstone, C. (2004) *A Biometric Study of Equids in the Roman World*. Unpublished PhD thesis. University of York.

Jones, R.T. (1977) *Animal Bone from Carn Euny*. Ancient Monuments Laboratory report OS 2301. London, English Heritage.

Kerridge, E. (1968) *The Agricultural Revolution*. New York, Augustus M Kelly Publishers.

King, A.C. (1976) A comparative survey of bone assemblages from Roman sites in Britain. *Bulletin of the Institute of Archaeology London* 15, 207–32.

King, A.C. (1999) Diet in the Roman world: a regional intersite comparison of the mammal bones. *Journal of Roman Archaeology* 12, 168–202.

Levitan, B. (1987) Medieval animal husbandry in South West England: a selective review and suggested approach. In N.D. Balaam, B. Levitan, and V. Straker (eds) *Studies in Palaeoeconomy and environment in South West England*, 51–80. British Archaeological Reports British Series 181. Oxford, Archaeopress.

Levitan, B. (1990) The vertebrate remains. In M. Bell (ed.) *Brean Down. Excavations 1983–1987*, 220–241. English Heritage Archaeological Report 15. London, English Heritage.

Lobley, M., Reed, M., Metcalf, R. and Stephens, J. (2006) *Food Production, Processing and Distribution in Cornwall and the Isles of Scilly*. Centre for Rural Research Report No. 20. Exeter, University of Exeter.

Locker, A. (1983) The bones (inc. mammal, bird and fish) (AM Report No. 3843). In D. Neal, Excavations on a settlement at Little Bay, Isles of Scilly, 66–70. *Cornish Archaeology* 22, 47–80.

Locker, A. (1996a) 6.2 The bird bones. In P. Ashbee, *Halangy Down, St Mary's, Isles of Scilly, Excavations 1964–1977*, 113–115. *Cornish Archaeology* 35, 9–201.

Locker, A. (1996b) 6.3 The fish bones. In P. Ashbee, *Halangy Down, St Mary's, Isles of Scilly, Excavations 1964–1977*, 115–118. *Cornish Archaeology* 35, 9–201.

Locker, A. (1996c) 6.4 The mammal bones. In P. Ashbee, *Halangy Down, St Mary's, Isles of Scilly, Excavations 1964–1977*, 118–26. *Cornish Archaeology* 35, 9–201.

Locker, A. (1996d) 8.5 Measurement archive. In P. Ashbee, *Halangy Down, St Mary's, Isles of Scilly, Excavations 1964–1977*, 176–183. *Cornish Archaeology* 35, 9–201.

Locker, A. (1997a) No title. In J. Ratcliffe, *British Telecom Trenching on St Martin's, Isles of Scilly Summer 1992. The Results of the Archaeological Watching Brief*. Unpublished document. Cornwall Archaeological Unit, Cornwall County Council.

Locker, A. (1997b) The mammal, bird and fish bones from excavations at Porth Killier 1996 Coastal Protection Scheme, St Agnes, Isles of Scilly'. Unpublished document.

Maltby, M. (1979) *The Animal Bones from Exeter 1971–1979*. Exeter Archaeological Reports 2. Sheffield, University of Sheffield.

Maltby, M. (2006) Salt and animal products: linking production and use in Iron Age Britain. In M. Maltby (ed.) *Integrating Zooarchaeology*, 117–22. Oxford, Oxbow Books.

Maltby, M. (2010) *Feeding a Roman Town. Environmental Evidence from Excavations in Winchester, 1972–1985*. Winchester, Winchester Museums and English Heritage.

Maltby, M. (2016a) The exploitation of animals and their contribution to urban food supply in Roman Southern England. In D. Bird (ed.) *Agriculture and Industry in Southeastern Roman Britain.*, 180–209. Oxford, Oxbow Books.

Maltby, M. (2016b) The exploitation of animals in Roman Britain. I M. Millett, L. Revell and A. Moore (eds), 791–806. *The Oxford Handbook of Roman Britain*. Oxford, Oxford University Press.

Maltby, M. and Hamilton-Dyer, S. (2012) Big fish and great auks: exploitation of birds and fish on the Isle of Portland, Dorset, during the Romano-British period. *Environmental Archaeology* 17(2), 168–176.

Mason, H. (1984) The excavation of a cottage on Samson, Isles of Scilly, 1977. *Cornish Studies* 11, 49–68.

Matolcsi, J., (1969) Historische Erforschung der Körpergrösse des Rindes auf Grund von Ungarischen Knochenmaterial. *Zeitschrift für Tierzüchtung und Züchtungsbiologie*, 87(2), 89–137.

Mulville, J. (2002) The role of Cetacea in prehistoric an historic Atlantic Scotland. *International Journal of Osteoarchaeology* 12, 34–48.

Nowakowski, J. (ed.) (2007) *Excavations of a Bronze Age Landscape and a Post-Roman Industrial Settlement 1953–1961, Gwithian, Cornwall. Assessments of Individual Key Datasets 2003–2006. Vol II*. Truro, Cornwall County Council.

Nowakowski, N. (2011) Appraising the bigger picture – Cornish Iron Age and Romano-British lives and settlements 25 years on. *Cornish Archaeology* 50, 241–61.

Nowakowski, J. and Johns, C. 2015 *Bypassing Indian Queens. Archaeological Excavations 1992–1994. Investigating Prehistoric and Romano-British Settlement and Landscapes in Cornwall*. Truro, Cornwall Council.

O'Neil, H.B. (1964) Excavation of a Celtic hermitage on St Helen's, Isles of Scilly, 1956–58. *Archaeological Journal* CXXI, 40–69.

Patchett, F. (1944) Cornish Bronze Age pottery. *Archaeological Journal*, 101, 17–49.

Payne, S. (1989) *Animal Bones from Excavations in 1987 at Trethellan Farm, Newquay, Cornwall*. Ancient Monuments Laboratory report 40/89. London, English Heritage.

Pearce, S. (2004) *South-Western Britain in the Early Middle Ages*. London and New York, Leicester University Press.

Pernetta, J.C. and Handford, P.T. (1970) Mammalian and avian remains from possible Bronze Age deposits on Nornour, Isles of Scilly. *Journal of Zoology* 162, 534–540.

Pollard, L.E. (1952) Ancient finds at St Martin's. *The Scillonian*, 27(3), 172–173.

Poole, K. (2011) *The Nature of Society in England c. AD 410–1066*. Unpublished PhD thesis, University of Nottingham.

Powell, A. and Serjeantson, D. (1995) Animal bone. In J. Ratcliffe (ed.) Duckpool, Morwenstow: A Romano-British and early medieval industrial site and harbour, 136–142. *Cornish Archaeology* 34, 81–171.

Preston-Jones, A. and Rose, P. (1986) Medieval Cornwall. *Cornish Archaeology* 25, 135–83.

Quinnell, H. (1986) Cornwall during the Iron Age and the Roman Period. *Cornish Archaeology* 25, 111–34.

Quinnell, H. (2004) *Trethurgy, Excavations at Trethurgy Round, St Austell: Community and Status in Roman and Post-Roman Cornwall*. Truro, Cornwall Archaeological Unit.

Randall, C. (2010) *Livestock and Landscape: Exploring Animal Exploitation in Later Prehistory in the Southwest of Britain*. Unpublished PhD thesis, Bournemouth University.

Ratcliffe, J. (1995) Duckpool, Morwenstow: A Romano-British and early medieval industrial site and harbour. *Cornish Archaeology* 34, 80–175.

Ratcliffe, J. and Straker, V. (1996) *The Early Environment of Scilly*. Truro, Cornwall Archaeological Unit.

Reed, S., Bidwell, P. and Allan, J. (2011) Excavation at Bantham, South Devon, and Post-Roman Trade in South-West England. *Medieval Archaeology* 55(1), 82–138. Doi: 10.1179/174581711X13103897378447.

Reynolds, A. (2001) *Newquay Sewage Treatment Works, Cornwall. Archaeological Mitigation. A Report for South West Water*. Truro, Cornwall Archaeological Unit.

Rose, P. and Preston-Jones, A. (1995) Changes in the Cornish countryside AD 400–1100. In D. Hooke and S. Burnell (eds) *Landscape and Settlement in Britain AD 400–1066*, 51–68. Exeter, University of Exeter Press.

Salvagno, L. (2020) *The Neglected Goat: a New Method to Assess the Role of the Goat in the English Middle Ages*. Oxford, Archaeopress.

Salvagno, L. and Albarella, U. (2019) Was the English medieval goat genuinely rare? A new morphometric approach provides the answer. *Archaeological and Anthropological Sciences* 11, 5095–5132. https://doi.org/10.1007/s12520-019-00843-2.

Serjeantson, D. (1988) Archaeological and ethnographic evidence for seabird exploitation in Scotland. *Archaeozoologia* II/1, 2, 209–224.

Serjeantson, D. (2011a) *Review of Animal Remains from the Neolithic and Early Bronze Age of Southern Britain (4000 BC–1500 BC)*. Research Report Series 29/2011. Portsmouth, Historic England.

Serjeantson, D. (2011b) *A Review of Animal Remains from the Neolithic and Early Bronze Age of Southern Britain*. [dataset]. York, Archaeology Data Service [distributor] https://doi.org/10.5284/1000396.

Straker, V. (2006) Overview of the results of environmental analyses. In C. Johns, *An Iron Age Sword and Mirror Cist Burial from Bryher, Isles of Scilly*, 58–60. *Cornish Archaeology* 41–42 (2002–3), 1–79.

Straker, V. (with contributions from P. Baker, A. Bayliss, C. Bronk Ramsay, R. Gale, G. Heathcote, J. Jones, M. Robinson and H. Tinsley) (2007) The ecofactual assemblages. In R.C. Barrowman, C.E. Batey and C.D. Morris (eds) *Excavations at Tintagel Castle, Cornwall, 1990–1999*, 280–301. London, The Society of Antiquaries of London.

Straker, V. (2011) Science and Archaeology. *Cornish Archaeology* 50, 175–186.

Summers, C.F. (1974) The grey seal (*Halichoerus grypus*) in Cornwall and the Isles of Scilly. *Biological Conservation* 6(4), 285–291.

Sykes, N. (2007) *The Norman Conquest: a Zooarchaeological Perspective*. British Archaeological Reports International Series 1656. Oxford, Archaeopress.

Teichert, M. nd in Driesch, A. von den and Boessneck, J. Osteometrische Untersuchungen zur Berechnung der Widerristhohe bei Schafen. Vortrag gehalten auf der Archaozoologischen Konferenz im April 1974 in Groningen.

Thomas, C. (1984) Appendix: The Occupation of May's Hill and it's dating – A Note. *Cornish Studies* 11, 78–80.

Thomas, R., Holmes, M. and Morris, J. (2013) 'So bigge as bigge may be': tracking size and shape change in domestic livestock in London (AD 1220–1900). *Journal of Archaeological Science* 40, 3309–3325.

Turk, F.A. (1965) Notes on Cornish mammals. Annual report – No. 7. *Royal Cornwall Polytechnic Society Annual Report* 132, 49–55.

Turk, F.A. (1967) Report on the animal remains from Nornour, Iles of Scilly. *Journal of the Royal Institute Cornwall* 5(3), 250–266.

Turk, F.A. (1968) Notes on Cornish mammals in prehistoric and historic times: 1. *Cornish Archaeology* 7, 73–79.

Turk, F.A. (1969) Notes on Cornish mammals in prehistoric and historic times: 2. *Cornish Archaeology* 8, 100–104.

Turk, F.A. (1970a) *Identification of Animal Bones from the 1969 Nor Nour Excavations*. Ancient Monuments Laboratory Report 1537. London, English Heritage.

Turk, F.A. (1970b) Notes on Cornish mammals in prehistoric and historic times: 3. *Cornish Archaeology* 9, 121–191.

Turk, F.A. (1971) Notes on Cornish mammals in prehistoric and historic times: 4. *Cornish Archaeology* 10, 79–91.

Turk, F.A. (1973a) Distribution patterns of the mammalian fauna of Cornwall. *Cornish Studies* 1, 5–33.

Turk, F.A. (1973b) *Notae de Ossibus in Cornubia Inventis: Manipulus* I. *Cornish Studies* 1, 49–52.

Turk, F.A. (1973c) *Preliminary Report of Animals Represented in Bone Material from Samson, Isles of Scilly (1970/1)*. Ancient Monuments Laboratory Report 1571. London, English Heritage.

Turk, F.A. (1978) The animal remains from Nornour: a synoptic view of the finds. In S.A. Butcher, Excavations at Nornour, Isles of Scilly, 1969–73: the pre-Roman settlement, 99–105. *Cornish Archaeology* 17.

Turk, F.A. (1982) The animal bones. In P.M. Carlyon, A Romano-British site at Kilhallon, Tywardreath: Excavation in 1975, 165. *Cornish Archaeology* 21, 155–170.

Turk, F.A. (1984a) *Notae de Ossibus in Cornubia Inventis: Manipulus* 2. *Cornish Studies* 12 (1986), 23–30.

Turk, F.A. (1984b) A study of the vertebrate remains from May's Hill, St Martins. *Cornish Studies* 11, 69–78.

Turk, F.A. (1991) 8. Faunal Remains. In Ratcliffe, J and Thorpe, C, *Lighting up the Past in Scilly. Archaeological Results from the 185 Electrification Project*, 101–6. Truro, Institute of Cornish Studies and Cornwall Archaeological Unit.

Turner, S. (2006) *Making a Christian Landscape. The Countryside in Early Medieval Cornwall, Devon and Wessex*. Exeter, University of Exeter Press.

Webster, C.J. (ed.) (2007) *The Archaeology of Southwest England. Southwest Archaeological Research Framework. Resource Assessment and Research Agenda*. Taunton, Somerset Heritage Service.

Wood, I. (2015) *Gunwalloe Through the Ages: Middle Bronze Age to the 12th century AD. Lizard Peninsula, Cornwall*. Evaluation Report for works carried out 2011–2012. Helston, National Trust Penrose Estate.

Worley, F. (2016) Seals in UK Zooarchaeology. Unpublished Presentation. Professional Zooarchaeology Group (PZG), Orkney, 3rd April 2016.

Yalden, D. and Albarella, U. (2009) *The History of British Birds*. Oxford, Oxford University Press.

*Appendix 16.1. Scientific and English names of mammal and bird species in Cornish assemblages. * only recorded at Launceston*

Order	Family	Scientific name	English name
Mammals			
Artiodactyla	Bovidae	*Bos taurus*	Cattle
Artiodactyla	Bovidae	*Capra hircus*	Goat
Artiodactyla	Bovidae	*Ovis aries*	Sheep
Artiodactyla	Cervidae	*Capreolus capreolus*	Roe deer
Artiodactyla	Cervidae	*Cervus elaphus*	Red deer
Artiodactyla	Cervidae	*Dama dama*	Fallow deer*
Artiodactyla	Suidae	*Sus domesticus*	Pig
Artiodactyla	Suidae	*Sus scrofa*	Wild boar
Carnivora	Canidae	*Canis familiaris*	Dog
Carnivora	Canidae	*Vulpes vulpes*	Fox
Carnivora	Mustelidae	*Lutra lutra*	Otter
Carnivora	Mustelidae	*Martes martes*	Pine marten
Carnivora	Mustelidae	*Meles meles*	Badger
Carnivora	Phocidae	*Halichoerus grypus*	Grey seal
Carnivora	Felidae	*Felis catus*	Cat
Carnivora	Felidae	*Felis silvestris*	Wildcat
Cetacea (large)	–	*Balaena/Balaenoptera size*	Balaena/Balaenoptera size
Cetacea	Phocoenidae	*Phocoena phocoena*	Harbour porpoise
Cetacea	Delphinidae	*Tursiops truncatus*	Bottlenose dolphin
Cetacea	Delphinidae	*Delphinus delphis*	Common dolphin
Cetacea	Delphinidae	*Grampus griseus*	Risso's dolphin
Cetacea (small)	–	*Delphinus/Phocoena size*	Delphinus/Phocoena size
Lagomorpha	Leporidae	*Lepus europaeus*	Brown hare
Lagomorpha	Leporidae	*Lepus timidus*	Mountain hare
Lagomorpha	Leporidae	*Oryctolagus cuniculus*	Rabbit
Perissodactyla	Equidae	*Equus asinus*	Donkey
Perissodactyla	Equidae	*Equus caballus*	Horse
Perissodactyla	Equidae	*Equus sp*	Equid
Rodentia	Castoridae	*Castor fiber*	Eurasian beaver
Birds			
Anseriformes	Anatidae	*Anser anser*	Greylag/domestic goose
Anseriformes	Anatidae	*Anser sp/Branta sp*	Goose
Anseriformes	Anatidae	*Cygnus sp*	Swan
Anseriformes	Anatidae	*Tadorna tadorna*	Shelduck
Anseriformes	Anatidae	*Anas platyrhynchos*	Mallard/domestic duck
Anseriformes	Anatidae	*Anas crecca*	Teal
Anseriformes	Anatidae	*Aythya ferina*	Pochard
Anseriformes	Anatidae	*Aythya marila*	Scaup
Anseriformes	Anatidae	*Clangula hyemalis*	Long-tailed duck
Anseriformes	Anatidae	*Bucephala clangula*	Goldeneye
Galliformes	Phasianidae	*Lyrurus tetrix*	Black grouse
Galliformes	Phasianidae	*Perdix perdix*	Grey partridge

(Continued)

Appendix 16.1. (Continued)

Order	*Family*	*Scientific name*	*English name*
Galliformes	Phasianidae	*Coturnix coturnix*	Quail*
Galliformes	Phasianidae	*Gallus gallus*	Domestic fowl
Gaviiformes	Gaviidae	*Gavia immer*	Great northern diver
Procellariiformes	Procellariidae	*Puffinus puffinus*	Manx shearwater
Ciconiiformes	Ciconiidae	*Ciconia ciconia*	White stork
Pelecaniformes	Ardeidae	*Botaurus stellaris*	Bittern
Pelecaniformes	Ardeidae	*Ardea cinerea/Ardea sp*	Grey heron/Heron
Suliformes	Sulidae	*Morus bassanus*	Gannet
Suliformes	Phalacrocoracidae	*Phalacrocorax aristotelis*	Shag
Suliformes	Phalacrocoracidae	*Phalacrocorax carbo*	Cormorant
Accipitriformes	Accipitridae	*Milvus milvus*	Red kite*
Accipitriformes	Accipitridae	*Buteo buteo*	Buzzard
Gruiformes	Rallidae	*Crex crex*	Corncrake
Gruiformes	Rallidae	*Gallinula chloropus*	Moorhen
Gruiformes	Rallidae	*Fulica atra*	Coot
Gruiformes	Gruidae	*Grus grus*	Crane*
Charadriiformes	Burhinidae	*Burhinus oedicnemus*	Stone-curlew
Charadriiformes	Charadriidae	*Vanellus vanellus*	Lapwing
Charadriiformes	Charadriidae	*Pluvialis apricaria/P squatarola*	Golden/grey plover*
Charadriiformes	Scolopacidae	*Scolopax rusticola*	Woodcock
Charadriiformes	Scolopacidae	*Gallinago gallinago*	Snipe
Charadriiformes	Scolopacidae	*Limosa limosa/Limosa sp*	Black-tailed godwit/Godwit
Charadriiformes	Scolopacidae	*Numenius arquata*	Curlew
Charadriiformes	Scolopacidae	*Tringa totanus*	Redshank
Charadriiformes	Scolopacidae	*Calidris canutus*	Knot (red knot)
Charadriiformes	Scolopacidae	*Philomachus pugnax*	Ruff
Charadriiformes	Laridae	*Larus marinus*	Great black-backed gull
Charadriiformes	Laridae	*Larus argentatus*	Herring gull
Charadriiformes	Alcidae	*Uria lomvia*	Brunnich's guillemot
Charadriiformes	Alcidae	*Uria aalge*	Common guillemot
Charadriiformes	Alcidae	*Alca torda*	Razorbill
Charadriiformes	Alcidae	*Alca impennis*	Great auk
Charadriiformes	Alcidae	*Cepphus grylle*	Black guillemot
Charadriiformes	Alcidae	*Fratercula arctica*	Puffin
Columbiformes	Columbidae	*Columba sp*	Pigeon
Cuculififormes	Cuculidae	*Cuculus canorus*	Cuckoo*
Strigiformes	Tytonidae	*Tyto alba*	Barn owl
Falconifomes	Falconidae	*Falco tinnunculus*	Kestrel*
Passeriformes	Laniidae	*Lanius collurio/sp*	Shrike*
Passeriformes	Corvidae	*Pyrrhocorax pyrrhocorax*	Chough
Passeriformes	Corvidae	*Corvus frugilegus*	Rook
Passeriformes	Corvidae small	*Garrulus. glandarius/Corvus monedula/Pica. pica*	Jay/Jackdaw/Magpie*

(Continued)

*Appendix 16.1. Scientific and English names of mammal and bird species in Cornish assemblages. * only recorded at Launceston (Continued)*

Order	*Family*	*Scientific name*	*English name*
Passeriformes	Corvidae	*Corvus corone*	Carrion crow
Passeriformes	Corvidae	*Corvus corax*	Raven
Passeriformes	Troglodytidae	*Troglodytes troglodytes*	Wren
Passeriformes	Turdidae	*Turdus merula*	Blackbird
Passeriformes	Turdidae	*Turdus philomelos*	Song thrush
Passeriformes	Turdidae	*Turdidae*	Turdid
Passeriformes	Passeridae	*Passer domesticus*	House sparrow
Passeriformes	Alaudidae	*Alauda arvensis*	Skylark
Passeriformes	Emberizidae	*Emberiza calandra*	Corn bunting

17

Fodder in the city: rye for animals in the 1755 earthquake in Lisbon

João Pedro Tereso and Lídia Fernandes

On the first of November – All Saints' Day – of 1755 a great Earthquake, followed by a tsunami and massive fires destroyed a great part of the city of Lisbon (Portugal). One of those fires led to the destruction of a housing area located where nowadays the Roman Theatre Museum is built. This destruction event was also responsible for the preservation by carbonisation of a great amount of plant remains in what is up to now an unprecedented context in Portuguese archaeology.

The archaeological excavations of this house allowed the recovery of a great amount of very well-preserved culms and spikes of rye (Secale cereale). *The straw was tied in sheaves and placed horizontally in the floor of the house, beneath a mezzanine. They were likely used as fodder. Evidence demonstrates these plant remains were sealed by the debris of the building and carbonized without direct contact with fire, allowing their preservation until today.*

Keywords: Archaeobotany, Lisbon, 1755 earthquake, rye

Introduction

This book pays a tribute to Simon Davis, one of the most brilliant and influential zooarchaeologists of the last decades. His influence derived not only of the impact his handbook *The Archaeology of Animals* and his many papers had among the scientific community but also of Simon's direct contact with many people working in zooarchaeology and archaeology in general. In Portugal, the impact of Simon's work is even more obvious and his arrival at Lisbon 20 years ago marks a shift in the zooarchaeological investigation in our country, for which we must be thankful.

This paper pays tribute to Simon not only by focusing on archaeobotanical material from Lisbon, such an important place in Simon's life, but also by addressing a particularly important topic for every zooarchaeologist, namely the identification of fodder. All animals need food and one of the main concerns in all communities who rely on the consumption of domestic animals and their secondary products, or on their working power, is the provision of feed (Jones 1998). Sometimes, strategies to cope with this necessity are such an important feature that end up determining the community's lifestyle or even its own identity, as in the case of transhumance.

The identification of archaeological fodder is not a simple task. The study of dung can provide precious little direct evidence of what the animals have eaten (Charles 1998; Derreumaux 2005), however, these findings haven't been identified regularly in Portuguese archaeology. In the Old World, this topic has also been addressed recurrently through the interpretation of archaeobotanical assemblages. Still, archaeobotanical evidence can hardly be called direct evidence since its interpretation derives mostly from deductions regarding the composition of plant assemblages. These assumptions can be rather problematic, particularly when it comes to actual crops, since the same species could be used to feed people and/or animals, by different communities, or even within the same community and within the same household, depending on cultural issues or simply on the yield of the following harvest (Jones 1998). An example of the difficulty in the archaeobotanical identification of fodder is the case of the Late Roman levels of Monte Mozinho (Penafiel, North Portugal). Here, a particularly diverse plant assemblage composed with multiple seeds of wild and domestic species, together with grains and abundant chaff mostly of rye (*Secale cereale*) and some bread wheat (*Triticum aestivum*) has been interpreted as grain-rich fodder (Tereso *et al.* 2013) but further investigation, including charcoal analysis of that same context and the surrounding buildings questioned that interpretation, suggesting the assemblage may simply be refuse from a nearby oven (Vaz *et al.* 2017).

The interpretation of plant assemblages must always include a careful analysis of the archaeological context where they were retrieved. This can a be a crucial aspect in the eventual identification of archaeological fodder. A particular case is that of the British medieval Bamburgh Castle's west ward, where oats (*Avena* sp.) were recovered in a building used to keep horses, as suggested by specific archaeological evidence, hence making the interpretation of these plant remains as fodder reasonable (Blakeney 2017).

Here we will present a peculiar case in Portuguese archaeology, the remarkable well-preserved plant remains recovered in the ground floor of a Lisbon house destroyed and sealed in the 1st November (All Saints Day) 1755 in the aftermath of the earthquake that destroyed a great part of the city. The archaeobotanical assemblage is composed of sheaves of straw and was promptly interpreted as fodder during the archaeological interventions. Here we will present the context and the plant remains in detail and discuss its possible interpretations.

Materials and methods

Archaeological context

The archaeobotanical materials presented here are, beyond any doubt, remains of the fires that followed the 1st of November 1755 earthquake that destroyed a great part of the city of Lisbon. They were detected in the 2011 archaeological intervention carried out in the Museum of Lisbon – Roman Theatre (Museu de Lisboa – Teatro Romano) (at the time known as 'Museum of the Roman Theatre').

The excavations at the reception area (Rua de São Mamede nº 3 A) and patio (Rua de São Mamede nº 3 B) of the abovementioned museum (Fig. 17.1) begun in 2001 and

Figure 17.1. Location of the site in the city of Lisbon and the Iberian Peninsula. Drawing by Carlos Loureiro (Museu de Lisboa/EGEAC, CML).

lasted for 10 years. In the 400 m^2 excavation area, a 10 m stratigraphic sequence was uncovered (Fernandes 2007), which included the context we present here.

The house where the plant remains were retrieved was located in what is now the reception area of the museum. At North it is limited by the huge structure of the *post scaenium* of the Roman theatre that in some areas is 9 m high (Fernandes 2013; 2017). This structure was reused throughout the years and during the 17th century and the first half of the 18th century was being used as a wall in a simple three- or four-storey house. The wall was covered by white plaster and a small niche and a cabinet with multiple nails, probably to keep tableware, were also found.

The botanical remains were recovered inside one of the house compartments. Only the ground floor was preserved with its pavement, a door and two windows which were opened to an alley called 'Beco do Aljube por Detrás do Celeiro da Mitra' (Fig. 17.2) that no longer exists. In this alley, people would have access to the first floor through exterior stairs, meaning there might be no direct connection between the first floor, probably used for habitation, and the lower floor, which might have had a different function (Fernandes 2015).

The name 'Beco do Aljube por Detrás do Celeiro da Mitra' is revealed in a document of that time (*Tombo da Cidade de Lisboa em 1755*). Through that document we know that the land and the house we are studying belonged to the Count of Valadares who rented them, while maintaining his own household elsewhere (Fernandes and Almeida 2012). Through the same document, it is possible to know that the other areas of the current building of the museum were part of a granary – Celeiro da Mitra – belonging to the collegiate

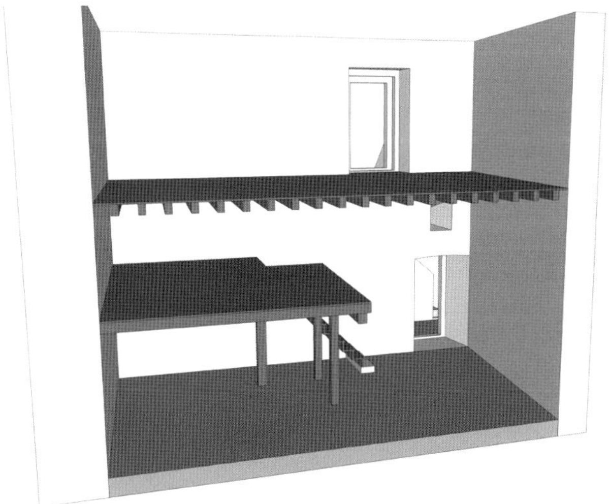

Figure 17.2. Reconstitution of the house from the 17th/first half of the 18th century, showing the ground floor, the first floor and, in between, the mezzanine that covered the cereals in the eastern part of the compartment. Reconstitution by Carlos Loureiro (Museu de Lisboa/EGEAC, CML).

of the city's cathedral, during the 17th century and the first half of the 18th century (Fernandes and Almeida 2012).

The archaeological excavations demonstrated that this house was partially filled with debris related to the reconstruction of that part of the city, after the earthquake, which took place only at the end of the 18th century, namely in April 1798, when the Roman Theatre was discovered. The disposition of the debris located at the western part of the compartment, next to the door, suggests some degree of intentionality possibly to create specific areas whose functionality is impossible to now nowadays. However, the remaining area was simply abandoned after the earthquake (Fernandes 2016).

In the eastern part of the ground floor, a great amount of mortar from the walls, with ceramics included, was found covering charred plant remains. The carpological remains we present here were revealed after the removal of this layer of construction material.

Beyond the layer of mortar, a great amount of straw was uncovered. The straw was disposed in several layers, each one with culms perpendicular to the ones of the preceding layer. The culms were tied in small bundles (Fig. 17.3) and a group of bundles would comprise a larger sheaf. A set of twined culms were used to tie each small bundle and each big sheaf. The sheaves were laid down side by side directly in the bedrock (Miocene marlstone) that was used as floor in this section of the house. Beam pockets in the eastern wall and pockets for stairs in the southeast wall demonstrate the existence of a mezzanine that would cover the remains we study here (see a reconstruction in Fig. 17.2).

Contrary to this area where the bedrock was used as floor, on the western area there was a pavement made with pebbles, demonstrating the existence of two different areas at ground level. These were also visible in the characteristics of the compartment's walls.

The functional interpretation of the ground level of the house is based on the archaeological and archaeobotanical data available. The area is thought to have been a stable for small-sized animals, possibly sheep or goats, to whom the straw was being given. Unfortunately, no zooarchaeological remains were found during fieldwork. Nevertheless, this compartment should also have been used for other purposes, as suggested by the presence of other archaeological materials. These included rings made of copper, probably belonging to barrels, which suggests this space was also used to keep wine. Harnesses and multiple metal buckles, some of which are decorated, were probably used to fasten leather straps, although some may have been clothing ornaments as other small metallic objects that probably fell from the upper floors.

Several pieces of faience were collected, mostly dishes made in Lisbon, however there is a predominance of common pottery, particularly small jugs and pots. It should be noted that the glaze in the faience fragments acquired

Figure 17.3. Excavation of debris that covered the cereals (left) and detail of the alternated bundles of cereals during the excavation (right). Photos by Lídia Fernandes.

a black colour due the high temperatures, over 150 ºC, caused by the fire.

Field and laboratorial methodology

The recovery and transportation of the plant material from the excavation area was a difficult task, considering its great size and fragility. As such, it was not possible for the field team to maintain the integrity of the sheaves. Thus the assemblages of plants that were recovered do not correspond to actual sheaves, rather they correspond to casual assemblages determined by the circumstances of the excavation.

Some of this material was incorporated in the permanent exhibition of the Museum of Lisbon – Roman Theatre and the remaining plant findings were sent to the laboratory of the Natural History and Science Museum of the University of Porto (MHNCUP) and the Research Centre in Biodiversity and Genetic Resources (CIBIO-UP) where they were analysed and provisionally deposited, until their future return to the Museum of Lisbon – Roman Theatre.

It is not possible to calculate accurately the amount of plant remains that were found in the excavation site but it was likely around 3 m^3 (*c.* 3000 l). The assemblages studied in the lab are composed of *c.* 0.15 m^3 (*c.* 150 l) of charred plant material.

Due to the brittleness of less dense material – culms and spikes – laboratory work was carried out in the least invasive way possible, reducing to a minimum the handling of the material. Thus detailed analyses were restricted to loose plant remains and those visible at the surface of the assemblages. Some material was extracted from the assemblages to allow the observation of informative details. As such, plant remains in the surface of the assemblages were observed with a stereomicroscope – Euromex, up to x45 magnification – in a long-reach articulated stand, to reduce handling. The remains extracted from the assemblages were observed with a stereomicroscope Nikon SMZ800, up to x126 magnification. It must be noticed that due to these methodological options, it was not possible to do an accurate quantification of the plant macroremains recovered.

Taxonomic analysis was carried out by comparing the archaeobotanical remains with the reference collections of CIBIO and MHNCUP and handbooks (*e.g.* Jacomet 2006).

Results

The analysed material included spikes, grains, culms and knots made with straw, characterised as follows:

- Spikes (inflorescences) of rye (*Secale cereale*), with 8 cm to 10 cm length. They are mostly intact, frequently flattened (Fig. 17.4). Spikelet bases are hairy and the spikelet-rachis disarticulation scar is quite short. Lemmas are pointed, pectinate-ciliate, keeled and awned. Spikes frequently do not have grains and when they do, spikelets without grains are still very common. Some spikes are not fully carbonized, showing a dark-brown colour.
- Grains without adherent glumes, slender and elongated; proximal end pointed; apex truncated; long scutellum and deep ventral furrow (Fig. 17.4). All grains were identified as rye (*Secale cereale*). Some grains were twisted. No biometric study was carried out, but a first analysis detected grains with length between 4 mm and 7 mm and width between 1.5 mm and 2.5 mm. Length is similar to other charred reference material (Jacomet 2006), but width is slightly lower, meaning these Lisbon grains are generally slender. Caution is necessary considering the plant assemblage was crushed by the collapse of the building, which may have influenced some measurements. Charring may have also affected grains' morphology and biometry but only slightly. Changes in cereal grains due to carbonisation have been well documented in experimental works (*e.g.* Braadbart 2008; Bonhomme *et al.* 2017) but these never included charring conditions similar to those studied here. Considering the exceptional preservation of remains far more fragile than grains in the same context, it is likely that heat exposure did not change the caryopses significantly.
- Knots composed by twined cereal culms (Fig. 17.5). Eight knots were recovered, all flattened.
- Culms of Poaceae at least up to 80 cm long, but since culms are fragmented true length is difficult to access. Culm perimeters vary between 2 mm and 10 mm, but since culms are flattened these measurements are also difficult to obtain. Still, these perimeter measurements fit well in the base-spike variability in single plants, when compared with whole plants of rye from the MHNCUP reference collection. Besides being crushed, culms frequently have a wavy aspect and some are not fully carbonised, showing a dark-brown colour. Although the proportion between spikes, grains and culms is impossible to quantify without severely damaging the plant remains, it is clear that culms are by far predominant.

Figure 17.4. Remains of rye: non-flattened and flattened spikes (left); detail of hairy rachises (centre, top); detail of lemma (centre, bottom); grains (right). Photos by João Tereso.

Discussion

The state of preservation of the plant remains is remarkable and some specific aspects are informative regarding the formation processes of the archaeobotanical assemblages. As mentioned before, several morphological features such as hairs from the apex of the grains and rachis are frequently documented in these assemblages, although they are seldom identified in archaeobotanical studies. Abundant culms and spikes are not fully carbonised, presenting a dark-brown colour, meaning that combustion was interrupted at an early stage, *i.e.* between dehydration and carbonisation (following Chabal *et al.* 1999; Allué 2002; Badal *et al.* 2003).

Charcoal formation implies the chemical degradation of cellulose and lignin, making plant remains almost inert and very brittle. As such, mechanical pressure is their main form of degradation. Less dense plant parts are not only more likely to disappear in direct contact with fire, but are also more prone to mechanical destruction. Thus, wood is more likely to survive a combustion than cereal culms and grains are more likely to survive than glumes, even in contexts where the latter were more abundant prior to the fire.

Considering these premises, it is not likely that the cereals here presented had direct contact with fire or else culms and spikes probably would not be so abundant. So, although the fire that contributed to the destruction of the house was the heat source that led to the conservation of the plant remains, it did not have a direct contact with the straw, probably because the plant remains were already covered with at least part of the debris of the structure. This also explains why many culms and spikes are flattened and wavy. If the plant material had been carbonized before the building collapsed, it would have become much more fragmented. It is clear that carbonized culms and spikes are brittle and

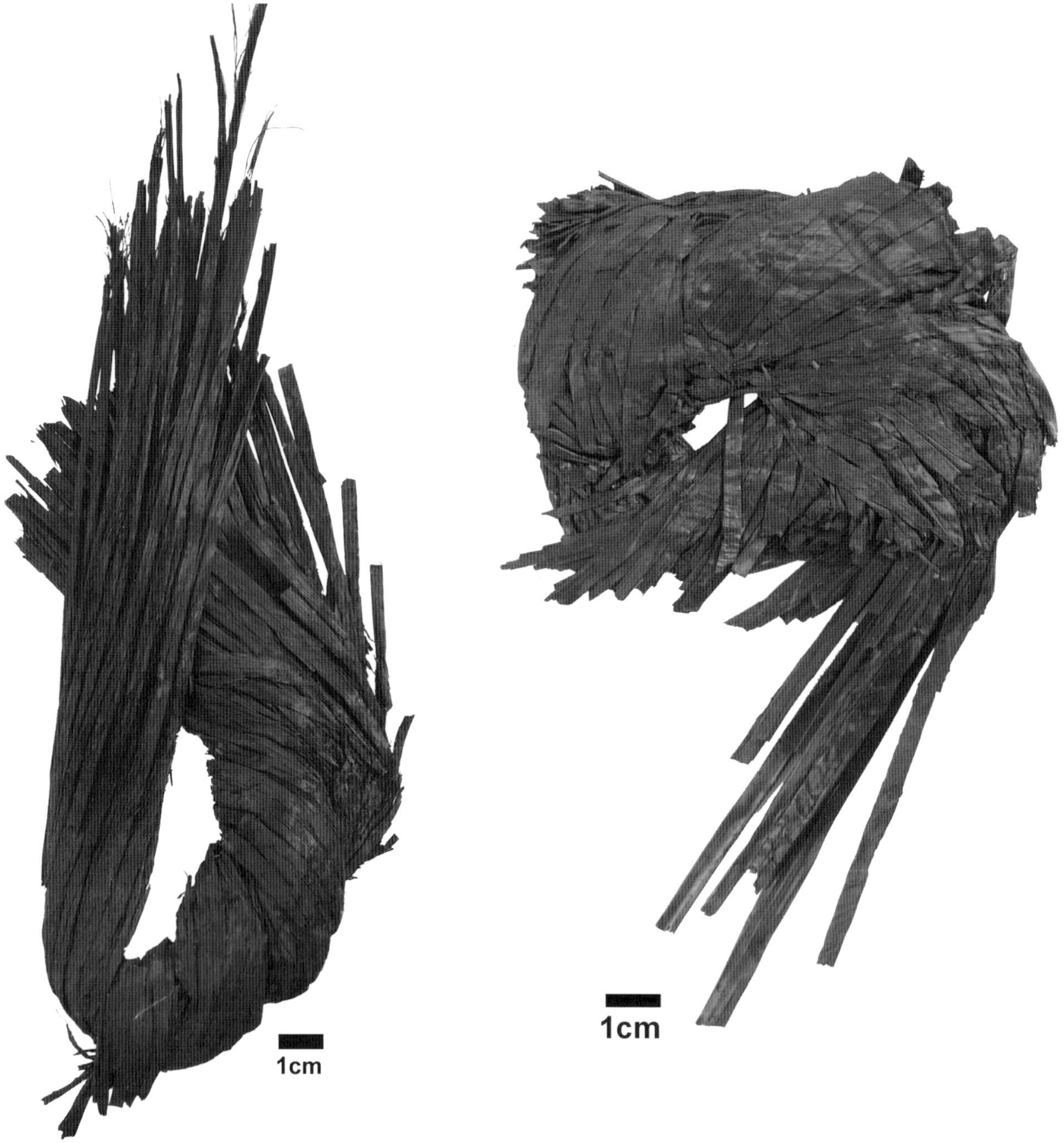

Figure 17.5. Knots made with twined cereal culms. Photos by João Tereso.

they do not bend or flatten. When submitted to mechanical pressure, they simply break.

Explaining the presence of such assemblages in the ground level of a house in Lisbon, in 1755, is not simple and several issues must be taken into consideration. The presence of rye, itself, is not surprising since this crop was introduced in western Iberia at least in 1st century BC (Seabra 2015; Seabra *et al.* 2018) and was widely cultivated in medieval times, particularly in central and northern Portugal (Marques 1978; Gonçalves 2012). The peculiarity relies in the composition of the assemblage. As mentioned before, most of the plant remains are culms of Poaceae as well as spikes and grains, although the latter are not very abundant, considering the great amount of spikes recovered. All spikes and grains were identified as rye. Thus, what was being kept below the mezzanine was mostly straw tied in sheaves and intentionally disposed in a horizontal position. It was not a chaotic concentration of plant remains and building material resulting from the collapse of the house.

It has been emphasised that although straw was tied in sheaves, the studied assemblages do not correspond to actual sheaves, hence when observing their outer surface, we are not in fact observing the outer surface of past sheaves that could hide different elements (plant parts or species) inside the bundles. As such, the non-invasive lab methodologies probably did not produce any sort of bias, namely some over- or underrepresentation of culms or spikes. However, when it comes to the grains further reasoning is necessary.

Rye is a free-threshing cereal and the grains, when ripe, fall easily from the spikelets. That can happen intentionally by processing the cereals after the harvest, or unintentionally during transportation or in the area where the plant parts are stored, if they are intensively manipulated. It is hard to evaluate if the collapse of the building could emulate such handling but, if so, would have originated a great accumulation of loose grains that would have been visible during the excavation. It is possible to sustain that the abundant spikes without grains did not have their grains removed after the carbonisation. In the lab work, extracting grains for observation always implied breaking the spikelets due to their extreme brittleness. Considering spikes and spikelets in the assemblages are well preserved and usually seem to be compact and apparently closed, it is clear they did not have grains at the time of the carbonisation.

As such, plants could have been harvested before most were fully ripened, coexisting spikes with and without grains, the latter being most abundant. Following this hypothesis, the plant was not processed to free the grain, leading to well-preserved culms that were not cut with a threshing sledge or beaten with a grain flail. However, it is also possible that plants were harvested ripe and processed to free the grain in a way that would generally maintain the integrity of the culms and spikes, in order to make possible their transport and use as a whole. This hypothesis could explain why we often found spikelets with and without grains in the same spike.

Either way, both hypotheses imply that what was kept in the ground floor of the house was straw intentionally without or with few grains.

Straw is a very relevant material that can be used for several purposes, such as fuel, animal bedding, fodder, thatching and even for making adobe but for some of these usages, waste products from crop processing – chaff – would be sufficient. We cannot exclude the possibility that the straw was stored awaiting further processing before being used. That would imply that such waste-producing task would be carried out in the city, instead of the threshing floor near the fields. Still, it does not explain the necessity to maintain the integrity of the plant to the level that was recorded.

It is possible that the straw was kept to be used for thatching, but the presence of grains would be a great disadvantage. Grains would invite birds to degrade the roof. So, the use of the straw for thatching would imply a further processing to remove the remaining grains.

Considering all of this, it is highly probable that the straw was kept in the ground floor of that house to be used as fodder, for which the presence of some grains would not be problematic, on the contrary, this good-quality fodder suitable for a variety of domestic animals would have enhanced nutritive value (Foxhall 1996; Derreumaux 2005; Tereso *et al.* 2013). The future use of this straw as animal bedding is also a possibility, but we must stress that the excavations did not reveal a specific area with a concentration of trampled straw or straw disposed in a disordered way.

Conclusion

Excavations at the Lisbon Museum – Roman Theatre led to the discovery of an extraordinary set of carbonised plant remains. The surprising preservation of whole culms and spikes and the recording of fragile morphological characters that usually do not survive the carbonisation process makes this a unique set of archaeobotanical remains. This assemblage, together with the remaining archaeological evidence recovered in the ground floor of a Lisbon house allows us to describe a small episode of the city's life, within a great historic and catastrophic event, the earthquake that destroyed a great part of the city on the 1st November 1755.

Following the earthquake, the house here studied collapsed and burned in the devastating fires that spread throughout the city. In the ground level, sheaves of rye were kept in the floor underneath a mezzanine, presumably to be used as fodder for small domestic animals that were kept in the compartment. The debris of the building covered the sheaves protecting them and allowing them to be carbonised without direct contact with fire, as suggested by the waving aspect and flattening of culms and spikes, only possible if the mechanical pressure over the straw occurred prior to charring. The building was never reconstructed after the earthquake, allowing the preservation of these botanical remains until the archaeological excavations, leading to their inclusion in the current museum.

Acknowledgements

Authors' contribution. JPT: archaeobotanical lab work. LF: coordination of the archaeological interventions. Both authors are responsible for the interpretation and writing of the manuscript.

The work of J. Tereso was supported by a Post-doctoral grant from Foundation for Science and Technology (FCT – Portugal) with the reference SFRH/BPD/88250/2012.

References

Cópia do Tombo da Cidade de Lisboa em 1755, que está no Arquivo Nacional da Torre do Tombo, feita sobre uma copia do mesmo tombo, da letra de José Valentim de Freitas; que está na Associação dos Arqueólogos, por João Marques da Silva, em Junho de 1894, Museu de Lisboa.

Planta da Baixa de Lisboa antes do terramoto, na zona das actuais freguesias da Sé, S. Tiago, S. Miguel, Castelo e S. Cristóvão e S. Lourenço [Cartographic material]. Drawing, indian ink on canvas. Museu da Cidade. Inventory: MC.DES.4295

Allué, E. (2002) *Dinámica de la vegetación y explotación del combustible leñoso durante el Pleistoceno Superior y el Holoceno del Noreste de la Península Ibérica a partir del análisis antracológico*. Unpublished PhD thesis, Universitat Rovira i Virgili, Tarragona.

Badal, E., Carrión, Y., Rivera, D. and Uzquiano, P. (2003) La Arqueobotânica en cuevas y abrigos: objectivos y métodos de muestreo. In R. Buxó and R. Piqué (eds) *La recogida de muestras en arqueobotánica: objectivos y propuestas metodológicas. La gestión de los recursos vegetales y la transformación del paleopaisaje en el Mediterráneo occidental*, 19–29. Barcelona, Museo d'Arqueologia de Catalunya.

Blakeney, R. (2017) *On Food and Fodder: Archaeobotanical Investigations of Bamburgh Castle's West Ward, 9th through 12th centuries*. Unpublished MSc thesis, University of Nottingham.

Bonhomme, V., Forster, E., Wallace, M., Stillman, E., Charles, M., Jones, G. (2017) Identification of inter- and intra-species variation in cereal grains through geometric morphometric analysis, and its resilience under experimental charring. *Journal of Archaeological Science* 86, 60–67.

Braadbaart, F. (2008) Carbonisation and morphological changes in modern dehusked and husked Triticum dicoccum and Triticum aestivum grains. *Vegetation History and Archaeobotany* 17, 155–166.

Chabal, L., Fabre, L., Terral, J.-F. and Théry-parisot, I. (1999) L'anthracologie. In C. Bourquin-Mignot, J.-E. Brochier, L. Chabal, S. Crozat, L. Fabre, F. Guibal, P. Marinval, H. Richard, J.-F. Terral and I. Théry-Parisot (eds) *La botanique*, 43–104. Paris, Editions Errance.

Charles, M. (1998) Fodder from dung: the recognition and interpretation of dung-derived plant material from archaeological sites. *Environmental Archaeology* 1(1), 111–122.

Derreumaux, M. (2005) How to detect fodder and litter? A case study from the Roman site Le Marais de Dourges, France. *Vegetation History and Archaeobotany* 14(4), 373–385.

Fernandes, L. (2007) Teatro romano de Lisboa – os caminhos da descoberta e os percursos da investigação arqueológica, *Al-madan* 15, 27–39.

Fernandes, L. (2013) Teatro romano de *Olisipo*: a marca do novo poder romano. In J. Arnaud, A. Martins and C. Neves (eds) *Arqueologia em Portugal. 150 anos. Lisbon, 21–24 November 2013*, 765–773. Lisbon, Associação dos Arqueólogos Portugueses.

Fernandes, L. (2015) Habitar em Lisboa nos séculos XV e XVI: modos de vida e de construir. O exemplo dos vestígios habitacionais na área de intervenção arqueológica do teatro romano. In *Lisboa 1415 Ceuta. História de Duas Cidades*, 54–56. Ceuta, Ciudad Autonoma de Ceuta – Consejería de Educación y Cultura; Câmara Municipal de Lisboa - Direção Municipal de Cultura/Departamento de Património Cultural.

Fernandes, L. (2016) Sinais de um quotidiano que o terramoto de 1755 interrompeu: os vestígios detectados no Museu de Lisboa – Teatro Romano. Colóquio Terramoto em Lisboa. *Arqueologia e História* 68, 89–102.

Fernandes, L. (2017) Aspetos construtivos do teatro romano de Lisboa: matérias primas e técnicas edificativas. In J. Arnaud and A. Martins (eds) *Arqueologia em Portugal – 2017 Estado da Questão*, 1265–1278. Lisbon, Associação dos Arqueólogos Portugueses.

Fernandes, L. and Almeida, R. (2012) Um Celeiro da Mitra no Teatro Romano de Lisboa: inércias e mutações de um espaço do séc. XVI à actualidade. Congresso Internacional de Arqueologia Moderna. Lisbon, 6–9 April. Lisbon, FCSH da Universidade Nova de Lisboa, 111–122.

Foxhall, L. (1996) Snapping up the unconsidered trifles: the use of agricultural residues in Ancient Greek and Roman farming. *Environmental Archaeology* 1, 35–40.

Gonçalves, I. (2012) *Por terras de Entre-Douro-e-Minho com as inquirições de D. Afonso III*. Porto, CITCEM/Edições Afrontamento.

Jacomet, S. (2006) *Identification of Cereal Remains from Archaeological Sites*. Archaeobotany Lab, IPAS, Basel University.

Jones, G. (1998) Distinguishing food from fodder in the archaeobotanical record. *Environmental Archaeology* 1(1) 95–98.

Marques, A.H.d.O. (1978) *Introdução à História da Agricultura em Portugal. A questão cerealífera durante a Idade Média*. 3rd edition. Lisbon, Edições Cosmos.

Seabra, L. (2015) *Estudo Paleoetnobotânico do Povoado da Idade do Ferro do Crastoeiro (Noroeste de Portugal)*. Unpublished MSc thesis, Universidade do Minho.

Seabra, L., Tereso, J.P., Bettencourt, A.M.S. and Dinis, A. (2018). Crop diversity and storage structures in the settlement of Crastoeiro (Northwest Iberia): new approaches. *Trabajos de Prehistoria* 75(2), 361–378.

Tereso, J.P., Ramil-Rego, P., Carvalho, T.P., Almeida-da-Silva, R. and Vaz, F.C. (2013) Crops and fodder: evidence for storage and processing activities in a functional area at the Roman settlement of Monte Mozinho (northern Portugal). *Vegetation History and Archaeobotany* 22, 479–492.

Vaz, F., Seabra, L., Tereso, J. and Carvalho, T. (2017) Combustível para um forno: dinâmicas de ocupação de um espaço em Monte Mozinho (Penafiel) a partir de novos dados arqueobotânicos. *Arqueologia em Portugal/2017 – Estado da Questão*, 1331–1345. Lisbon, Associação dos Arqueólogos Portugueses.